MUSIC IN VIVO

MUSIC IN VIVO

THE MUSIC OF BIOLOGY AND THE BIOLOGY OF MUSIC

Robert Labaree

THE BOYDELL PRESS

First published 2026
The Boydell Press, Woodbridge

ISBN 978 1 83765 003 3

The Boydell Press is an imprint of Boydell & Brewer Ltd
and of Boydell & Brewer Inc.
website: www.boydellandbrewer.com

Our Authorised Representative for product safety in the EU is
Easy Access System Europe – Mustamäe tee 50, 10621 Tallinn, Estonia,
gpsr.requests@easproject.com

A CIP catalogue record for this book is available
from the British Library

To Frances, Aaron, Hannah and Sophie

CONTENTS

FIGURES

Chapter 1

Chapter 2

Chapter 3

Chapter 4

Chapter 5

Chapter 6

Chapter 7

Chapter 9

Chapter 10

Chapter 15

Chapter 16

Chapter 17

Chapter 18

Chapter 20

MUSIC EXAMPLES

Chapter 3

Chapter 4

Chapter 8

Chapter 11

Chapter 15

Chapter 18

Chapter 19

Chapter 20

Chapter 21

PREFACE: IT'S THE SONG, NOT THE SINGER

To tell a story about the music of biology and the biology of music, I have to begin in nature. Admittedly, for anyone opening this book because music is on the cover, starting off in an unfamiliar terrain inhabited by epigenesis, autopoiesis, and lateral gene transfer may be off-putting. However, by expecting, right off the bat, that we will be applying our native musical curiosity directly to nature, I am more likely to accomplish the goal I have set for this project – let me call it an attempt to define a biology-centered approach to the making of music – than if I opened in the comfort zone of songs and sonatas.

In 2017 the molecular biologist W. Ford Doolittle was speaking to other biologists, not to musicians, when he claimed that "it's the song, not the singer."[1] He was stating a biological truth by inverting The Rolling Stones' more famous claim, "it's the singer not the song," an appeal to a lover to see the individuals submerged in their habitual patterns – "the same old places and the same old songs." Doolittle was interpreting "songs" broadly, of course, not as individual ditties but as structures, models and patterns which are themselves not so much things that we have made (though it sometimes helps us to speak of songs that way) as much as they are events in which we participate. We are, as individuals, crucial in song-making. Without us, there would be no songs at all, not even ideas of songs. However, Doolitle's perspective proposes to those of us in the world of professional music-making a shift in emphasis from individual artists and individual musical objects toward processes and systems that are self-making and self-sustaining. A better understanding of this shift is one of the possible outcomes of this biology-centered story about music.

Of course, on these pages the beginning in biology will only lead to more biology, and still more after that, and even though the biology will always be read alongside more and more music (the goal here is

1 Doolittle, W.F. and Booth, A. (2017). The Rolling Stones: "The Singer not the Song" (Official video with lyrics). https://www.youtube.com/watch?v=wtlenQW63NM (accessed 1/29/26).

integration), music-oriented readers may also wonder at times what they have gotten themselves in for. I believe that any amount of time devoted to the non-musical science of this story is well-spent. But in saying this, I am asking for an act of faith, since at this point in my argument no evidence has been offered for why all this biology matters to music, beyond a biologist's risky trespass into the sanctuary of rock 'n' roll. My goal is to refresh the very old conversation about music and nature, not to end it. It is my attempt to re-calibrate ways of thinking about music accumulated over at least two centuries in the public sphere and in education. It relies heavily on the work of other musicians, scholars and observers of music.

What makes the argument here a bit different from these precedents in music is that it directs all that musical thought and experience toward the natural order from which music has sprung as an indispensable feature of human civilization. This book calls on those of us who care deeply about music to reconsider our long-established preference for thinking about music primarily in relation to music itself. Nature, civilization and music are nested systems, fitting inside of each other approximately (very approximately) like Russian dolls, with nature as the largest doll and art the smallest. It is telling that in the world of music, at least music as it is talked about among professionals in the west, these positions tend to be conceptually and practically reversed. Nature, the largest of the dolls, is treated in the music world as nearly invisible, as if it was the smallest doll contained inside of, and obscured by, human artifice, a footnote to what humans have made of the "ecosystem services" nature has provided. In this view, music, indeed, the arts in general, are big enough to incorporate into themselves – or not – pastoral symphonies and bucolic landscapes but this has remained the artist's prerogative. In this view, nature has been an option, just one of the many colors of the aesthetic palette.

What I am suggesting is an orientation that is not exactly new to music conversations, but in the practice rooms, classrooms and performance venues where music is worked and taught it is heard only rarely. To get to a place where we can feel music's connectedness to nature in our daily work we will need more tools than this one book can provide. But now, in the spring of 2026, after four decades during which the material of this book assembled itself, I am looking back on a period marked by global heating, global pandemics and global extinctions. The urgency in civilization's dislocation from nature is no longer ignorable. It is humanity's problem to figure out, which makes it also music's problem. But I also expect that in a book that claims to be about music, references to warming oceans and species decline may feel like intrusions.

One final warning. As with so much science and with so much art, novelty often emerges in the course of accidental or opportunistic boundary crossings based on a hunch. This book is a story of a journey from music to biology and back again, but it is also a story about hunches and their role in discovering what's true, as much in music as in science. A guiding hunch in what follows is that there is a wholeness to the life on this planet from which music cannot be extracted as if it were simply a toy, a clever object made on some other planet, one without a history of evolving life forms making a home in all its minute crannies.

ACKNOWLEDGEMENTS

This book is mostly made of questions. Luckily for me, the questions rolled out gradually over decades, allowing me the time to slowly wake up to how much music there is in nature. Essential assistance was provided to me by a village of people. I mention a few here.

Music in Vivo only managed to get this far because of the patient oversight of editor Michael Middeke of Boydell and Brewer. From the very beginning, Michael never meddled, even as he encouraged, guided and corrected. He has been a very special presence, forgiving each broken deadline and looking forward to the next. The meticulous and sensitive copyediting of Ailén Cruz is the main reason the text has achieved its current level of clarity. Crispin Peet shepherded it through the final production process.

The scientific observations here are my responsibility alone, but they have been disciplined by two biologists whose words appear on these pages as often as those of Darwin, Margulis, Gould and Varela: Stuart A. Newman, Professor of Cell and Molecular Physiology at New York Medical College and Scott F. Gilbert, Howard A. Schneiderman '48 Professor Emeritus in Biology at Swarthmore College. Their interest and comments, dating back more than a decade, and their corrections to my moments of confusion and overreach, were always accompanied by yet more recommendations of research I should know about. They even suggested that biology might have something to learn about its own work from music.

My colleagues in ethnomusicology, James Cowdery, Mark Slobin, Michael Tenzer, Jeff Titon and Aaron Allen played different roles in shaping the multi-decade path of this project. Jim and Mark oversaw my doctoral work in the 1980s on the intersections of oral and written tradition. Jim's refreshed look at the tune family concept became, decades later, a formative insight into what musicians have long intuited about the continuity of music and nature. Michael's generous comments and criticisms about an early version of *Music in Vivo* ended up informing the chapter which Jeff and Aaron suggested I write for their 2023 volume for Oxford University Press, *Sounds, Ecologies, Musics*. That chapter was a bite-size first pass through the material that became this book which has retained the imprint of Jeff's and Aaron's thoughtful comments and edits.

My colleagues in the Music History Department of New England Conservatory since 1984 – Sean Gallagher, Helen Greenwald, Anne Hallmark, Mathew Cron, Katarina Markovic, Peter Row (1944–2018) and Greg Smith – modelled for me what musical practice looks like as history. Some of the material here was first presented in conservatory classes in musicology, and benefitted from being filtered through the responses of students.

Thanks to my friend, musicologist and cellist Lewis Lockwood, for his comments on the chapter featuring his well-known work with the Juilliard Quartet. As an authority on the music of Ludwig van Beethoven and the wisdom of Yogi Berra, and with his long experience with the writing of books about music, he was a quiet and steady voice over the years.

Thanks also to Adam Sacks (1947–2026), Jim Laurie, Christopher Haines and Frederick Jennings, my colleagues at Biodiversity for a Liveable Climate in Cambridge, Massachusetts. Between 2018 and 2023 they made up a hospitable seminar where a naturopath, a chemical engineer, an architect, an economist and a musician could each week dive deeply into the science behind the warming of our world. It was up to me to apply that science to music.

The presence of the Ottoman Turkish tradition in this book can be traced back forty years to my teachers in Istanbul, the multi-instrumentalist İhsan Ozgen (1942–2023) and the revered neyzen Niyazi Sayın (1927–2025). They embodied self-organization and co-creation at every moment, though it took years before I could appreciate it. My colleagues in *The Eurasia Ensemble* over the years, Feridun Özgören (tanbur) (1942–2024), Frederick Stubbs (ney), Beth Bahia Cohen (yaylı tanbur), Christos Govetas (voice, ud) and Reinmar Siedler (cello) were participants with me in an ongoing *meşk* (lesson) in Turkish makam. The *meşk* continued with udi, neyzen, jazz pianist, composer and scholar Mehmet Ali Sanlıkol, my partner for two decades in *Dünya*, a musicians' collective through which the global musical perspective of this book was explored and deepened in every concert and recording.

Cellist, composer and bibliographer Jason Coleman engraved most of the musical notations on these pages. His informed interest in the continuity of music and biology went well beyond staves and notes. Choral conductor Holly Druckman provided additional notations.

Pianist Jean Butler (stage name Maria Rodriguez) (1926–1998), my teacher of piano improvisation for dancers in Washington, D.C. in the 1970s, used to say, "you've got to feed yourself," and would then open another score for me to devour. This was my earliest experience of western art music repertoires as ecosystems.

I am grateful to historian Philip Burnham and my brother David Labaree, Emeritus Professor in the School of Education at Stanford University, who read portions of the book's early drafts. More experienced than I in the ways of book-writing, they each, in distinct ways, provided support for the ecological view of culture on these pages.

Finally, I am grateful to my wife Frances Lang, a writer, editor and psychoanalyst, who put up with my preoccupation with this project for too many years. She read and commented on most of it at various stages, and even read much of it out loud to me during periods when I was incapacitated because of eye surgeries. In the end, it was just another of the many things that we have done together for 50 years, although it still doesn't come close to what we have created in our children.

Brookline, Massachusetts
February, 2026

Preparing a bio-musical workplace

Mental states do not exist, any more than flame states.
Mind and flame are both processes.
Mark Bickhard[1]

The two opening chapters of this book are intended to prepare a space in our musical thinking where ideas about natural processes and ideas about musical processes can pass freely in both directions across disciplinary boundaries.

The storyline which begins in Chapter 1 is my own wayward path from music to biology. But the personal thread in this story is only important as a way to track the thinking of specific biologists and music scholars I have encountered along the way whose contributions suggest how the work of those separate fields converge and even interact. Chapter 2 sketches a preliminary definition of what I am calling *music in vivo,* a biology-centered approach to music-making, followed by an outline of the book's chapters. The chapter ends by returning to the convergence of music and biology introduced in Chapter 1 with a few preliminary examples of how those convergences show up in the everyday shoptalk of biologists and musicians.

1 Bickhard, M.H. (2005), 205–218. Quoted in. Deacon, T.W. (2012), 175.

1

Coronavirus is playing our song. Are we listening?

Several decades ago, for no reason that I could have articulated to my music colleagues at the time, I found myself drawn to the work of nineteenth-century naturalists. Beginning with Charles Darwin, I moved on to his predecessor in France, Jean-Baptiste Lamarck, his contemporary in Moravia, the scholar-monk Gregor Mendel, and his nemesis, the self-appointed promoter of "Darwinism," the Englishman Herbert Spencer. These classic writers drew me, in part, because their biographies intersected with the biographies of figures well known in the music world. (Jacques Barzun's famous exploration of a few of these overlaps in 1941, *Darwin, Marx, Wagner,* comes to mind.) But inevitably this interest led to the work of others in the life sciences who, in addressing selected communities of readers, in some cases, as much through widely accessed lectures and interviews on YouTube as through their writings, also speak to the likes of me. That is, they speak to non-scientists who are ready to consider biology as the place to start any inquiry into how all the processes of the planet work.[1]

Over the years, my interest gradually turned to the fields of cell biology, developmental biology, genetics and immunology. It turned out to be a good time to be a beginner in the life sciences. I was finding a rich literature by specialists who were also intelligible, even compelling writers, people like Erwin Schrödinger (a physicist), Steven J. Gould (a paleontologist), Francisco Varela (a biologist and neuroscientist) and Lynn Margulis (a

1 A short list of studies written by and for professionals in biology which are also accessible for non-biologists would include three collections: Müller, G. and Newman, S.A. (2003); Caporale, L.H. (2006); and Gissis, S.B. and Jablonka, E. (2011).

microbiologist). As time went on, I discovered a host of other specialists in a wide variety of disciplines, like the physical chemist Ilya Prigogine, the team of Fritjof Capra and Pier Luigi Luisi (a physicist and a biochemist), cardiologist/guitarist Denis Noble, neuroscientist Terrence W. Deacon, developmental biologists Scott F. Gilbert and Stuart Newman, science writer David Quammen, literature-scholar-and-rock-'n'-roll-bass-player-turned-bio-cyberneticist Bruce Clarke, and science-saturated novelist Richard Power.[2]

From the beginning, time-lapse photography of developing organisms and electron microscope videography played an important role in my seduction into life forms (Figure 1.1). These images deliver to people like me a glimpse into the ceaseless and otherwise invisibly small actions of living systems which subtly disrupt my habitual view of nature's surface – nature as a collection of bulky, more-or-less beautiful and more-or-less threatening objects. In the same way, the famous whole earth photos taken from Apollo 8 in 1968 had disrupted humanity's view of itself in a single snapshot of our planet as a shiny blue-green ball, constantly changing, against a black background stretching out on an otherwise invisibly grand scale. I believe that these are developments with implications for music which test our ability to make room for a fresh view of musical creativity, one which looks more like the creativity of nature than we have up to now been able to see.

Nature's tangled bank

A musical interest in all this biology can be both mundane and speculative. The mundane part is all about muscles and memory and experience and inputs from the current moment intersecting to create the flow of musical ideas. This is the everyday, journeyman level of music-making which is focused on execution of musical tasks, on teaching and on mastering long-established musical practices. So much of the moment-by-moment tasking of music and the moment-by-moment tasking within living cells is mundane – repetitive and utilitarian. Interest becomes more speculative when these musical tasks are juxtaposed with the insights of biologists who are scrutinizing both the split-second variations in an organism's body and the variations in a species of organisms over tens of millions

2 Representative works by these writers: Schrödinger, E. (1945); Gould, S.J. (1989); Varela, F.J. (2017); Margulis, L. (2000); Prigogine , I. (1996); Capra, F. and Luisi, P.L. (2014); Noble, D. (2006); Gilbert, S.F. et al. (2012); Newman, S.A. (1995); Clarke, B. (2020); Power, R. (2018).

Figure 1.1 Five vegetation stages of a bean plant growing. Photo credit: *Pryzmate*. Licensed by Istock.

of years. The biologists' insights into these variations have themselves evolved, as has the very concept of evolution. With the unfolding of many generations of scientific thought to be aware of, the speculative part can overtake the mundane, and the prospect of a deeper integration of music and nature can seem remote.

It is at this point, when a musician decides "I should just stick to music," that the perspective of poets can matter most. Here is how William Butler Yeats approached what is one of the recurring questions of this book: how can we understand the individual human imagination as a natural phenomenon?

> Like a long-legged fly upon the stream
> His mind moves upon silence[3]

Yeats offers no musical example, but he does describe Caesar in his tent (his eyes "fixed upon nothing a hand under his head"), Michaelangelo

3 Yeats, W.B. (1937).

on his scaffold (his hand moving to and fro "with no more sound than a mouse makes"), even a young woman executing a simple "tinker's shuffle picked up on a street." All these individual humans make room in their own internal spaces for creative processes that float precariously above a ceaseless stream of their own experience, experience which merges with that of other people, places and times. In Yeats' view, human creators ride the current with practiced dexterity, watching, attentive to what bits can be pulled to the surface before they disappear. It turns out that the eddies and ripples and water-walking fly of Yeats' stream fit the current thought ways of biology and might therefore persuade the musician and music lover that it is both desirable and necessary to think of nature and musical practice, different as they are, in a single glance, as a continuity. The poet's willingness to see a human actor as an insect and to think of the individual human imagination as a participant in the shared stream of human experience provides a solid platform for beginning to think about music as continuous with the natural world.

Here is another simple idea about natural processes that at first glance might be easily dismissed as simplistic, a layman's dumbed-down version of science: *There are no straight lines in nature.* The idea also has a corollary: *neither are there circles in nature, movements which simply repeat.* So many of nature's circular movements turn out to be more like spirals that never quite get back to where they started, and tend to corkscrew, often unpredictably, into new territory. It sounds contradictory to say it, but replication – astonishingly precise replication – can be seen everywhere in nature. But the replication of essential structures in living bodies is decidedly not the replication of the assembly line or the computer. Nor is it the replication of the printed text or the musical score, where multiple copies of identical ink shapes on a page might persuade us that they stand for identical events and meanings. Nature's curlicues are never solitary, but are always interlocked with other curlicues in the staggeringly complex webs of energy exchange and communication we see everywhere, if we choose to look: in the eddies of a stream, in the nearly invisible webs of root hairs in the soil's mycelium, in neural networks and in the pathways of interdependence among species in a habitat. Musical habitats, as well, are made of tangles just as dense, with curlicues of excited airwaves and delicate strands of memory embedded in human brains, vocal cords and muscles.

In 1859, all the circuitous repetitiveness of lifeforms doing the same things over and over again, while still tending toward difference, was encapsulated in Charles Darwin's famous untidy image of the complexity of nature in the closing pages of his *The Origin of Species*:

> It is interesting to contemplate a tangled bank, clothed with many plants of many kinds, with birds singing on the bushes, with various insects flitting about, and with worms crawling through the damp earth, and to reflect that all these elaborately constructed forms, so different from each other, and dependent upon each other in so complex a manner, have all been produced by laws acting around us.[4]

Darwin's homely image is also paradoxical in its embrace of both the disorder and the order of nature. By implication, it accepts the interdependence of chaos and lawfulness, or even, in one of many paradoxes we face in recent science, the lawfulness of chaos. This is a chaos which, in the words of Fritjof Capra and Pier Luigi Luisi, "only appears to be random but in reality shows a deeper level of patterned order."[5] The tangled bank is not at all the stately tree with a straight trunk, tap root and symmetrical branching that is the common icon of the unidirectionality of natural growth. This is the tree which has long been the image humans have tended to apply to the developmental history of families, languages, ideas and artistic styles. But, though Darwin at times referenced the tree which was the consensus image of natural development in his own time, in the end he favored the tangled bank.[6]

Non-scientists who are reading this should take heart in the fact that Darwin arrived at his preference for the tangled bank – and, for that matter, at his many other insights into the evolution of species – as much by instinct and reasoning as by experiment and calculation, without the aid of the astonishingly precise tools we now take for granted. His image of the tangled bank is like an informal snapshot which can in an instant freeze the frenzied movement of a crowd leaving a stadium – the movement itself appears to be random, and yet almost everyone heads in the same direction, usually without injury. Somehow, perhaps through unseen laws governing large swarms of individuals, the chaos is ordered. I make no apology for the "somehow" cropping up repeatedly in this book, but I also recognize that it may be unsettling for some looking for what they regard as sturdier proofs. Darwin's work, we should remember, is full of somehows, but this has not kept the models he established from enduring,

4 Darwin, C. (1859/1993), 648.

5 Capra, F. and Luisi, P.L. (2014), 105.

6 Others have seen in Darwin's image of entanglement an important insight into the complexity of living systems. See, for example, two recent treatments of quite different topics in biology: David Quammen's *The Tangled Tree* (2018), an account of the changing views of inheritance in genetics over the past century, and Merlin Sheldrake's *Entangled Life* (2020), a detailed portrait of the fungus kingdom.

even into our own age of computer simulations and dizzyingly detailed data streams.

The tangle in Darwin's image is also surprisingly consonant with terms like *recursive, coupled, self-referential, reticulate, rhizomatic,* and of course, *complex* in the terminology of scientists who came after Darwin. Latter day scientific terms like recursive and reticulate are also consonant with my own preference for spirals and corkscrews in the description of both music and biology. Applied to music, these biological terms bring the added sense of a whole system of looping feedbacks which is *self-sustaining* and even *self-organizing.* Remarkably, life provides for its own provisioning. Life manufactures or catalyzes the very ingredients and processes which drive the whole. This self-provisioning is key to understanding how life, once established on earth between three and four billion years ago, has never been absent on our planet. In that time, multiple extinction events triggered by drastic atmospheric changes, volcanic eruptions and meteor strikes have wiped out millions of species, proving over and over that life is fragile. And yet, the basic mechanisms of energy production, reproduction and growth common to all life forms have continued uninterrupted, mainly in the form of microbial life too small for the human eye to see.

The tangled bank in music

In this pattern of persistence in the face of extinction, self-provisioning is a curlicued process which can strike a musician as familiar: it does what music does. Musical events provide the fuel for other musical events. Musical actors feed on each other, pulling on threads of energy which are as mental as they are auditory and muscular, and the threads continue to be pulled long after a Ravi Shankar or a Franz Joseph Haydn or a Bessie Smith is deceased. An organism contributes essential elements to the chemical exchanges in its habitat which follow a looping path out and back, pulling threads of communication which tie each organism to others, even as the threads are connected to still other organisms deceased thousands of millions of years ago.

The understanding of how *biome* or *biosphere* make and sustain themselves has continued to deepen since Darwin's day, especially in the past half century. The presence of life is the identifying feature of our unique blue-green planet, the tangled web embedded in almost every cubic millimeter of the earth's geology and atmosphere. The planet's citric acid cycle, also known as the Krebs cycle after the pioneering work led by Hans Adolf Krebs in the 1930s, is one of the defining components of the biosphere. Figure 1.2 is a simple diagram of the Krebs cycle by itself. If we were able to view it in context, its circular form would quickly disappear in a tangle of still more circles. Inputs and outputs link the Krebs cycle to other cycles of chemical reactions, which in turn feed back into the very cycle. Together, they drive the metabolism of the cell, any cell of anything living.

From the perspective of the arts, such self-generating and self-sustaining processes would suggest a potential alternative view of human creativity. In the common view in the arts, we are free-standing, willful beings who elect, or not, to work with others in order to fashion the works of civilization. But as the details of the self-sustaining biome have sharpened over the past century, those details have come to look more like the tangle of journeyman creativity in which musicians also assume that they are immersed every day. My teacher, my ensemble partner, my compositional models, my audience

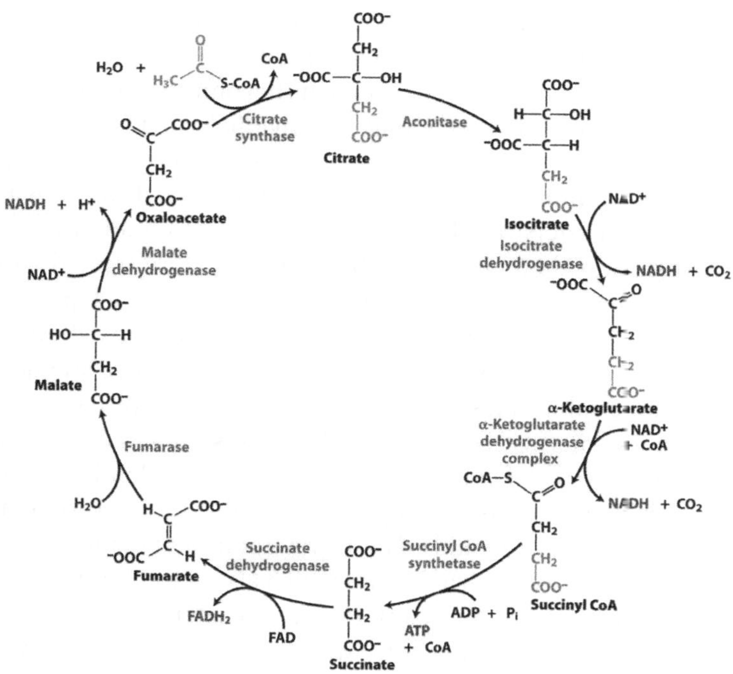

Figure 1.2 A schematic diagram of the Krebs cycle or citric acid cycle which is essential for creating the energy of metabolism in the cells of all oxygen-breathing organisms. Diagram from Dawson, J. (2021).

(adoring and critical), my student, my student's student, the teacher of my teacher's teacher – their words and sounds and muscular disciplines – are all threads connecting my own music-making to that of others. The threads are tangible, but ephemeral and we are each dwarfed by them, just as the long-legged fly is dwarfed by the stream.

Even without these latter-day scientific models, at certain moments of insight humans have been able to look past their own membranes and see themselves as conduits, as one of countless nodes which connect the sounds, habits, insights and products of others manifesting themselves in us at a particular moment in time. *Soli Deo Gloria* ("to God alone the glory") which J.S. Bach and other composers wrote at the bottom of many of their scores, and the medieval iconography of the heavenly dove dictating the thousands of melodies of Gregorian Chant to Pope Gregory I both express a common understanding of musical "inspiration" and musical "tradition." With everyday words like these we attempt to fill the gap between our perception of our personal musical agency – "*I* made this, it is *mine*" – and the mostly invisible web of agencies in which we are embedded and to which we are indebted.

As our own musical experience broadens beyond what we learned at our mother's knee, we have to accept that as individual experiences, inspiration and tradition do not necessarily translate universally. For example, the repertoire of the *mbira* (the so-called "thumb piano") of the Shona people of Zimbabwe is understood by Shona musicians to be the music created by ancestors. Individual pieces of mbira repertoire are treated with great reverence but they are also transformed by musicians in each performance.

I once tested the profound difference in assumptions about art and artist hinted at in these ancestor stories by asking, before an audience, a seasoned Shona mbira master whether he had ever composed a piece of his own and was met with confusion. The confusion was understandable. Everything he played, of course, was *his* and he confidently accepted the respect of his peers for his accomplishments. But if you were to ask him what he was playing, he would call it by its traditional name, for example, "Nhemamusasa" or "Bangidza," names which also stand for distinct tuning systems.[7] Though these compositions are not attributed to a particular creator, my mbira master would no more claim them as his own than I would put my name to a Bach Prelude and Fugue which I had just played or claim the major and minor scales as my personal creation. What, in fact, had I meant by my question, *have you ever composed a piece of your own?*

7 Berliner, P. (1978).

The Shona mbira master's confusion at a naïve question about musical creation – and my own confusion at his confusion – brings us a step closer to seeing how musical thought and biological thought might converge. Concepts like systemic creation and self-organization, now common in the natural sciences, offer a more comfortable place to the Shona view of creation than to the view which inspired my question. The Shona view also fits more comfortably into that of the geneticist W. Ford Doolittle. His ironic reversal of The Rolling Stones' line mentioned in my preface proposes a potentially disturbing alternative to my own individualistic assumptions. His preference for "it's the song, not the singer" over "it's the singer, not the song" could strike a musician less like a *bon mot* and more like a symptom of an existential shift. Some may find Doolittle's emphasis on patterns and systems (songs) disturbing because it appears to diminish the importance of the individual creator. Whatever we discover about the laws of nature acting in music, by the end of this book, I believe uniqueness and genius will remain, only re-contexted. All the same, I cannot promise that with a deeper understanding of nature's laws there will be no disturbance. Disturbance is to be expected when musicians start thinking – that is, *thinking as musicians* – about bacteria and slime molds, about the intertwining of order and chaos, and about individual creation and systemic creation.

The great derangement and the nature-music divide

Darwin's tangled bank is also a reasonable starting point for understanding the presence of a lethal pandemic in the same sentence as music. The title of this chapter "Coronavirus is playing our song," is intended to be more than a figure of speech. These words fit comfortably into a growing, decades-old conversation about how the processes which create human culture and the processes which create the biome are continuous. In this wider conversation, systemic life-like behavior is found not just in viruses, redwoods and whooping cranes but also in global markets, educational institutions and Broadway musicals. One corollary of this growing trend (which clearly demands more explanation) is that nature's way of working through ceaseless, impersonal processes of variation and interaction cannot be ignored in markets, education and music any more than in biology. Another corollary (which will demand still more explanation) is that many centuries of denial of the seamless continuity of human and natural creativity is a large part of the reason that we find ourselves in our current environmental predicament, what Amitav Ghosh has called "the great derangement."[8]

With these thoughts, we have somehow passed, in a few short pages, from the creativity of biology to the creativity of music, and from there to the very un-musical terrors of melting glaciers, uncontrolled fire, species die-off, dwindling resources, desertification and sea level rise. For the music-oriented reader, where in this thought sequence can we find a place for the diverse practice of sound art prized for millennia by humans on every continent? What has the daily work of musicians to do with these natural disturbances and the social disorder which we know will follow from them?

At this late date, it is undeniable that the chief agent of the growing disorder has been one species out of many millions, a species which arrived at the last minute to the 3.8 billion year story of life on this planet[9] and which, even now, represents less than one one-hundredth of a percent of the total mass of all living things. I can now say with certainty that my growing awareness of this circumstance is the source of the increasing urgency I have felt in this music in vivo effort over decades. That urgency has helped push this project out of the hermetically sealed world of professional music-making into the anxiety-ridden public sphere.

For all the times my music colleagues and I have worried together about the trending topics of "environment" and "global warming," we have also shared a deep, yet barely conscious conviction that there is no *real* connection between the impersonal events of nature and the individual heroic act of pulling beautiful sounds from an instrument. The anxiety we carried with us that we and our beautiful work might not be separate from nature would linger, unspoken, following our conversations. It was unspoken because we lacked the confidence to affirm it and we were, like everyone else, intimidated by the alien complexity of the science we needed in order to grapple with what was happening to the natural system. Behind the daily headlines about the biosphere's progressive distress was a growing recognition that its cause was not simply the burning of fossil fuels or the excesses of consumption. Rather, the results of ten thousand years of human manipulation of nature were leaking into the would-be monastic setting of professional music-making. This was a crack in the

8 Ghosh, A. (2016).

9 In his novel *Overstory*, awarded the Pulitzer Prize for fiction in 2019, Richard Power sketches a 13.8-billion-year timeline of life on earth from the Big Bang to the present as if it were one day of twenty-four hours. Anatomically modern humans appear four seconds before midnight.

wall of the artistic cloister that our civilization had built around humans and therefore around artists.

Perhaps the present time is a Copernican moment for music. It just might be an opportunity for those of us in the music world, like the opportunity which Galileo faced in the seventeenth century. That was when the consensus within the Christian world about humanity's place in the universe began to erode. The conviction that humans were the goal of God's creation, and therefore of history, and that the planet holding human civilization was the gravitational center holding all the other planets and the sun itself in place was being challenged. Galileo, confirming what Nicolaus Copernicus had said a generation earlier, had used his own observations and mathematical calculations to confront the people of his time with an unimaginable morsel of self-awareness – that it was the sun at the center of our solar system, not the earth. What would a Copernican perspective on music look like?

2

What is music in vivo?

The hybrid term *music in vivo* in the title of this book is a mashup of terms from the fields of music and biology, reflecting a biology-centered approach to music. Music in vivo is a made-up term that attempts to locate a hypothetical region where musical practices and the workings of nature intersect. I am happy to call it a metaphor. Like all metaphors, music in vivo (like its adjectival equivalent, bio-musical) can be elusive and dependent on context, at times feeling more like poetry than like factual description. The following concise explanations of the term will require further elaboration in coming chapters. I offer them here as a guidebook to the terrain ahead.

Music in vivo is a perspective on musical experience derived from late twentieth- and early twenty-first-century biology and modelled on the study of living organisms.

The expression *in vivo* (in the living) is a standard term in science denoting the study of moment-by-moment changes in living organisms while they are alive and still connected to the ecology which sustains them. In vivo study is different from the study of living organisms or tissues either isolated in a test tube (*in vitro:* in glass) or after they are no longer living, as in the case of the butterfly, stilled in formaldehyde, pinned to a lab tray. Music in vivo pays special attention to music and living organisms as processes acting in real time. It foregrounds the inherent variability of musical events, regardless of whether they occurred at the moment we perceive them, in the distant past, or in the imagining of some future event.

Music in vivo is an ecological perspective on musical practice and experience.

In an open-ended spirit of experimentation modelled on biology, musical compositions will be approached here as participants in a complex interactive web of forces and agents which resembles an ecosystem. Each event in a sonata ecosystem will be like other sonatas, even though each one will also be unique in its details. Within the constraints of the sonata ecosystem other sonatas can be expected to emerge, as it were, effortlessly, while non-sonata traits will be excluded or inhibited. In current biology, this emergent behavior is referred to as self-organization. As in all natural ecosystems, in music the self-organizing behavior of individual musical events and the human habitat supporting them engage in continual co-creation – that is, they create, sustain and renew each other.

Music in vivo, based on biological models, attempts to envision musical processes as a continuity with natural processes.

Putting this another way, a re-orienting of musical thought to include biological benchmarks alongside artistic benchmarks is a single step toward the de-centering of individual human creativity within the creativity of the biome. By implication, the de-centering of individual human music-making is a step toward the de-centering of human civilization within the living systems of this planet.

Finally, music in vivo does not refer to a special class of repertoire, but is a proposal for a biology-oriented approach to any musical practice.

It is a perspective on any human sound art in any cultural or historical context, regardless of whether it is sounded in the air, silently remembered or imagined, created in performance or contemplated as a page of a score Music in vivo is intended to supplement, not replace, the conventional focus on unique sonic objects by individual creators.

Preparing a bio-musical workplace

The view that music is continuous with nature is where a musician's curiosity about biology can lead. I want to call this musical re-orientation to nature *bio-musical* (hyphenated), a term which distinguishes itself from the more common term *biomusical* (unhyphenated) which for much of the past 100 years has focused more narrowly on the relevance of neuroscience, psychology, anthropology and genetics to the evolution of music.[1] It is under the caption "biomusic" or "biomusical" that the songs of whales and birds, music in brain science, the music of plants and music as a shaping force of human evolution are studied in laboratories and in the field. These important bodies of research all participate in the ideas presented here, and will find their way from time to time into this story.

To limit the potentially unmanageable scope of what I am presenting, I have chosen not to give these areas of research pride of place. Instead, the hyphenated combination of *bio-* and *music* is meant to suggest less-plowed ground where the real time life-sustaining processes of cells and habitats and the real time processes of music intersect. In the attempt to describe the workings of living organisms and the making of music with the same language, commonplace words stressing action taking place in time will recur on every page – words like *event, practice, happening, occurrence* and *behavior. Musical behavior* and *musical event,* for example, are as acceptable in musical parlance as *natural behavior* and *natural event* are in biology. These words invite the reader to appreciate, as I proposed in Chapter 1, the normalcy of music, not just its normalcy in human society, buts its normalcy in nature itself. From here on, the hyphenated form will be used in place of the unhyphenated and will serve as a more self-explanatory synonym for music in vivo.

Part 1 of this book, "Finding music in biology," introduces four examples of organisms which serve to expose us without delay to the insights and challenges which can arise when we choose to view nature through a musician's eyes.

Part 2, "Finding biology in music," covers the same terrain as Part 1, but proceeds from music to biology and looks at musical cases for moments when we see behavior which reminds us of what is described in biology. Part 2 concludes with an examination of the great Gregorian chant project, the founding effort of the new science of musicology in the nineteenth century during the lifetime of Darwin, when the modern science of biology was emerging. By the end of Parts 1 and 2, my hope is

1 For the conventional (unhyphenated) use of the term biomusical see: Wallin, N.L. (1991); McDermott, J. and Hauser, M.D. (2001). "The Music of Nature and the Nature of Music," *Science*, January 2001, is a summary of presentations at the BioMusic Symposium held as part of the American Association for the Advancement of Science Annual Meeting (February 2000, Washington, DC); The BioMusic Program of National Musical Arts (NMA), which sees itself as "a unique conduit between art and science, as it seeks to examine music in all species and to explore and understand its powerful role in all living things."

that these ten chapters will have allowed the main lines of *Music in Vivo*'s story to unravel in a form readable in a few sittings.

Parts 3, 4 and 5 revisit and update the conversation between biology and music which began in earnest in the age of Darwin, the period when the developmental ideas embedded in the evolutionary tree of life were absorbed into thinking about the history of societies, cultures and the arts.

Part 3, "The self-organization of everyday life," features the key twenty-first-century biological concept of self-organization and its related concepts which are likely to strike musicians as oxymoronic. (Art that makes itself?). Self-organization provides a platform for exploring possible alternatives to current intellectual trends derived from biological determinism. Self-organization to the western art world's individualistic belief in the artist as the answer to any question about how music happens.

Part 4 begins with the macroevolutionary ideas of the nineteenth century which found their way into European musical thought. Two key ideas in these chapters are common metaphors that should be familiar to music-oriented readers: the image of biological evolution as a tree of life and the image of history as a single road leading from primitive to advanced along which all civilizations must pass. The burden of Part 4 is the gradual refashioning of these metaphors that has occurred over the past two hundred years in light of biological discovery.

Part 5, "Emergent order in two musical ecosystems," applies the accumulated experience of the preceding chapters to a detailed comparison of two song repertoires, one rooted in Istanbul and the other in Vienna and separated by 150 years.

Throughout the book, a modest lexicon of scientific terms, as well as a modest gallery of thinkers in the sciences, social sciences and the arts, will accumulate and will hopefully link up into a plausible outline of what I believe amounts to a global intellectual and social movement with biology at its core, a movement which far exceeds the scope of music. If this scope, projected against a backdrop of evolving life forms over billions of years, seems daunting to a music-oriented reader, I would urge patience. The very fact that the boundaries of discussion in this book are bigger than music can help us imagine fresh ways of thinking about where music belongs in the noisy, diverse and conflictual environment of our time. I believe that any disorientation which this discussion may provoke in music-oriented readers can, in the end, lead to an enhanced sense of belonging for both music and musicians. By belonging I do not mean the more familiar sense – belonging within the protected corner of culture conventionally reserved for the arts. Instead, I am referring to *music's belonging within the natural order.* As will be said repeatedly on these pages, the natural order as it has come to be described in the life sciences of the past fifty years has made it possible to talk meaningfully, as never before, about music's place in nature.

By beginning, as I do in Part 1, with a search for the signs of music in the behavior of fish, birds and photosynthetic life, requires a leap across a divide. Leaps are the business of art, so I am choosing to begin there, in the region of metaphor. If my own experience is any guide, it is likely to take time to find how the lexicon of melody, harmony and rhythm, of artist, composition and style might translate to the lexicon of species, phenotypes, metabolism and metastability. The metaphors of all kinds resorted to by scientists within their own disciplines and by non-scientists attempting to evoke nature in their descriptions of the arts, do not hold still. We should not be surprised to find that the usefulness of bio-musical images based on apparent likenesses can easily convert to something else – to conjectures, propositions or rules of thumb. In Parts 1 and 2, for example, the casual simile "a repertoire is like an ecosystem" in Parts 3 and 4 becomes a proposition – "let's say a repertoire *is* an ecosystem." From there, the next step is to take that proposition based on an image and set "ecosystem" to work as an experimental tool for exploring and ultimately teaching musical practices, as in the extended musical examples in Part 5. *Metaphor to heuristic to pedagogy*: this is one way to understand the spectrum of meaning through which bio-musical terms, including "music in vivo" itself, can be expected to move in what lies ahead.

Before there was an Adam there had to be an Eden

Musical phenomena are like natural phenomena – they occur in time. Regardless of the form in which we happen to encounter them, whether as auditory experience, or as imagined, remembered, notated or recorded experience, their meaning to us as music is not as an abstraction, but as something which is *happening now*. At each moment in a musical event, music emerges from what has taken place in the moment before, and no matter how well-known or conventional the event is, we don't experience it until it happens. It is always, to some degree, a surprise. Biological and musical artifacts are both sites where something is happening. The snapshot of the moment is also a freeze-frame from a movie, like the time-lapse photos of a growing plant in Figure 1.1. Music in vivo invites us to understand variation the way we understand a flame, not only as an object but as something which happens. Flame, the very phenomenon of combustion, is an example of a recurring metaphor habitually resorted to by scientists studying life forms. The motion of the flame and its flickering light are its proof of a dynamic process. The flame – the process of burning – resists being frozen in time, measured and broken down into parts or reverse-engineered, since once combustion has stopped, what is left is no longer the thing we set out to understand. This is a circumstance which I believe musicians and music lovers will recognize.

Fritjof Capra and Pier Luigi Luisi declared in their *Systems View of Life,* that the general direction of twenty-first-century science is toward *ecology* – toward the relationships among individual agents and phenomena in a system. In this view, it is not the individual organism which is "the basic ecological unit," rather, "it is the ecosystem."[2] This is also the conclusion of many researchers – both physicists and biologists – who are seeking the origins of life. The Santa Fe Institute physicist Eric Smith stated in 2017 that, contrary to the training in physics and biology he received, ecosystems are "more fundamental than organisms." In other words, to borrow a biblical image, the origin of human life is not in a presumptive Adam, a particular individual progenitor, but rather in the tangled garden itself. This Eden is the fecund habitat where, as developmental biologist Stuart A. Newman puts it, living tissues are *inevitable.*[3] Newman's claim (which will be discussed more fully in Chapter 14) is the kind of statement which has always created a zone of discomfort between science and religious faith, the same zone that Darwin had to navigate with extreme caution, only occasionally venturing into public or joining the polemics over evolution going on around him in which, for some, his name was anathema. If life, given the right conditions, turns out to be inevitable, then what is the role of The Master Gardener in Genesis? On the music side, the question is similarly risky: if music comes alive (in some sense) inevitably, if rhythms, melodies and even whole pieces (somehow) self-organize in the ears or throats or hands of humans, then what is the role of The Creative Executive, the artist?

Life: something edible, lovable or lethal

Switching continually between the morphologies of organisms and the morphologies of human sound art is illuminating, but in my experience, one of its possible side effects is vertigo. James Lovelock, a former NASA engineer, first proposed in the 1970s that the unique life-sustaining environment of earth over more than three billion years was actually

2 Capra, F. and Luisi, P.L. (2014), 342.
3 Newman, S.A. (2003b), 221.

created by the presence of life itself.[4] This is a vertiginous idea which not only updates but extends the old puzzle, "which came first, the chicken or the egg?" Lovelock captioned his proposition "Gaia" after the goddess of earth in Greek mythology and went on to develop and amplify it over the following decades through his collaboration with microbiologist Lynn Margulis. Now, fifty years later, in the current scientific view of life, for approximately 3.8 billion years our planet has been approaching, but never achieving, the state of a *perpetuum mobile*. It is a self-making, integrated and self-renewing system, a whole which is greater than the sum of its myriad parts. It falls short of being a perfect perpetual motion machine because it requires the constant input of energy from outside the machine – primarily from the sun. Lovelock's proposition is mind-bending but has become one of science's staple topics of debate, a still controversial oxymoron which is captured in the mind-bending image by the artist M.C. Escher of two human hands drawing each other (Figure 2.1). The hands are fully fleshed out and three dimensional, but they are scratched from the flat page as simple outlines by the hands as they are being drawn, as mysteriously as life was first self-created from the inert chemical substrate of the early earth. Chicken or egg, which came first, no longer feels like the point.

In 1988 Lovelock characterized the system of lifeforms on earth in a way which may be recognizable to musicians. I offer Lovelock's pithy summary below as an introductory contribution to an emerging bio-musical lexicon. Note that Lovelock's words carefully avoid insider science talk, and so they are more easily understood – at least, at this early stage of this story – as a string of metaphors. These kinds of statements which are equally ambiguous as metaphor or as science should not be dismissed as a layman's vulgarizations of true science. A surprising amount of talk among scientists takes the form of this kind of street level informal discourse. Science talk is not all about data and mathematics, any more than music talk among musicians is all about music theory. I believe these are the kinds of scientific statements which have a reasonable chance of being usefully translated to music and put to work.

Life, Lovelock said, can be expected to be "something edible, lovable, or lethal."[5] These are words which squeeze the trophic (feeding) cycles, the reproductive cycles and the energy cycles present in all regions of the biosphere into everyday concepts which barely hint at either biology or art. If, as I believe, it turns out that human sound art emerges from processes shared with bacterial colonies and tropical rain forests, then according to

4 Lovelock, J. and Epton, S. (1975).
5 Lovelock, J. (1988), 16.

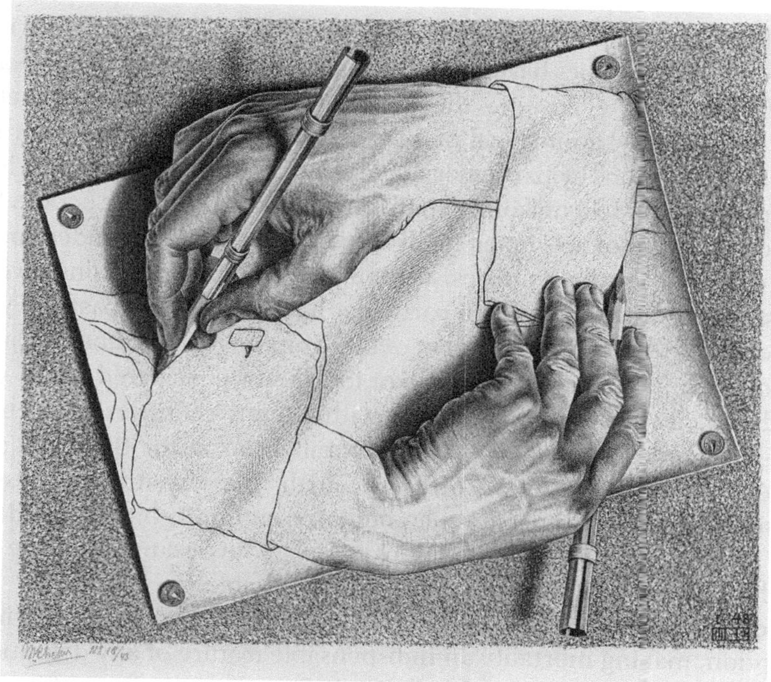

Figure 2.1 "Drawing Hands." (M.C. Escher) Lithograph (1948). M.C. Escher works © Cordon Art-Baarn-the Netherlands. Used with permission.

Lovelock, there should be no surprise that we find music delectable and enticing, an experience evoking both hunger and eros in every culture. The physicist Nick Lane refers to the second law of thermodynamics which describes the instability of energy flows through living systems as "the science of 'desire.'" The existence of atoms and molecules is dominated by 'attractions,' 'repulsions,' 'wants' and 'discharges,'" he says. "Molecules 'want' to lose or gain electrons; attract opposite charges; repulse similar charges; or cohabit with molecules of similar character."[6]

Taking Lovelock's point of view, the human "appetite" for music – as universal as the human need for food – could easily stand without the scare quotes marking the presence of a metaphorical leap. Music's potent and habitat-dependent attractiveness to humans qualifies as the kind of lawful behavior Lane describes in nature. Looking more closely at specific musical behaviors, what has traditionally struck European musicians since

6 Lane, N. (2009), 13–14.

the Middle Ages as the natural attractions (and repulsions) of pitches in the unfolding of harmonies have been expressed as musical laws of *consonance* and *dissonance*. The tension – the repulsive force – created by two melodic lines separated by an interval of a second or seventh was considered to require "resolution" by particular movements of one of the melodic line – either downward to a unison or sixth, or upward to a third or octave. Such musical patterns of action, as in all forms of life, are now understood to have evolved over the centuries, as for example in the increasing diversity of dissonance treatment within particular habitats over the past 150 years. In the Middle Ages, the movement of two or more melodic lines in parallel motion separated by an octave, fifth or fourth was considered a God-given standard, a form of natural law. But within a few centuries, the sense of God-givenness was being invested in parallelisms at the third and sixth, to the exclusion of the octave, fifth and fourth. The very attractiveness and digestibility of music by a particular human community have depended on such specific qualities as these. They are just a few examples of the ways by which music, like life itself, is edible and lovable.

Lovelock's term "lethal" reminds us that all living things die, without exception, making mortality an indispensable feature of any definition of life. Music is also experienced as living, in large part, because of its variability – the form a piece or gesture took a moment ago in a performance is extinguished in the variations of it, however small, which immediately follow. But we are glad of the precariousness of this circumstance because it is one of the surest proofs we have that music is alive. Any musical moment which somehow lacked this sense of constant change intuitively strikes us as mechanical and robotic – that is, lifeless. The continual revision of musical events in time parallels the continual replacement and adaptation of tissues in an organism. The vast majority of the cells making up the human being we see in the mirror, bearing our name, are recycled approximately every 24 hours, creating the gradual changes we all accept as signs of life.

We have always celebrated music for the way it provokes human emotions and makes human bodies move. These dynamic qualities combine with music's tendency toward variation to create music's inherently disruptive power, its ability to threaten stasis and equilibrium. What is more, in its edibility and lovability it is irresistible and therefore an attractive hazard for civilization, a potential threat. One proof of its destabilizing potency, provided by history, is that human religions and governments have so often felt a need to hold music in check and even at times to prohibit it. The puritanism of Savonarola's Florence in the fifteenth century, of Calvin's

Geneva in the seventeenth and of the reactions against rock 'n' roll in the United States in the twentieth come to mind.

Edible, lovable and lethal: the first of many attempts on these pages to translate the intimidating complexity of living systems into terms useful for thinking musically about biology.

This book asks: What if our human-centered view of music were replaced with a nature-centered view? What if we believed that the act of drawing a musical gesture from a string or from our vocal cords – even the act of silently imagining that gesture – was (in some sense yet to be defined) the same as the acts which enliven a bacteria colony or the moss on a rotting log for ten thousand times the duration of human civilization? And what if this belief were to become as assured as our unspoken assumption that things fall down, that water is wet and that fire burns? What if the music world believed that any discussion of music which lacked any reference to nature was incomplete? Could we get over the shock of accepting this direct continuity of music with nature, its *normalcy*? Would we then be able to allow that new musical awareness to play a part in the daily work of music-making and even in the civilizational course correction that we now need?

If we were to embrace the normalcy of music in human life along with the one-of-a-kind exceptionalism of its products, could music still be precious, though normal, as self-perpetuating and as reliable as a field of dandelions, a human embryo, or a beehive?

Part One
Finding music in biology

Life is a property of planets rather than of individual organisms.
Harold Morowitz[1]

Imagine that all the creativity of invisible microbes and of bulky plants and animals – including the creativity of humans – gets done in a common well of activity. Then imagine that no living creature in the well operates free of the well's tangled mutuality. The intertwined pathways of energy exchange and signaling in the well would resemble a maze operating on a planetary scale which is both larger and smaller than the vocabularies of either science or music could encompass. In this book, the main difference between the music of biology in Part 1 and the biology of music in Part 2 is where they allow us to enter that maze.

In the following five chapters, the entry point into the maze is a small selection of just four species which invite us to consider whether we find signs of musical experience in them. The selection is admittedly random and personal, as random and personal as the examples of global repertoires chosen for discussion throughout this book. Over years of exploring biology as a layman, these four families of organisms happened to be the ones which most compellingly embodied such alien concepts as epigenesis, symbiosis, self-organization and metastability. These alien terms percolated to the surface in the stories of the finch, starling, stickleback and lichen and gradually persuaded me that they deserved a place in my musical lexicon.

1 Morowitz, H.J. (2004), 8.

3

Finches on adaptive radiation

In 1839, following his five-year globe-encircling voyage aboard the "Beagle," and twenty years before the publication of his famous *On the Origin of Species,* Charles Darwin devoted part of his published report on the voyage to a comparison of the subtle variations in the beaks of a particular bird species.[1] His conclusions were based on his study of the finch specimens he had collected on the Galapagos Islands off the coast of Peru and on the Cocos islands in the Indian Ocean. This finch study helped to establish the unknown naturalist as an important new voice in science. It also provided lasting insights into avian history, and even the history of life itself, a trial run on his yet unarticulated theory of natural selection. At this early stage of his thinking, Darwin hazarded a novel theory: different food sources available to finches in different habitats – seeds and insects, leaves and grasses, as well as different nest-building materials – were a factor in the evolutionary changes in the form of the finch's beak over centuries. Darwin argued that if a finch with a seed-cracking beak migrates to a region where insects are the most plentiful source of nutrition an adaptation to an insect-grabbing beak would enhance its survival. With time, this adaptation could be passed to finch offspring because the birds with the adaptation were more likely to survive and reproduce. Later generations of biologists called Darwin's 1839 finch theory *adaptive radiation,* and it can now be seen as one piece of the puzzle that would eventually become the central argument of his later book, the full title of which is *The Origin of Species by Means of Natural Selection* (Figure 3.1).

The theory of natural selection states that any new trait arising in a species (note that in Darwin's view, new traits arise mostly randomly

1 *Journal and Remarks* (1839), published in 1905 under the title by which it is better known, *Voyage of the 'Beagle.'*

Figure 3.1 The Finch beak in adaptive radiation, as collected by Charles Darwin during his trip around the world from 1831 to 1836. Darwin, C. (1845).

and by chance) will be perpetuated if it enhances the survival of that individual. Having survived, the individual will have a chance to pass on that trait to its progeny. Both natural selection and adaptive radiation would later be supplemented and amended by the understanding of inheritance through the genes which became standard knowledge in the generations after Darwin.[2] These finches also provided one of my first encounters with a biological phenomenon where I immediately felt that music and nature might have more in common than my musical education had led me to believe.

What happens to music in new habitats?

Keeping our musical sensibilities intact as we observe nature through Darwin's eyes, this famous finch story offers something immediately

2 While the principle of adaptive radiation remains a staple of modern biology, discussion continues about the details in Darwin's data, including the exact taxonomy of Darwin's finches. See Abzhanov, A. (2010).

familiar, as if it were a piece of the history of European music as it is often told. The material culture of music – the evolution of musical instruments, for example – comes most easily to mind as the equivalent of differently-shaped animal structures such as a bird's beak. In a violin, as in a seed-adapted beak, a vital function adapts to local materials and conditions. For the bird, the vital function is the ability to feed on particular food sources, and for the musician, the ability to evoke a pleasing sound from an inanimate object. In nature, a feeding structure capable of cracking seeds is less suitable for pecking insects from a tree trunk or building a nest from local grasses. In the case of musical instruments built on a common principle, there is the example of the very large groups of so-called *bowed chordophones,*[3] consisting of one or more strings stretched tightly over a sound box and activated by a bow, usually made of animal hair. In one region, the sound box is fashioned of thin strips of wood, and in another region, from the hollowed burl of a tree or a tortoise shell covered with animal hide. Each of these adaptations of the box and string structure must adapt to the materials, conditions and social preferences in their habitat. In this way, instruments fit comfortably as examples of the material model of adaptive radiation in music.

What the examples of musical instruments do not address are those aspects of sound art in which musicians tend to be most deeply invested, that is, in the practices that come to inhabit the box and bow. The diverse acts of conceiving, remembering, hearing, appreciating and teaching the music emanating from the box are as much immaterial as material. Once the finch model is pursued further into these mental realms, we discover other ways that adaptive radiation is already at home in musical thought. Consider the many regional and cultural transformations of the familiar triadic structure of European polyphonic music during the period between the late fifteenth century and the present. It is commonplace in music histories to refer to such diverse forms as the aria, concerto, symphony, trio sonata, dance suite, popular song and requiem mass as variants of a centuries-old polyphonic paradigm in which stacked thirds and fifths are the controlling features of the musical texture. In the expanding universe of popular music spearheaded by jazz in the early twentieth century, musicians built an array of repertoires on many of the same chord structures as did composers of cantatas and operas two centuries before. Among music scholars, it would not be unusual to acknowledge these developments as responses to the differing demands and resources

3 In its modern usage, the Hornbostel-Sachs system consists of five classes of instruments: *idiophones, membranophones, chordophones, aerophones and electrophone.*

of stage, court, tavern and church, although it would be less usual if the biologist's vocabulary of habitat, ecology and adaptation were used as musical descriptions.

Like the finch's beak which attracted Darwin's attention during his travels, the demands of habitat shape how a common musical structure is expressed within that habitat. That shaping process is not optional. In music, as in nature, the cost of unsuccessful adaptation to local conditions is failure to thrive and ultimately, extinction. The rightness and normalcy of the triadic texture can seem to those operating within those conditions to be natural, as natural as the green color of vegetation in spring, an expression of the natural order. The musicologist Rob Wegman pointed out in 2014 that the matrix of thirds and fifths underpinning music in the European Renaissance was the only cultural expression besides Gregorian chant which qualified at the time as music. Any other music, he said, "had no claim to being music at all."[4]

In its simplest form, the model which evolved as the common framework of all these triad variants was transmitted between the seventeenth and nineteenth century by music teachers as a figured bass line with melody, the so-called *basso continuo.* A complex polyphony-generating practice, centuries in the making, could be triggered in the mind of the trained musician with just a few standardized jottings of numbers under a notated bass line. Though basso continuo or *figured bass* now tends to be read and taught to students as essentially blocks of sound, this written shorthand was originally a tool for disciplining the dynamic interplay of separate moving voices. The numbers 1, 3, and 5, standing for the intervals of the unison, third and fifth above the bass, were so established as a norm by the end of the sixteenth century that they were commonly not even written, but presumed. In most cases, a single digit – a 2, 4 or 7, for instance – might be all that was needed to suggest, not just the particular triad intended, but its exceptions and complications. The presumed normalcy of the triadic structure, together with the composer's bass line and melody line, left to the seventeenth- and eighteenth-century keyboardist or lutenist to realize the particular forms of the triad which the composer had chosen, but in a wide variety of ways, adapted to the moment.

Over more than two hundred years, this efficient model embedded itself in the hands, voices, ears and imaginations of countless individual composers and performers. In Figure 3.2 below, the notation in Mozart's own hand of the "Dies Irae" movement of his unfinished *Requiem Mass,*

4 Wegman, R.C. (2014), 15.

K. 626 from 1791 illustrates how the pedagogical model of the basso continuo could become second nature in the composition of a complex structure of eleven staves of independent vocal and instrumental parts (Figure 3.2). The numbers written below the bass line on the bottom staff (indicated with boxes) show that the composer could quickly sketch in the desired harmonies above before filling in the moving parts on each staff, even at a later time.

Long before Mozart, this model – itself an adaptation of the layered consonances based on the interval of the third and fifth created by countless vocalists and instrumentalists in the Renaissance – had by the seventeenth century become a default practice, taking the typical written form in Figure 3.2. It was as unconscious, "natural" and "God-given" as the default feeding structure of finches found on islands as remote from each other as the Galapagos and Cocos, in so many more variants than the four general types shown in Figure 3.1. Biologists, confronted with a multi-century phenomenon like basso continuo would be inclined to

Figure 3.2 The first page of the "Dies Irae" movement of the (unfinished) Requiem Mass in D minor (K. 626, 1791) in Mozart's handwriting, showing eleven staves for orchestra and chorus. (See figured bass in boxes at the bottom of the page.) Austrian National Library, Codex 17561a, folio 1 (recto). *Wikimedia*.

say that what might have begun as a unique solution to a local need had gradually hardened over many repetitions and over many generations through natural selection. As will be taken up in more detail later, the so-called "soft matter" of multi-cellular organisms in very early stages of evolution gave way over millions of years to standardized body types and organs – for example, the membranes and organelles of plant and animal cells, the jointed limbs of crustaceans and vertebrates, and the eyes of a many types and sizes – which could provide stable, proven, repeatable structures on which later adaptations could be built.

Teaching musical reflexes

To test how a biologist's view of life forms might be adapted to the development of Europe's polyphonic musical practice, it helps to unburden that practice of a strictly written definition. This is not hard to do. Basso continuo was a written process, a technique that was spelled out in treatises and teaching manuals. But its written form is also a record of the music-making habits of countless individual musicians. From its beginnings in the middle ages, the making of multi-voiced music was a singer's art which began as singing "over the book" *(cantare super librum)* – inventing a melody to accompany a notated chant. In the European Renaissance, cantare super librum evolved into what came to be called in the following centuries "counterpoint." The musicologist Phillippe Canguillem tells us how musicians auditioned for positions as music director and head teacher for large cathedrals in Spain and Italy in the sixteenth and seventeenth centuries. The audition would usually include rigorous examinations in a range of written skills and practical skills in cantare super librum. In some cases, candidates were even expected to improvise a melody to accompany a sung chant melody while at the same time indicating for another singer the pitches of a third melody through conventional hand signs.[5]

Unlike the textbook descriptions of counterpoint in western art music since the nineteenth century, studies of oral tradition practices in our own era depict multipart singing in social terms. Jean-Jacques Castéret describes twentieth-century polyphonic folk singing in the Pyrénées as a combination of individual and communal impulses – "a dialectic between horizontal and vertical musical thought." connecting person to person. "An entire society and its cultural values," he says, "are expressed throughout

5 Canguilhem, P. (2011), 98.

the construction process."[6] Viewing polyphony in all these instances as a form of human behavior, as social interaction, draws it closer to the ecological perspective of the finch.

In the widespread eighteenth-century pedagogical method of *partimento,* the teaching of basso continuo relied heavily on imitation of concise written models of multi-voiced textures which singers and instrumentalists were expected to master and then freely adapt.[7] They were, as notated musical schemata, potential compositions, indeterminate in all but their basic structure. The partimento pedagogy developed in music conservatories in eighteenth-century Naples became the required early keyboard regimen of most German, French, and eventually English composers of opera, symphony and sonata well into the nineteenth century. In 2012, Giovanni Sanguinetti described partimento as widespread but also hidden from view to all but the musical practitioners themselves, for whom it was "an esoteric doctrine," an "oral tradition [by a] circumscribed circle of initiates." The training began at the keyboard as early as the age of twelve and was inseparable from direct imitation of a master teacher.[8]

The so-called *rule of the octave* in Music Example 3.1a illustrates one of the most basic partimento exercises in which students were given an ascending or descending scale in any key in the left hand and a set of model harmonies in the right. The famous *Goldberg Variations* by J.S. Bach published in 1741 was called *Klavierübung* or "keyboard exercise" on its title page, and each of the thirty variations of the opening aria (Music Example 3.1b) were given a title describing the contrapuntal technique being applied to the aria's bass line. Variation 7, for example, was a quick dance in triple meter, while Variation 9 was canon at the third.

From a present-day perspective, it would be easy to treat these chordal models in partimento collections by famous teachers like Fedele Fenaroli (1730–1818) and Giovanni Paisiello (1740–1816) as if they were hard, pre-formed blocks of sound, like Lego pieces. But they are actually more than static handfuls of intervals dominated by thirds and fifths. For all that partimento had become focused on instrumentalists, and especially on keyboard players, we should not forget that partimento grew out of the contrapuntal discipline of singers of preceding centuries, for whom polyphony was not composed of chords but independent melodic lines

6 Castéret, J.-J. (2013), 58. See also ICTU Study Group for links to sound files, symposia and study groups of living multipart music.

7 For fuller accounts of the singer-centered view of contrapuntal science, see: Labaree, R. (2023), 304; Wegman, R.C. (2014); Cumming, J.E. (2013); Canguilhem, P. (2011).

8 Sanguinetti, G. (2012), 10.

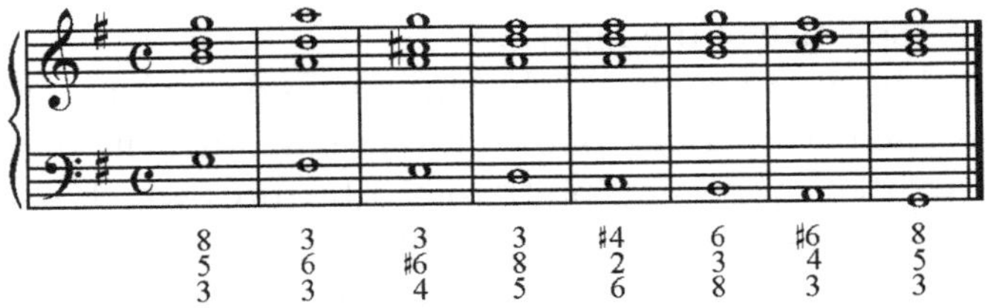

Music Example 3.1a The "rule of the octave," a basic paradigm in the partimento pedagogy of eighteenth century Italian conservatories which provided models to be mastered and then varied by converting the limited model of chords to an infinite array of melodies, contrapuntal lines and accompaniment patterns full of moving parts. Sanguinetti, G. (2012), 121.

Music Example 3.1b The first four measures of the *Goldberg Variations* (BWV 998, 1741) by J.S. Bach illustrate how the so-called "rule of the octave" in Italian partimento collections could be employed in a formal composition, a 32 measure aria, followed by 30 variations of the same bass line.

which exhibited dynamic tendencies of attraction and repulsion, of activity and rest. Looking closely at the actions of individual musicians and birds over many generations, adaptive radiation and the realization of partimento models are both behaviors – actions occurring in the cells of a bird's body and in the mind and muscle of a human being. In both cases, adaptation of the materials of the moment is not optional, but is a systemic demand on species and musical practices according to the rules of their respective habitats. They are both operations which appear very different from each other within their respective ecosystems – in music, they are recognizable as habit and reflex, and in nature, they are recognizable as bio-chemical pathways. But both finch and created what they needed by processes which they inherited. In Mozart's case, the inheritance was transmitted through long hours of exercises in partimento as a child under the watchful eye of his father Leopold. Bird and musician were co-evolving with ecologies they did not create. Neither fish nor musician created a feeding structure or a requiem mass movement out of whole cloth, as the God of Genesis is depicted fashioning Adam out of clay.

We have never been individuals

In the life sciences, variations in phenomena are more often than not the main event. What researcher in biology would engage in a study of atmospheric carbon dioxide, or a bacterial culture, or a chicken embryo which presumed that they were not in a constant state of change? What is more, biologists do not tend to put much energy into locating, as is so often the habit in music, the responsibility for variation in a particular agent – as if the changes were all the doing of a particular carbon dioxide molecule, or a particular bacteria, or a particular chicken. In music, we commonly treat an individual example of a musical phenomenon as if it were a thing – a distinct piece, a work, a performance, a recording, a notation. When we rehearse for a performance, we tend to shut everything else out and focus on getting it right. *Practice* – repetition of whatever we planned to present – *makes perfect.* "Play it again" expresses what the listener hears, as well. It is a *thing* we are experiencing at the moment, and if it is pleasing, it is worth repeating. These are the kinds of musical moments when we will tend to emphasize the uniqueness and replicability of musical art. At such moments, musical processes and musical behavior can take a place at the margins of our attention while we wrestle this beautiful *thing* into its optimal form.

And yet, adaptive radiation is a story which is already familiar in music, re-told in countless music history textbooks, program notes, concert reviews and professional conversations. The many historical accounts of how the basso continuo model has been adapted to the demands of different habitats over centuries, are examples of the way that the adaptive radiation of musical traditions is transmitted through anecdote and history. This kind of contextual thinking comes as naturally to musicians as to biologists, but the differences between the two professions are instructive and go beyond the degree of objectness we happen to be feeling about this or that musical moment. The stories we tell about music tend to foreground individuals, not systems. In practice, systemic behavior in music is acknowledged among music historians and theorists as a kind of insider knowledge, while in everyday teaching and rehearsal, or in communications with lay audiences, the systemic perspective is more often than not relegated to the background, leaving the unique variants contributed by individual musicians as the center of attention. The difference between the seventeenth- and eighteenth-century German and Italian Baroque versions of the basso continuo model, for example, are commonly epitomized in the works of giants like J.S. Bach in Germany and his contemporary Antonio Vivaldi in Italy. Students and audiences tend to be left with the message that the individual artist, and the accumulated effect of many such artists, is most of what needs to be known to understand the two regional practices, as if any variability in Baroque music-making was simply a result of individual authorial choice. "Historical style" and "tradition" becomes the colorful scenery against which the high drama of individual artistic achievement can be projected.

In 2016, Scott F. Gilbert, a developmental biologist, and Alfred I. Tauber, an immunologist, described biology as it is now viewed by biologists in a way which may be startling to musical sensibilities and will therefore require revisiting in future chapters. It can alert us to what the study of life forms might contribute in the long run to the study of musical forms: "We have never been individuals." During the eighteenth and nineteenth centuries, they said, the rise of biology in the modern period paralleled the rise of the idea of the individual and for a time "biology simply followed suit." Individuality, is "a remnant of a lapsed semantics."[9] For the biologist, even the study of plant and animal morphologies from specimens which are no longer living, and therefore no longer changing, is usually approached comparatively. Those separate specimens, like the array of time-lapse photographs in Figure 1.1 pull the observer away from an easy fixation on stand-alone objects toward their continuities.

Darwin's finches, the young naturalist's first serious foray into the question of how species change, are examples of what we can find when we turn our musical attention to the beak of a bird.

9 Gilbert, S.F. and Tauber, A.I. (2016), 845.

4

Starlings on co-creation

Turning now from bird beaks to bird behavior, consider a different and more radical bio-musical model, an example of how it is possible to look directly at nature with a musician's eyes and find musical principles acting in it: an English folk song collector's descriptions more than a century ago of *murmuration,* the swarming behavior of starlings.

The behavior of birds and melodies

Cecil Sharp was born in 1859, the year of the publication of Darwin's *Origin of Species,* so it is not surprising that 48 years later his book *English Folksong: Some Conclusions* would exhale a bit of the Darwinian air he had been breathing during his entire life. His chapter entitled "Evolution" opens with the claim that folk songs evolve in culture through "continuity, variation and selection."[1] He made no mention of Darwin or even biology, nor was this necessary, it seems. In 1907 he had reason to expect that his readers would not have been puzzled by the reference. His claim leaves us today to wonder if these same ideas based in biology might also apply, not just to anonymously composed Anglophone folk music, but to music in general.

By the time he wrote his book, Sharp had been researching and notating the performances of singers in peasant communities and writing about them for a decade, first in Australia, but then in Somerset in England. He was recognized in the English-speaking world as a major authority on folk music. In a few years, he would tour the mountains of Kentucky and North Carolina with the American collector Maud Karpeles to eventually produce a similar volume of tunes, *English folk songs from the southern*

1 Sharp, C. (1907), 21–41.

Figure 4.1 Starling murmuration, Shell Bay, Studland. Original photo in color. Copyright Tanya Hart, Creative Commons License.

Appalachians (1932), where he found that "the people are just English of the late eighteenth or early nineteenth century...and their manners are old-fashioned English."[2] Finding the qualities of one group of humans reflected in another group of humans at a different time and location is commonplace in the social sciences.

At the time, Sharp was contending with two different scholarly trends that purported to explain how folk music was made, one trend favoring individual creators and one favoring "communal creation." These were concepts rooted in the work of previous generations of folksong scholars and enthusiasts. In 1853, for example, the folk song collector William Stenhouse accounted for any especially worthy specimens he found in various villages in the English countryside by proposing that they were the work of anonymous creators resembling the composer of art music: "Some early minstrel, or musical shepherd, composed the simple air; which being picked up by the learned musician, took the improved form it bears..."[3] At first glance, this individualistic interpretation also appears to find expression in Cecil Sharp's outspoken contempt for the scholarly trend

2 Yates, M. (1979), xi.
3 Stenhouse, W. (1853), cited in Shapiro, A.D. (1975), 13.

favoring communal creation. "Corporate action has originated nothing and can originate nothing," he said dismissively. "Communal composition is unthinkable." But then he went on to say:

> There are few things in nature more wonderful and more incomprehensible than the ordered flight of a flock of starlings. Many thousands of these birds will fly together in a compact mass; they wheel about in the air…with a precision which argues complete unanimity of purpose. If attention be concentrated upon the bounding lines of the moving and living mass, it will be noticed that…[t]he edges, instead of being smooth and even, are rough and jagged. Further observation will show that these irregularities are due to the aberrations of flight on the part of individual birds, who are constantly separating themselves from their fellows, darting out at acute angles to the line of flight, and then swiftly returning to the flock. Every now and again, however, it will be seen that one of these birds is followed by the rest, and the course of flight of the whole mass is immediately changed.[4]

Stepping away from the communal creation theory and from the model of the art music composer, Sharp saw in starling murmuration something of both, and he did this with a musical reference. The birds' swarming behavior, he concluded, was a faithful picture "of the way in which a folk song is evolved. Of the innumerable changes made by individual singers, only those that win general approval are perpetuated; the rest, being ignored, pass into oblivion." Especially interesting is Sharp's claim that each starling's variations, and therefore the collective murmuration phenomenon itself, are less signs of intentionality than of accident or reflex – for example, the result of this or that bird's peculiar "waywardness" or "search for food." He likens this to song-making in peasant communities in a way which suggests a potential disruption of the special intentionality usually attributed to art, and the special importance of the individual artist: "the changes which singers introduce into the words or melodies of their songs proceed from many causes – forgetfulness, chance, accident and what not; but very rarely, if ever, from a definite and conscious desire to improve." Then, in the next paragraph following his emphasis on creation which is unintentional he articulates a view of folk song as a systemic interaction of individual and group. His summary has a distinctly Darwinian sound: "The individual, then, invents; the community selects."[5] He goes on from there:

4 Sharp, C. (1907), 39.
5 Ibid., 21–41.

> Every line, every word of the ballad sprang in the first instance from the head of some individual, reciter, minstrel, or peasant; just as every note, every phrase of a folk-tune proceeded originally from the mouth of a solitary singer...The community plays a part, it is true, but it is at a later stage after and not before the individual has done his work and manufactured the material.[6]

The artist and the creatorless fecundity of nature

The relationship which Sharp saw between the behavior he observed in folk song performance and the behavior of birds he observed in the evening sky suggests that there may be larger principles at work running through, not just biological and musical thought, but biological and musical phenomena. At first, we may find a musical perspective more compatible with Sharp's observations about music and nature when we consider his words simply to be an extended simile – song-making is *like* starling murmuration – a statement no different from the poet's "O my love is like a red, red rose,"[7] Sharp's comments on murmuration did not get more detailed than those quoted above. More importantly, they appeared in a chapter entitled "Evolution," so whether we choose to interpret his words as poetic metaphor or as objective description, we would also have to take into account ambiguities of the word "evolution" itself. This is a term which served in Sharp's time (as in our own) as both a specific term of science and as a general synonym for the changes or developments which occur over time.

Whether we take his words as simile or as science, Sharp's comments suggest that music and nature do not merely *resemble* each other in some general way – for example, in the way the four movements of Antonio Vivaldi's "The Four Seasons," composed in 1743, are recognizable musical representations of the contrasting emotional qualities of each of those times of year. Instead, Sharp simply proclaimed that a starling swarm and a folk song *develop* in the same way. He draws our attention, not to how biology and music *appear* to be similar or how they register on human emotions similarly, but to the possibility that *nature and music depend on the same processes.* Sharp was no scientist, though he spoke of needing to do his work according to scientific standards. He was known in his own day as the father of the English folk song revival, and his views

6 Ibid., 31.

7 A traditional Scottish song made famous by Robert Burns. The first quatrain reads: My luve is like a red, red rose / That's newly sprung in June / O my luve is like the melody / that's sweetly played in tune."

of culture were sometimes colored by the class and race prejudices of his time.[8] But in his fascination with the mysterious creativity of oral tradition among ordinary people, he intuitively saw both song-making and natural processes as *activities*. This is perhaps his most disruptive legacy – his audacious suggestion that natural creation and human creation operate by the same principles. His observations are scientifically naïve and non-specific, and they suggest more than they argue. But the principles he alludes to are revealed concretely in observable patterns of behavior among both birds and humans. He is presenting a view of a local microcosm which is consistent with Darwin's far grander view that laws were responsible for the richness of nature's tangled bank over millions of years of evolution.

Sharp's suggestion that nature and music are continuous in this way, inadequate as it is to current natural science and musical science, nevertheless provides us with a layman's model for how a musical perspective can provide legitimate insight into nature. He leaves us to ponder the possibility that *music as music and nature as nature do overlap*. What if Sharp's intuition is consistent with current science? If this is the case, how and to what extent might the celebrated, individual will of the artist be maintained in a separate domain from the creator-less fecundity of nature?

The behavior of birds and "boids"

Sharp's starling example is one which many life scientists today would assign to the general category of *self-organization*, a concept which will recur throughout this book. The concept may be expressed simply as the tendency of complex systems to spontaneously create order.[9] In the case of the starling, the components of the system are some 500 to 5000 birds moving in a densely packed flock in the early evening sky before settling down for the night. We will return to self-organization in its many manifestations in Part Three, where more will be said about how the concept is now at the center of controversies in science over the origins of life itself. The non-directedness of cell division, of the human immune system, of mangrove forests and of colonies of bacteria will also offer further examples. But for now, it will be enough to simply pay close attention to the self-organization which Sharp, the scholar of song, saw in the starling's swarming behavior. It is behavior which suggested to him the

8 See: Bearman, C.J. (2000); Boyes, G. (2010).

9 This simple definition of self-organization is based on Shultz, T.R. (2001).

mysterious way songs are born, evolve and die. Self-organization was not in his vocabulary, nor in the vocabulary of his contemporaries. But what he observed convinced him that murmuration did not happen by executive agency – by a lead bird, for example – but by interactive behavior which he felt was a as mysteriously self-governing as the making of folk song.[10]

Looking back from our position more than a century after Sharp, the behavior he saw in the sky was convincing as music and is now also convincing as science. Generations after Sharp, in the age of technologies unknown to earlier scholars, studies[11] of starling murmuration have harnessed computer animation tools to create virtual birds, or "boids." These digital entities were made to mimic starling choreography by programming into each boid just three basic rules: *avoid collisions, match velocity, and stay close.* Researcher-programmers endowed each boid with skills which each bird must bring to the ballet, including the ability to respond to gravity, velocity and wind drag, the ability to turn on a dime in any of three dimensions, and sensitivity to the locations and direction of the other boids in the flock. These requirements greatly taxed the capacities of their computer programs.

These were just the programmers' minimal selection of what they considered the most important of the requirements placed on each bird in the aerial ecology where the murmuration plays out. Individual birds which do not conform to these requirements wander out of the swirling pattern – out of the reach of the aerial habitat's requirements – into isolation from the flock. An isolated starling loses the protection of the group from falcons and severs its connection to the pool of potential mates. When Sharp observed the starlings at work doing what folk singers do, he explained how coherence could emerge from such feverish variability by referring to "community." Ecology and habitat, like self-organization, were not in his vocabulary. His summary drew, consciously or unconsciously, from the natural selection theory of the biological sciences of his time: "the individual invents and the community selects."

This turns out to be not far from the conclusions of the computer animators in our own day. They found that, in order to produce a convincing murmuration, they had to create a field of pseudo-starlings, each of them capable of inventing its own unique path through the sky by locating itself within the path of the group. The gamer-scientists ran

10 An alternate theory for the swarming of starlings conforms more closely to the logic of straight-line causation: predators, mainly falcons, are the choreographers of murmuration. See Hunt, G. (2013).

11 Young, G.F. et al. (2013); Reynolds, C.W. (1987).

their codes and studied the behavior of their on-screen avatars. Here is my own paraphrase of what they took home from their examination of the complex phenomena they were observing: *Boid flocks behaved most like bird flocks when each boid was able to relate to at least six or seven of the boids nearest them. The whole group is shaped instantaneously and continuously by thousands of individual choices, while at the same time the group is shaping each of those separate choices.*

Variants of something – but of what?

It was the dazzling variability of starling flight which, more than a century ago, suggested to Cecil Sharp the creation and transmission of folksong. Still photographs of murmuration would have shown, then as now, gorgeous painterly images like Figure 4.1 which would appeal to human standards of beauty. But multiple photos taken even once every few seconds would have immediately revealed that the images were more than separate pictures suitable for framing. All that furious movement of thousands of individual birds, with no apparent leader, somehow coheres into a graceful and shapely movement that at one moment suggests to human onlookers the strokes of a paintbrush and at another, the perfect synchrony of a ballet. Sharp noted the impression of unanimity in the birds' swarming patterns, but he also kept his eye on the infinite variability of their idiosyncratic flight paths. Variants beyond counting were subsumed into the murmuration's collective oneness. From a musical perspective, it was as if each bird was singing the same melody, but each in its own way.

Earlier generations of folk song collectors had also observed that there was a generous spectrum of variability among versions of what they considered to be the same song. The eighteenth-century philosopher and inventor of the term "folksong," Johann Gottfried von Herder (1744–1803), embraced the tendency toward variability as the antithesis of the products of his own "paper civilization." One of the folk song's chief attributes, he declared, was its lack of fixity: "all folksongs on this earth are altered; no province sings its own without variation."[12] The folk song scholar Samuel Bayard echoed this sentiment in 1939 when he extolled "the recreative individualism of singers."[13] However, for many collectors, this variability was an annoyance to be expected of unlettered farmers in the countryside, implying that literate musicians performing art music would never play so

12 Herder, quoted in Wiora, W. (1966), 26.
13 Bayard, S. (1939), 124.

fast and loose with their material. The problem, these collectors declared, was best disposed of by choosing for their published collections the one variant they considered the best, the oldest or most popular, and then ignoring the rest. For example, William Chappell (1809–1888), the English folk song collector and music publisher, included examples of tune variants in some of his more than a dozen published collections. But we see in his work some of Stenhouse's standard of the art music composer. Speaking as a music publisher, Chappell declared that examples of song variants were included in his publications as demonstrations of "the mode in which airs become corrupted and altered by untutored singers."[14] By 1810, Thomas Morley had already described the simplicity and purity which was assumed to be the natural state of folk song and which it was the collector's duty to restore: "...the chief corruptions of which we have to complain arise from the unskilled performance of our own traditional musicians, from whom, too frequently, the airs are noted down, encumbered by their tasteless decorations, and responsible for all their ignorant anomalies... the most delicate and difficult duty of a compiler is to endeavor, as much as possible...to restore the regularity of its form, and the chaste simplicity of its character."[15]

For Sharp, we should remember, the variability of folk song transmission was "rarely, if ever, from a definite and conscious desire to improve" and more likely to be from "forgetfulness, chance, accident and what not." In the 1980s, the ethnomusicologist James Cowdery proposed a correction to Sharp's view by offering proof that both chance and conscious variation were factors in the making of folk songs. In a conversation about melodic variants with the famous Conamara baritone Joe Heaney (1919–1984), Cowdery asked him about the kinds of variants he might apply to a single line of a song. Heaney replied:

> ...the old people used to say about songs: 'you take a song; you divide it into parts.' And each verse of a song varies, you know. You put the emphasis on the main parts of the line. 'Going to Mass', that's the main part of that line... you've got to decorate that particular spot – it means a lot to you, you want to hold on to it as long as you can, but you can't forever, so you decorate and go on to the next line...[16]

In a moment rare in folk song collecting, Heaney then, responding to Cowdery's request, came up with no less than eight different versions of

14 Quoted in Shapiro, A.D. (1975), 26.
15 Ibid., 27.
16 Quoted in Cowdery, J.R. (1990), 43.

a single line in the song "Going to mass last Sunday." His answer, recorded and transcribed by Cowdery, (Music Example 4.1), placed him more in the camp of Bayard and Herder than Sharp.

The Scottish scholar Gavin Greig, writing in 1906, took the question of folk song variants a step further. The variants, he said, offered clues to

Music Example 4.1 Joe Heaney's eight versions of one line from "Going to mass last Sunday" Based on the transcription in Cowdery, J.R. (1990).

a creative principle which was common to all the many versions, though usually the principle remained elusive. "These endless variants," Greig said, "are apt to become the despair of the conscientious collector."

> He cannot take one version and say he has got the tune, so long as there are other versions in vogue. He must take all he can get. But then where is the tune? What is it? He finds that he must revise his definitions. A folk-tune is certainly no individual entity. It is hardly even a generic conception. After much comparative study of his brood of variants he may think he has discovered something common to all. He can trace a framework on which they are all built. It is a skeleton affair at best, rickety and evasive, but it is something and he is thankful for it – thankful until he comes across another variant which knocks a plank out of his poor structure...In despair he falls back on a kind

> of intuition, through the exercise of which he feels that his ruck of records are all variants of *something* which is at once the air and in it.[17]

Looking beyond what seem to be variants of the same tune within a particular English county, other scholars discovered striking similarities among tunes with different texts which were otherwise unrelated by region, time or culture, as in the three continental examples of what the musicologist Walter Wiora called a shared *tune type,* which he named after a well-known English ballade "Lord Randall," which they resembled (Music Example 4.2).

Music Example 4.2 Three different songs of the "Lord Randall" tune-type, according to Walter Wiora. W. Wiora (1966), quoted in Shapiro, v. II, (1975), 50.

Chronologically, the items shown here in Wiora's tune type span six centuries. Geographically, they extend from France to Hungary, and in genre they include love songs, hymns and village songs. It is a spread which may seem more out of place in music than in biology, where for example, the mammalian type in Linneaus's scheme includes animals as distinct and diverse in habitat, behavior and morphology as the anteater and the human, the Beluga whale and the shrew.

Like Sharp's reflections on the coordinated, yet infinitely variegated patterns of birds in the sky, Grieg stepped back from his "ruck of records," looking for something common to them all and asked, "Where is the tune?

17 Greig, G. (1963), 58. Quoted in Cowdery. J.R. (1985), 4.

What is it?" In answer to his questions, he turns to intuition, but concedes that he is still confronted by a mystery: they are all variants of something which is at once the tune and in it. Without reference to the life sciences, Greig's mysterious *something* did intersect with Sharp's Darwin-tinged perspective of group behavior. Sharp didn't say as much, but he suggested that the mysterious something uniting the variants was not a particular version, but a pattern of *action*, a way of *behaving*. Instinctive and general as it is, this idea is enough to open a window for musicians into further levels of potential intersection between music and biology.

The compossible in nature and music

From the starlings' point of view, individual actions and collective actions shape each other in the swarm as they are happening in real time. We can say that the gorgeous coherent form of their movement is *co-created*. This is biology's word, but the idea of co-creation, like adaptive radiation, is also commonplace in music. When a reviewer in *Downbeat* finds important differences between Tony Bennett's renditions of popular songs in a duo with jazz master Bill Evans and Bennet's renditions of the same songs with full orchestra at Carnegie Hall,[18] the critic is affirming music's instinctive embrace of the principles of co-creation, and by implication, co-evolution and co-development. In music these principles are even so common as to be banal: soloists (of course) modify their behavior in the groups they perform with, even as the group's sound is (of course) an adaptation to the soloist, and this mutuality is routinely celebrated on all sides.

Concepts like co-creation, co-evolution and co-development may also be considered examples of *non-linear causation,*[19] a term in the sciences which could be used to support my assertion in the introduction of this book that there are no straight lines in nature. Darwin's preferred image of natural processes as a tangled bank rather than as a straight symmetrical tree also suggest a non-linear understanding of causation. (More will be said about this in chapters ahead.) When a single cause is thought to result in a single effect, it is considered linear. Hurricane force winds sweep through a town, and the trees that fall can be explained linearly – the wind made them fall. But when we notice that some trees fell and others did not, we are forced to allow a place for other, more indirect, or *non-linear* causes, as well – for example, the age of the fallen tree, its

18 Lutz, P. (2023).

19 Gilbert, S.F. and Bard, J. (2014), 130.

history of disease, parasite infestation and damage, as well as changing local conditions of soil and water over decades.

The readiest examples of linearity in twentieth-century biology have commonly been selected from genetics, that is, genetics in its narrowest interpretation – genes as "the law code and executive power" and "the architect's plan and builder's craft in one,"[20] as the physicist Erwin Schrödinger put it in 1945. Recent biology has had to adapt to the growing evidence of regulatory elements in the genome which describe something more interactive. By 1992, the science writer Natalie Angier was likening the understanding of DNA emerging from recent research to "a certain kind of politician, surrounded by a flock of protein handlers and advisors that must vigorously massage it, twist it, and on occasion, reinvent it before the grand blueprint of the body can make any sense at all."[21]

In the late twentieth century, following the cooling of several generations of enthusiasm for linear interpretations of genetic evidence – what Stuart Newman called "the twentieth century's gene bender"[22] – non-linearity seems to have won an equal place with linearity in conversations within genetics and the life sciences more generally. Writing in 2011, cell biologist and cancer researcher Lynn Helena Caporale described the dramatic shift in biological thought in this way: "For over half a century we have been in the thrall of the double-helical structure of DNA, which, in an instant, revealed that information can be transferred between generations by a simple rule: A pairs with T, G pairs with C." (Here she is referring to the protein molecules which are considered the building blocks of the genome of any organism, the so-called "base pairs" *adenine-thymine,* and *guanine-cytosine.*[23]*)* "This structure," she says, using language which invites a comparison to music-writing, "had entranced us into believing that we can fully understand the information content of a DNA sequence simply by treating it as a text that is read in a linear fashion. While we have learned much based on this assumption, there is much that we have missed." The DNA molecule has not turned out to be, as first thought in the 1950s – an inert code "like a bag of letters" operating as if it were "a passive tape running through a reader."[24] The result of these reinterpretations of genetics, says Caporale, is that "at the beginning of

20 Schrödinger, E. (1945), 21, quoted in Newman, S.A. (1995), 217.

21 Quoted in Gilbert, S.F. and Barresi, M.J. (2016), 61.

22 Newman, S.A. (2003a), 8–12.

23 The number of base pairs in the human genome is currently considered to be ~3.2 billion. The number of actual genes in the human genome is currently estimated to be between 20 and 25 thousand.

24 Caporale (2006), 20–21.

the twenty-first century, with entire genome sequences appearing before us, biologists can appreciate the feeling experienced by physicists at the beginning of the twentieth century, of the firm ground slipping out from under their feet. We are indeed in a 'new world.'"[25] It is this vision of a new world, one shaped not only through a central command, but also by complex interactions of large numbers of individual actors, that I believe makes a continuity of nature and music more plausible.

The oldest version of the linear command-and-control model in nature is the assumption in many cultures that whatever we see in nature was put there at the beginning by a deity. In this version of linearity, causation acts *downward* on a vertical axis from god to planet, as in the Genesis account of God's purposeful work beginning on the third day of creation. On the same vertical axis, much of twenty-first-century biology has turned to the *upward* causation and even what might be called *outward causation* that occurs when ecologies of living organisms catalyze and cocreate more life. This is the model of the garden without which an Adam or an Eve, a worm or dinosaur, a lichen or a sunflower cannot come into being. Neither the downward causation model nor the upward causation model is understandable in the absence of the other. This is a view which long pre-dates Cecil Sharp and the reviewer of Tony Bennett. The developmental biologist Scott F. Gilbert describes this phenomenon with the term *compossible* as it was first used by the mathematician and philosopher Gottfried Wilhelm Leibniz in 1697. "While all permutations may be possible, very few will be compossible," says Gilbert. "Not all parts can function together to make coherent wholes. Ecosystems are examples of compossible systems: a squirrel and a whale are both possible, but not compossible in the same habitat."[26] A handful of starlings in a flock of pigeons cannot be expected to survive for long, let alone produce murmuration, nor will Tony Bennett be able to find a place for his voice and musicianship in *p'ansori,* Korean folk opera. Whales cannot survive in forests, nor squirrels in oceans. Compossibility is a characteristic of ecosystems, which limit the repertoire of qualities of organisms in their domains. At the same time, these limits of ecosystem also enhance the survival of birds adapted to swarming. In Gilbert's example, the squirrel cannot be expected to adapt to a marine habitat, but it has shown adaptability to other terrestrial settings, including cities.

In the midst of all the chaotic variability of thousands of individual actors in the swarming behavior of starlings, there is no choreographer,

25 Ibid., 11.
26 Gilbert, S.F. and Bard, J. (2014), 131.

no director, no controlling program, only behavior which organizes itself through hyper-sensitive interaction among individual birds who are individually directed. Each murmuration resembles other murmurations. And yet, each occurrence is new and unique, a performance which, for all that it may suggest "complete unanimity of purpose," is not designed or directed in any of the ways that humans imagine the making of complex things, especially those that we consider beautiful.

5

Sticklebacks on the epigenetics of performance

Now consider the case of the three-spined stickleback, known in one of its many forms as *Gasterosteus aculeatus,* a family of fish averaging no more than seven centimeters in length at maturity. This prickly little fish has been the subject of hundreds of studies since the 1970s in the extremely wide range of its habitats distributed across the higher latitudes in North America, Europe and Asia. Sticklebacks have attracted scientific attention, in part, because of the elaborate choreography of their courting and nesting rituals, but more recently, studies have increasingly focused on their nimble adaptability to changing environments and on the genetic basis of these adaptations.[1]

Selecting from the researchers' findings, I want to offer three of my own bio-musical observations, inspired by Sharp's musical interpretation of the real time swarming behavior of birds. Looking through musical eyes as Sharp did, I see in the scientific studies, images and videos of the stickleback what I would like to call an *ensemble sensibility* enabled by a combination of *ecological memory* and *adaptive flexibility.* These three terms, scientific though they may sound, have never been used as descriptors in either biology or music. They are a musician's idiosyncratic application of biological terminology to species behavior that strikes me as musical. With these hybrid terms as a starting point, it remains to be seen what stickleback evolution offers us about the experience of small ensembles performing jazz repertoire and classical western chamber music.

[1] See: Nelson, T.C. and Cresko, W.A. (2018), 21; Laurentino and Moser (2020); Heckwolf et al. (2020); Maier, E. (2022).

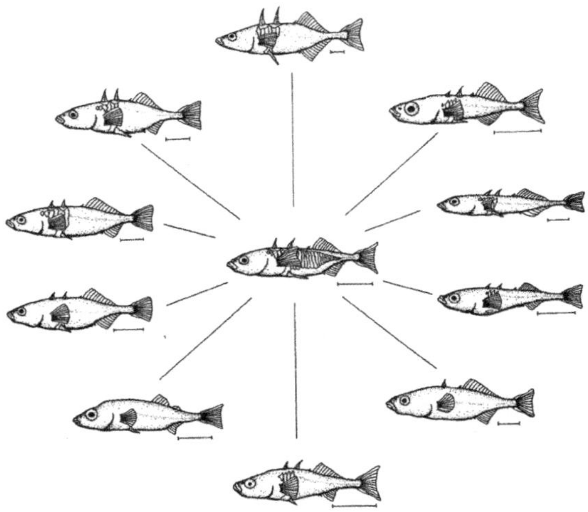

Figure 5.1 *Gasterosteus aculeatus,* the three-spined stickleback. The image at the center is the ocean-dwelling variety. Additional images are freshwater forms showing modification in armor and other changes in morphology. Bell, M.A. and Foster, S.A. (1994). Copyright open.

The stickleback's adaptability has earned it a minor celebrity status among geneticists and ecologists as the field equivalent of those laboratory test favorites, the common rat and the fruit fly.[2] Since the Pleistocene age 2.5 million to 12,000 years ago, the stickleback has managed to adapt its mouth, gills, spine, body armor and body chemistry to both salt and fresh water, to ocean, stream bed and lake, to running water and still water, to sub-arctic and temperate climates, and more recently, to unusually rapid vacillations in those conditions because of climate warming and the resulting pressure to migrate. In recent studies in Eastern Europe, for example, over the past 150 years the movement of stickleback populations from the shallow running water of a stream to the deep and still water of a lake has stimulated adaptations in the fish's body within a decade or two – a mere moment in the scale of evolutionary time. As recent as these stickleback studies are, they also participate in a long tradition of controversy which began before Darwin. Behind the mountains of accumulating biochemical, genetic, behavioral and ecological data is an enduring debate over how evolution happens, and how fast it happens, topics with potential relevance for music.

Lying dormant in the stickleback's genes for generations are coded recipes for mutations acquired over thousands of years, representing

2 "Stickleback are known for their propensity to rapidly diversify through habitat dependent divergent selection, and this makes them a particularly useful model for addressing questions about how ecological divergence occurs in invasive species..." Hudson, C.M. et al. (2021).

adaptations to a wide range of conditions, what biologists refer to as *standing genetic variations* (SGV).

From a musical perspective, we might think of this archive of options as the stickleback's memory which can be triggered by current conditions. For instance, when a formerly saltwater sea or a bay full of life became landlocked and lost its salination, sticklebacks in that region managed to alter the process of *osmosis* taking place in their body to adapt. That is, it modified the way its body regulated the flow of water across the membrane of each of its cells, a process dependent on a particular concentration of salt (sodium chloride) in the water making up the body's cells. These adaptations enabled the sticklebacks to either transform the salt-tolerance of their body chemistry to adapt to an exclusively salty or an exclusively fresh medium, or by becoming *euryhaline* like the salmon, striped bass and sturgeon – capable of living part of their lives in each habitat.[3]

Epigenesis in nature and in music

The ability to alter body chemistry and morphology in response to surrounding conditions and then to pass those alterations on to their offspring is not unique to the stickleback. Most other organisms, both the bulky and the microscopic, are similarly nimble in many of their interactions with habitat, making them all potential examples of *epigenesis,* a concept with a long history of competing meanings dating back to Aristotle in the fourth century BCE. The precise meaning of the term has varied, in part, in accordance with the contrasting perspectives of specialists genetics, bio-chemistry and developmental biology. But as a general concept standing for the context-dependence of all life forms, the term epigenesis has participated over the centuries in an evolving understanding of how living systems of all kinds develop and adapt interactively rather than according to a pre-determined structure built into the organism, the so-called *preformationist* view. In his *On the Generation of Animals,*[4] Aristotle had first argued against preformationism by observing that the development of animals from embryos did not consist simply of the rolling out of a structure already in place in the body of the embryo. Rather, Aristotle used examples of the embryos of chickens to argue

3 Recent studies tell us that this adaptability of the cell's osmosis can allow rapid adjustments to changes in salt content by as much as 400 ppm (parts per million). See: Bonzi, L.C. et al. (2021).

4 Aristotle, *On the Generation of Animals,* trans. by T. Pratt. https://penelope.uchicago.edu/aristotle/genanimals5.html

that development happens in stages, the structures in each stage making possible the transformations in the next stage. During the twentieth century, a substantial body of biological thought gradually coalesced around the view that interactions within an animal's body and with its surrounding habitat determine which of many possible developmental pathways of a given cell are put into play during each stage of development.

The twenty-first-century concept of the *stem cell* found in all organisms has picked up where Aristotle left off. In a human embryo, for example, the development of cells in stages from generic to specialized is dependent on the cells and systems forming around them. The fertilized egg is unspecialized, but as it divides, the cells are increasingly restricted toward a particular function by signals from surrounding conditions. Gilbert and Barresi refer to "the microenvironment that surrounds a stem cell" as the *stem cell niche.* This is the field of mechanisms and inputs outside the cell that create changes inside the cell, determining its "fate" (their word, not mine). What begins as an undifferentiated embryonic cell is increasingly restricted, for example, to brain or spinal cord, then next to brain proper, next to a particular brain region and finally, to a specific brain area. None of these crucial distinctions of function and cell chemistry were observable in the fertilized egg. Each of them arose in response to signals from environmental factors in real time, including signals from the immune system and from the unique physical properties of the tissues surrounding the stem cell.[5]

Competing with this growing view which left room for epigenesis was an updated preformationism, given new life during the first phase of modern genetics at the end of the nineteenth century. The growing enthusiasm over research on genetic controls of inheritance, driven by improved microscope technology, gave rise to an exclusively gene-centered understanding of how organisms develop and adapt. During the first half of the twentieth century,[6] a so-called *neo-Darwinian* consensus came together which argued that all species changes happen exclusively through randomly-generated mutations in a species' *genome,* the suite of coded recipes written in the protein chains of its genes. The neo-Darwinists also embraced a version of Darwin's own view of the speed of species change which supported their gene-centered view. Evolution was, as Steven J. Gould characterized the neo-Darwinian view, "a stately and orderly

5 Gilbert, S.F. and Barresi, M.J. (2016), 145–147.

6 Neo-Darwinism is also called "the modern synthesis" because it brought together Darwin's theory of natural selection with the genetic science that he never knew.

process, working at a speed so slow that no person could hope to observe it in a lifetime."[7]

The gene-centered tenet of neo-Darwinism was first expressed in the 1890s by the father of modern genetics, August Weisman (1834–1914), who declared that the sperm cells and egg cells which carried inheritance in sexual reproduction were transmitted in an entirely separate line from the cells making up an organism's body. While this fact may strike non-biologists as a bit arcane, the implications have proven in the long run to be far-reaching. From Weisman's claim it followed that no environmental impact on the body of any organism will ever be passed on to offspring by way of the genes. Weisman himself demonstrated his theory by cutting off the tails of multiple generations of mice. When each new generation was born fully tailed, he offered this as evidence[8] against what neo-Darwinists called "soft inheritance," a derogatory term for any theory that traits could be passed on by means other than through the genes. As the distinguished biologist Ernst Mayr declared as late as 1980, "It was perhaps the greatest contribution of the young science of genetics, to show that soft inheritance does not exist."[9] Erwin Schrödinger's mid-century claim that the genes were the "law code and executive power" and the "architect's plan and builder's craft in one" provided for two generations of researchers a one-size-fits-all explanation for how organisms both stay the same and evolve.

But a growing number of cases like the stickleback were suggesting something different. The variability of a species – its adaptability – was not describable as a simple linear cause and effect in the genetic code, a case of one-way-command-and-control, like flipping a switch on or off. Rather, the adaptability of a species – its capacity for generating life-enhancing modifications to its own body – was the product of a dynamic swirl of interactions and influences coming and going in all directions, 360 degrees around the organism and within it. This new contextual view of evolutionary change did not discount the role of the genes in the stickleback's adaptive agility. But the triggering of those stored traits by influences from outside the genes was now being seen as part of an ongoing conversation

7 Gould, S.J. (2006), 261.

8 This so-called "germ plasm" theory was adopted as neo-Darwinian doctrine for much of the twentieth century more rigidly than Weisman himself may have intended. Weisman also continued to believe, as did Darwin, in the importance of environment in the expression of traits through the germ plasm.

9 Mayr, E. and Provine, W.B. (1980), 17, quoted in Gissis, S.B and Jablonka, E. (2011), 105.

between the stickleback's genetic archive and its environment.[10] As the studies of a range of species like the stickleback accumulated in the late twentieth century, the deterministic models of control in orthodox genetics professed by Weisman, Schrödinger and Mayr were looking increasingly simplistic and mechanistic. Linear causation, command-and-control, and flipping a switch no longer matched what biologists were observing in their studies of living systems.[11]

Now, looking back on the past two generations of scientific work, it can be said that the cries of heresy at epigenesis and soft inheritance among life scientists have subsided, and the stickleback and the other three species featured in the opening chapters of this book have each played a small role in this change. Whatever the musical lessons we draw from these life forms, we would do well to remember that the lessons are inseparable from decades of sometimes career-threatening debates roiling the life sciences. If epigenesis proves as useful in describing the making of music as in describing the life of a fish, we have at least taken a significant step toward an integrated bio-musical view of how music works. It would also be no surprise if this line of thinking were to provoke musical controversies on a par with those experienced by our scientific colleagues.

Ensemble, ecology and memory

The combination of the long-term memory seated in the stickleback's genetic code and its extremely sensitive response to habitat are biological traits that I want to call musical. I propose that these traits add up to a general quality which musicians – especially those performing in small groups – would recognize as an ensemble sensibility. Ensemble performance requires that each player respond continuously to extremely small and fast-moving inputs from his or her playing environment. In the act of musical performance, input comes in at every moment from ensemble partners, from the limits set by the compositional task of the moment, from the audience, and from the musician's own internalized habits and memories. Not all these inputs originate within the performer but they are experienced and processed personally and uniquely, blurring the conventional distinction between "internal" and "external" in the subjective experience of the musician. Observing nature through the musician's eyes, we can perhaps recognize a kindred experience in the continual interactive

10 This was expressed in the original Greek: *epi-* + *gene*, meaning "outside of" or "in addition to" the genes.

11 See, for example: Gould, S.J. (2006), 444–466; Newman, S.A. (2003a), 8–12.

relationship between the stickleback's body chemistry and its ecology, a relationship viewed by biologists as epigenetic.

Ecological memory is the name I am giving to the situation-sensitive learning curve necessary for good music-making and for a thriving stickleback population. This made-up term suggests that time and habitat intersect in all living ecologies, both the aquatic and the musical. In a fish and in a musical performance, actions in the present are not invented new at each moment, but always emerge from remembered interactions with surroundings. A musician in the heat of performance draws on lessons learned from decades of practice, rehearsal and performance, as well as from the additional years of experience acquired from teachers, colleagues, reading and study. In order to preserve continuity and unity in the face of challenges in rhythm, pitch and tone presented at every moment, a musician cannot stop to think, but allows all these currents of experience to flow, normally saving reflection for later. It could be said that the stickleback's memory is longer, extending back millions of years, while its lifespan of perhaps three years is just 1/23rd of the human's three score and ten.

No less musical is the *adaptive flexibility* of the stickleback's responses, the organism's tendency to preserve many options in response to surrounding conditions. Adaptive flexibility means that, in the adjustments which an individual stickleback makes in the course of a lifetime, and in the adjustments a musician makes in the performance in an ensemble, fixity seems to be less the point than flexibility. As the multiple options stored in the stickleback's genetic code demonstrate, the stickleback species has had to adapt to many changes of habitat over the millennia. This stored experience waits, temporarily dormant, until the moment that a change in habitat evokes it. After years of playing with many musicians in a variety of ensembles and compositional contexts, performers accumulate a growing repertoire of options for how to adjust to the ever-changing circumstances of music-making. It does not seem like much of a stretch to conclude that, for musicians as for fish, it has always been more important to learn *how* to adapt than to set in place a particular *ideal* adaptation.

A stickleback view of jazz performance

Paul Berliner's imposing 800-page study of the jazz performer's learning process[12] is based on lengthy interviews conducted over sixteen years

[12] Berliner, P. (1994).

with more than fifty musicians whose experience spans the 1920s to the 1990s. The interviews provide many examples of ecological memory and adaptive flexibility – that is, how musicians learn to adapt, how they build on their memory of prior musical experiences, and how they apply those memories to new circumstances. The path from novice to master is marked by the gradual acquisition of tools for surviving in a widening range of contexts. The contexts are defined to some degree by the types of performance venues – small and large clubs, theaters, recording sessions, jam sessions – but more significantly, by the different bands and their leaders with whom an individual plays. From the stickleback's point of view, the band and even the personality of a particular star soloist, are shaping forces of a habitat in which an epigenetic relationship with environment is a condition of life.

Trombonist Melba Liston told Berliner that each band, with its different leader, had its own "sound," a term implying an array of musical behaviors to which she must adapt her own musical habits in order to match the prevailing texture and mood. "You have to go that way," she said, "because if you don't, you don't fit in." Playing with Dizzy Gillespie, "when you're on the bandstand, you're ready to burn," while with Billy Holiday, "you're not going to holler, stomp and carry on like you do with Dizzy's band." With Count Basie, says Liston, the "tone colors and feelings" were "more organized and routine. You're going to stay about the same way all night long, whereas with other bands, you reach greater highs and lows."[13] Liston's adjustments to a band's sound suggests a range of adaptations over time which are more osmotic and unconscious than explicit and scripted.

Moving among such different habitats over a lifetime requires that a musician learn to adapt sensitively and rapidly, shaping his or her musical vocabulary, tone color and style. Saxophonist Lee Konitz recalled how, after years of playing in the band led by pianist Lenny Tristano, he recorded with another band. A fellow musician, having heard the recent recording, asked Konitz – with a hint of criticism – why he no longer played the way he used to play with Tristano, to which Konitz answered – with a hint of frustration – "because I wasn't playing with Tristano [of course]!"[14] Adaptation can be explicit as well as osmotic. The kind of specific communications that alter behavior within any ensemble was described by pianist Fred Hersch when he spoke of how he "signals" a change to the group with "a loud rhythmic figure or a complex cross rhythm, or certain chords."

13 Ibid., 445.
14 Ibid., 436.

In Hersch's case, he sees the signal not as a command, but as a request to adapt to new conditions: "Let's get out of our present format and take the music somewhere else."[15]

When habits are imprinted deeply over time through gradual osmosis, their impacts can become vanishingly subtle. Many of Berliner's interviewees speak of the moment-by-moment communication in performance as so rapid as to be noticeable only after it has happened, for example, when listening to a recording of a performance. This is the quality which was captured by a reviewer in *Downbeat* when he described the collaborations between pianist Bill Evans and bassist Eddie Gomez as "almost-telepathic."[16] These words match the pianist Roland Hanna's description of mature ensemble behavior using himself and the bass player Richard Davis as examples: "I've heard him enough to know *how* he makes his lines," Hanna says, "so I may not know exactly what note he's going to play, but I know in general the kind of statement he would make...We train ourselves over a period of years to hear rhythms and anticipate combinations of sounds before they actually happen."[17]

The almost telepathic sensitivity to ensemble partners aptly describes how an organism like a fish interacts with its web of relationships in its environment. It is an interaction on which the organism depends to survive and thrive. As much as artistic performance may feel to the artist like a solitary, willful act, dependent entirely on the artist's own energy and discipline, the ecological nature of music learning and transmission is at times identified by a musician in language which conveys a network of inputs and communications, both conscious and unconscious, built up over many years. The virtuosic jazz instrumentalist Arthur Rhames, known for the intensity of his rehearsal and practice discipline, used the concept of intuition to stand for the tangled ecological web of inputs into his own performance. It is a web of awareness and skill built up through a combination of osmosis and conscious acquisition which I am calling epigenetic: "Improvisation is an intuitive process for me now...I'm calling upon the resources of all the years of my playing at once: my academic understanding of the music, my historical understanding of the music, and my technical understanding of the instrument that I'm playing. All these thing are going into one concentrated effort to produce something that is indicative of what I'm feeling at the time I'm performing."[18]

15 Ibid., 366.
16 Kalin, A. (2019).
17 Berliner, P. (1994), 365.
18 Ibid., 16.

A stickleback view of string quartet performance

Looking back on his collaboration with the Juilliard String Quartet to produce a book on the performance of three Beethoven quartet movements,[19] the musicologist Lewis Lockwood recalled that rehearsals and conversations about these pieces would provoke "ferocious arguments about details."[20] Any given passage of a composed score, in other words, will have room in it for multiple versions. Disagreements aside, what the members could all agree on was that the activity they were engaged in was "interpretation," and that the variables which were the focus of their arguments were confined to details of dynamics, tempo, phrasing, accent and bowing. Lockwood's scholarship provided another layer of variables recorded in Beethoven's own drafts, sketches and revisions. The transcript of the group's discussions of the scores and of Lockwood's documented variants represented still another layer of variants, completing a picture – like that of Arthur Rhames – of the multiple timescales of listening, thought and memory converging on any moment when a bow is put to string.

In contrast to Berliner's testimonies, the discussions in The Juilliard Quartet are dominated by references to notation, both to the printed score and to the markings the group agreed to add to the score representing the consensus outcome of their arguments. Much thought and effort go into deciding how the group will execute the conventional Italian markings for gradations of dynamics and emphasis. At one point in a discussion of Op. 59 no. 1, the composer's *sforzando* is interpreted as "quick sharp accent" (m. 53), but a few measures later as "longer, more expressive accent." (m. 57)[21] Their conversations veer between extremes: from one member's interpretation of a marking to consideration of how it is used by Beethoven in other pieces, or to its changing use in the composer's revisions, or to its use by other composers. Every topic is connected to other topics streaming into the rehearsal room – from the advice of teachers, from the many years of performing the piece now on the music stand.

All decisions about how best to play a passage remain fluid. As the cellist Joel Krosnick put it, "if you play expressively the same way all the way through, then you indeed haven't moved anywhere."[22] The interpretation of dynamics, rather than a certain setting on the volume controls,

19 Lockwood, L. (2008). The book examines the first movements of just three quartets: op. 18 no.1, op. 59 no. 1, and op. 130.

20 Lockwood, L., personal communication, December 2023.

21 Lockwood, L. (2008), 152.

22 Ibid., 125.

might require a complex negotiation with the harmonic ambiguity of a passage, provoking an extended dialogue. To someone other than his ensemble partners, Krosnick's attempts to describe these negotiations sound tortured. They are a puzzling mix of abstract and hyper-specific: "that *forte* is played with a different sort of sound quality, it's not solid and thick, it's still a little bow speed, still a little air, and an air of uncertainty."[23]

Taking a stickleback point of view of the Juilliard Quartet's life-long engagement with these three Beethoven movements, we catch hints of the moment-by-moment challenges any organism faces in its environment. Lockwood, as historian, continually calls attention to the layered record of the composer's sketches and drafts, hidden for 200 years from the view of all but a few scholars the way geneticists make visible the layers of changes in the stickleback over millennia.

In the composer's earliest quartet, Op.18 No.1 in F major, Lockwood notes how, in the composition process over several months in 1798–1799, the theme "undergoes a lot of change from its primordial state." At first, it appears in the composer's notebooks in 3/4 meter. It then "entered into a 4/4 phase and then went out again."[24] What is gained from accessing the memory of these variations in the notated record is the same as what is gained in the stickleback's continual accessing of its own genetic record. The goal in both fish and quartet is not to arrive at an ideal solution, but to create a full range of options for how to behave now and in the future. In both cases, the solution reached might be called optimal in the sense that, in the negotiation between memory and environment at the moment, it is the most workable. In both cases, we should not be surprised if at another time other options are favored and new options exposed. As the Juilliard's second violinist Ronald Copes observed, the purpose of adding the much-discussed markings in their scores was "to capture the most common experience that we have." But, as Krosnick had observed about the interpretation of expressive markings, Copes was also uncomfortable with the impression their own markings gave "that there is only one way we play this."[25] Arguments, conundrums and final decisions aside, preserving the potential variability of the composer's immortal composition is still fiercely defended.

In his memoir entitled *Indivisible by Four*,[26] violinist Arnold Steinhardt of the Guarneri Quartet, "interpretation" is also a key word as he describes the

23 Ibid., 132.
24 Ibid., 50–51; 32–34.
25 Ibid., 63.
26 Steinhardt, A. (1998).

quartet's multi-decade immersion in the 200 years of repertoire which they have rehearsed, performed and heard thousands of times. Here too, are signs of stickleback epigenetics, viewed through different lenses: in Lockwood's performer-scholar collaboration, the musical text and its history are the focus, while Steinhardt's memoir is a story of individual musicians, their interactions and examples of moments from particular performances.

With Franz Schubert's famous Death and the Maiden Quartet (No. 14 in D minor, D 810) written in 1827, Steinhardt could look back over 32 years at the quartet's many versions of it. Their first recording of the movement in the 1970s feels different to him, so he launches, stopwatch in hand, into a comparison of the quartet's performance durations, then and now. This is his attempt, recounted more than once in the memoir, to locate an objective anchor for his sense that the ensemble's performances of this or that piece felt to him so different from each other. But comparing two performances of the Schubert spaced more than twenty years apart, the stopwatch gives him little to go on: out of a total playing time of 41 minutes, the variability in tempos added up to a difference of a mere 21 seconds for the middle two movements, while the first and last movements were identical.[27]

Another, more revealing approach to understanding these differences is Steinhardt's attempt to create a blow-by-blow recreation of a performance as it is happening on a concert stage in Albuquerque in 1997. His account follows minute details of dynamics, bowing, articulation, fingering, and more, as they are happening. We experience with him the surprises and the risky moments of unexpected behavior which threaten to shatter the ensemble. In the second movement which features the melody of Schubert's song "Death and the Maiden," all surprise changes are registered in nano-gradations, almost imperceptible, but his experience of them is huge. His voice-over narration of this performance begins in the opening measure: "In four-part harmony we begin the solemn theme softly. Decent blend. Together. Good. John [second violin] and Michael [viola] now have the moving voices…" and continues in this way, filling the final four pages of his book.[28]

Steinhardt's blow-by-blow monologue is a stickleback's view of Schubert's composition. The piece is a fish, always in motion, and we are in the water with it and with the quartet, without which the piece would not, at that moment, exist. Schubert's art and the Guarneri ensemble

27 Ibid., 300.
28 Ibid., 302–306.

are, at this moment, one, and Steinhardt is speaking for them both. He is articulating the ensemble's intimate communication with everything presented to it by the habitat of this performance. As with the Juilliard's Beethoven quartet conversations, this habitat includes not just the score, but also the group's stored memory of decades of interacting with the score. Steinhardt's attention, like the Juilliard's, is constantly shifting from dynamics ("we recede – barely audible. Softer than usual") to bow ("more bow, less sound on the sixteenth notes"),[29] and then to unexpected tempi and shifts of emphasis. He is a participant in these events, and yet, as the measures flash by, he communicates a sense that he is being driven by a current somehow not of his own making. Suddenly, his eyes are on the cellist's third finger, and he recalls the revered violinist Sascha Schneider's advice many years before to "eat the fingers" of his partners. The graphic physicality of the phrase leaves no room for misunderstanding the total intimacy of their connection. The threads of the web that binds them in this habitat are made of sight as well as sound, though they are so delicate as to be imperceptible to all but the participants. They are the inter-cellular interactions which are invisible to the naked eye, but still consequential.

There are constant surprises which, though subtle to the listener, they are experienced as full of tension and uncertainty – modulations of tempo and tone and articulation by the four musicians which are so reflexive and instinctive that they are, as the jazz reviewer put it, almost-telepathic. They are only noticeable to Steinhardt a split second after they happen. At times, the surprises within this nano-world almost derail the ensemble altogether, as in the cello's "flight of fancy" in the third movement and his own playing which at one point is "too free." His fellow violinist rescues the tempo and the unity is preserved. ("Sorry, John. Thanks, John.")[30] At the finale each of them is "hurling thunderbolts" with their bows, but he knows that the second violin's will be "different every night."[31]

The boring dependability of living systems and musical practices

Common to the cultures of jazz and western art music is a tendency to presume the central importance of the individual actor – the composer, performer or improviser. This actor provides the ready answer to any

29 Ibid., 303.
30 Ibid., 303.
31 Ibid., 304.

question about any elements of structure or any moments of beauty or surprise or originality. At the same time, both sets of musicians habitually speak through references which may be called epigenetic. The Berliner interviews and the Juilliard and Guarneri testimonies are equally dense with references to context, to history, to tradition, and to the web of relationships with colleagues and teachers which make a particular musical moment possible. For musicians in both arenas it is commonplace to insist that whatever happens in an ensemble performance transcends mere individual achievement. Accumulated musical experience, preserved in written documents, in recordings and in memory, is no more separable from a performance of Beethoven's Op.18 No.1 than are Bill Evans and Eddie Gomez's almost telepathic interactions in a performance of a jazz standard, or in the Guarneri's hurled thunderbolts in the performance of Schubert's "Death and the Maiden" quartet.

For jazz musicians and classical musicians, a bio-musical perspective which compares their artistic labors with the living processes of a bird or fish would be, at best, a colorful metaphor, perhaps recalling Eduard Hanslick's description in 1854 of Beethoven's *Prometheus* overture which "develops itself in organically distinct gradations, like sumptuous blossoming from a bud."[32] At worst, to describe the works of Handel or Coltrane in the same terms as one discusses a species of fish might strike the musicians as irreverent, if not demeaning. Both fish and sound art can be said to have evolved "naturally," an expression often resorted to in order to describe a phenomenon which in some mysterious way transcends mere individual causation. "Natural" stands for what seems indisputably right, appropriate to time and place. In the end, the application to music of expressions like ensemble sensibility, ecological memory and adaptive flexibility suggests what is the real take-home of all this hyphenated bio-musical interdisciplinarity. Simply put: *This is how nature works. If these natural processes resemble musical performance in any way, then perhaps this is how music works as well.*

In a range of musical circumstances, the musician's newly acquired knowledge within a specific habitat is not welded into a rigid performance outcome, at least not ideally. Using musical eyes to view the life of a fish the way that Cecil Sharp's folk song experience shaped his view of starling murmuration, we can recognize in stickleback epigenetics the musician's ensemble sensibility, a familiar quality which over time becomes an increasingly refined tool in a player's toolkit for acting within

32 Hanslick (1986), 81, quoted in Watkins, H. (2018), 50.

a performance – *any* performance. One of the marks of more evolved musicianship in any genre is an accumulated repertoire of options which are the basis of confident interactive behavior now and in the future.

My emphasis on the epigenetic lessons learned about music from a fish does not diminish the individual achievement or rare artistic genius. But with a focus on the systemic creativity of nature, the deeply cut journeyman habits of musical language, the commonly accepted everyday musical practices, provide as much to think about as individual genius. These musical habits are like the habits of spoken language which are as often uninteresting in their normalcy as they are inspiring. Nature's dependable and boring ability to crank out life, to pull off the intricate variations in the chemistry of an organism's cells under constantly changing conditions, all in the absence of a command-and-control center, seems out of place in the world of the arts, with its nearly exclusive focus on individual and unique human achievements. But the contexts from which the Jelly Roll Mortons and Franz Liszts arise were packed with serviceable, but relatively predictable and mundane musical creations, which were therefore, for the most part, beyond our notice. This is the microbial, fungal and molecular level of musical life taking place in the soil and atmosphere of a culture. These variations can arise with an apparent effortlessness that suggests an agency transcending the individual. This life is rich and tumultuous and yet pedestrian and normal. It exists for its own sake, requiring no other justification than its own continuance. In the process, all this liveliness also makes possible the emergence of the occasional Great Barrier Reef, sequoia or hummingbird, for which we are all grateful.

6

Lichens on the normalcy of chimeras

The lives of flying and swimming species, besides the interest they have for us on their own, provide a window into the workings of biology since the eighteenth century. It was Swedish naturalist Carl Linnaeus (1707–1778) who first proposed to divide the catalogue of all known organisms into two parts: animals (L. *anima:* breathe), those which have breath and can move, and plants (L. *plantae: plantare,* to fix in one place), those which do not breathe and are stationary. Those two large categories of organisms, along with the tree of life on which the familial relationships among Linnaeus' categories were eventually displayed in the nineteenth century, have undergone considerable revision over the years. More will be said in Part Four about the drastic editing of that arboreal image in the century and a half after Darwin.

But we should remember that Darwin himself, and Linnaeus before him, had depended almost exclusively on observations of the gross features of living things, that is, those features which were, as a rule, visible to the naked eye. Remarkably, this pattern of dependence on gross morphology has persisted over the years, even though innovations in microscope technology in Linnaeus's own day had already begun to open up observable nature to include the vast populations of formerly invisible life forms. Microscopic organisms were first observed within the lifetimes of G.F. Handel and J.S. Bach by Antonie van Leeuvenhoek (1632–1723) and Robert Hooke (1635–1703) and were eventually incorporated into the germ theory of disease during Darwin's lifetime by Louis Pasteur (1822–1895).[1] Through modern microscopy, we now know that the pool of microscopic life forms actually dwarfs in both number and sheer heft the planetary pool of much bulkier creatures. Microscopic bacteria and

1 Gest, H. (2004). 187–201.

archaea – single-celled life forms lacking a nucleus, known collectively as *prokaryotes* – nearly equal the weight of all trees and other vegetation on the planet.[2] In support of the view that microbes were the planet's "unseen majority," a 2011 editorial in a microbiology journal placed some "back of the envelope calculations" into more familiar contexts: "If all the 1 x 1031 [1 + 31 zeros] viruses on earth were laid end to end, they would stretch 100 million light years," and "the number of microorganisms in a teaspoon of soil (1 x 109 [1 + nine zeros or one billion]) is the same as the number of humans currently living in Africa."[3] Nevertheless, as the editors point out, microbial life remains a footnote in the story of life on this planet. The larger life forms of plants and animals are still the examples which have continued to be presented in natural history exhibits and educational curricula as the poster children of all life forms.

Composite identities

Straddling the visible world of Linnaeus and Darwin and the microscopic world of Leeuvenhoek, Hooke and Pasteur are *lichens* – those ubiquitous, not-exactly-plant-like growths we see every day appearing spontaneously on tree trunks and stone walls. Lichens may also be considered poster children, but they are the poster children of a new, more challenging biology than Darwin could have known. I believe that it is this new biology which makes a bio-musical view more possible and even more necessary. Current biology's renewed terms and assumptions about our planet's life forms encourage us to re-examine the boundaries of science and art which for centuries have been kept comfortably in place by focusing on the appearance of nature's bulkiness and on the non-bulkiness – the immateriality – of artistic experience. Between these two domains of science and art is a space which bio-musical metaphors have traditionally filled. Hanslick's blossom to bud characterization of Beethoven's music is an example of a bio-musical metaphor which asserts by way of a simile a connection between the gross external likenesses of tangible nature and musical experience.

Looking at lichens as they are commonly found on every continent and in virtually every climate on the planet, what appears to the naked eye to be leaf-like, moss-like or fungus-like among the 18,000-odd catalogued lichen species is deceptive. Readily observable likenesses, science has learned, can

2 "The total carbon of prokaryotes on earth is enormous, approximately 60–100% of the total carbon found in plants." Whitman, W.B. et al. (1998).

3 "Microbiology by numbers" (2011), 628.

turn out to be contradicted by the discovery of mostly invisible relationships and differences among them.[4] In the case of lichens, it is probably safe to say that biologists of neither an eighteenth- nor a twenty-first-century persuasion would be inclined to place any of these species on the animal side of the ledger of life. But beyond this simple distinction, lichens are found in so many different forms that they require in some cases extreme contortions in the plant column if they are to fit onto the ledger at all. On the one hand, they are part of the planetary biome which are considered *autotrophs,* or self-feeders – that is, they produce their own nutrition as plants do, through photosynthesis, which harnesses the energy of light to produce organic molecules from the inorganic elements in the soil and air. In this way, lichens are different from animals *(heterotrophs),* which cannot self-feed and must consume other organic substances for nutrition. However, just because they are photosynthetic like a plant doesn't mean that lichens always, or even mostly, possess leafy structures or even look green. Red, yellow, black, brown, grey, orange and white are just as much a part of the lichen color palette. The drawings of the three most common lichen body types in figure 6.1 hint at how hard it is to begin to capture the diversity of lichen morphologies.

But the diversity of their body structures is not the main reason for considering the lichen here alongside the finch, starling and stickleback, species which, by comparison with the lichen's diversity, can look drably conformist. The reason for bringing up lichens at the beginning of this book is the complexity of lichen identity, an identity which is, in contrast to how we commonly think of birds and fish, *composite.* The austere yet gorgeous, furry and leafy presence of lichens in forbidding environments first triggered my own curiosity. They were a dependable part of the muted coloring of New England winters when nothing seemed alive, especially on miserable, cold and wet days when suddenly the lichen's customary grey turns pale blue-green against a rain-soaked tree trunk. But my musical curiosity was triggered only later when I realized that lichens did more than what I was able to see: they also defied my everyday understanding of the inviolable *singularity* of a species. They emphasized just how much my sense of genre itself – even of genre as a category in my own field – might be a kind of reductionism which distilled the rampant variability of living phenomena into essences more comfortable than they were descriptive.

The lichen's composite identity is a peculiar mashup of living types in a single organism which in classical mythology would have been pictured

4 For example, see Schulz, M. et al. (2022).

as a *chimera,* an animal which at one end was a lion, at midsection a goat, and at the tail a serpent, What to the Greeks was a chimera, in current biology is an example of *symbiosis.*[5] Each lichen is composed of a photosynthetic algae (also called *cyanobacteria*) which is of aquatic origin, and a moisture-gathering and nutrient-gathering fungus whose natural habitat is terrestrial – two distinct species adapted to two distinct habitats which persistently function side-by-side as one. Each of these collaborating species or *symbionts* contributes to the composite organism life-sustaining processes which the other lacks, and each brings to the bargain its own independently evolved package of genetic code.

Symbiosis itself, like the epigenesis so apparent in the stickleback's shape-shifting conversations with its environment, is another example of a nineteenth-century concept which went through 150 years of marginalization before returning to the scientific mainstream in the late twentieth century. By 2012, the immunologist and philosopher of science Thomas Pradeu was reflecting a consensus definition of symbiosis when he defined it as "a lasting relationship between organisms belonging to different species that is beneficial to one, neutral or beneficial to the other."[6] In the period some 400 million years ago, when the waters of the earth were the only viable habitat for life, in the regions between water and the planet's solid rock surface of recently-cooled lava, aquatic blue-green algae and the microscopic spores of terrestrial fungi made a union and, quite literally, put down roots. The fungi provided the system of hair-like *hyphae* which anchored the two symbionts to the hard archaic earth and enabled the breaking down of life-supporting minerals like calcium and phosphorus hidden in the rocks. These chemical processes were powered by solar energy through the photosynthesis of the algae, and the combination gradually contributed to the processes by which the planet's hard mineral surfaces were transformed into the rich and loose soil – the medium of most vegetative life on land. This planet-altering collaboration of distinct species over a long term joined with, or gave rise to, an array of new species – the leafy plant life so familiar to humans – which for all time established photosynthetic life on the barren earth.[7] From the workshops

5 Among developmental biologists, the term chimera tends to be reserved for special circumstances, as when a person receiving a bone marrow transplant ends up with blood cells with two distinct genomes, their own and that of the bone marrow donor. See Khan, K. et al. (2004).

6 Pradeu, T. (2012), 117.

7 There continues to be controversy about just how much lichens contributed to the establishment of vegetative life on earth. See: Nelsen, M.P. et al. (2019).

within the leaves of plants of all sorts – from pine needles to oak leaves and grasses, palm fronds, seaweeds and mosses – oxygen, an essential ingredient for most multicellular life, is pumped into the atmosphere as a byproduct of photosynthesis (Figure 6.1).

And yet, despite their lengthy cooperation, in most lichens the two symbionts persist as co-habitants in a shared space. They remain genetically distinct, which distinguishes them from what is referred to by biologists as a *hybrid,* a combination which, perhaps surprisingly, is far less common in nature than symbiosis. A hybrid exhibits intermediate features which are the result of the mingling of the genetic codes of its constituent species during sexual reproduction. The classic example of a hybrid is the mule, the horse-like and donkey-like offspring of a male horse and female donkey. Popular speech today confuses the distinction between hybridity and symbiosis by using hybrid as the general label for any entity which combines the qualities of two different things – a hybrid car with both electrical and gas-powered engines, for instance. In symbiosis, the genetic codes of two symbiotic partners do not merge as they do in the hybrid, but they continue to replicate and evolve independently in each generation of offspring. The two symbionts and their distinct genetic packages not only persist, but in most cases do not surrender their ability to exist separately from the symbiotic union. For these reasons, lichens tempt us to make the all-too-frequent mistake of viewing the union anthropomorphically, as "voluntary," or as a well-considered "marriage" of convenience and purpose.

Symbiosis everywhere

The sheer size of the lichen group compared to other groups of organisms speaks for symbiosis as more than an evolutionary freak. If symbiotic unions were as exceptional in nature as hybrids – that is, if the number of cases like the lichen and the number of cases like the mule were reversed – there would probably be little reason to bother wondering whether symbiosis was a process we could expect to observe in music. But compared to the 6400 mammalian species, 10,900 bird species and 34,000 fish species which have been catalogued, the 18,000 or so catalogued lichen species (with an estimated total number of as much as 250,000[8]) establishes the normalcy of symbiotic unions in nature overall.

8 Recent studies estimate that between 17% and 30% of all fungus species are capable of becoming lichens. Since at least 1.5 million fungus species are believed to exist, then the number of lichen species could be on the order of 250,000. See Fertig, W. (2025).

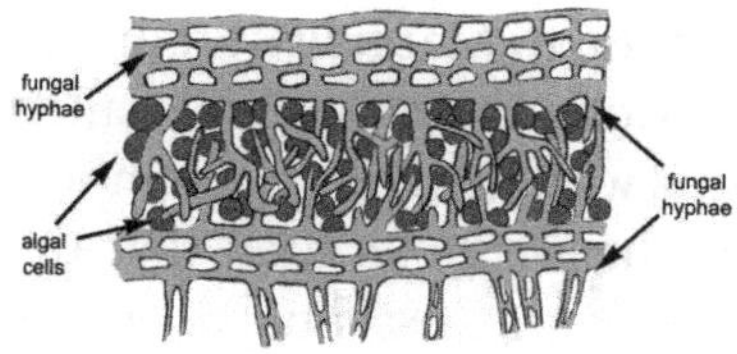

Figure 6.1 A *lichen* is a partnership of a photosynthetic algae (*Cyanobacteria* or blue-green algae) and a fungus. Homeomeric Lichen Thallus. In Yavuz, M. (2018). © www.suboptimist.wordpress.com.

Just as important for a bio-musical perspective, symbiosis also offers itself as an important counterargument to the twentieth century's Neo-Darwinian emphasis on glacial change in species over millennia through random generation of traits and the modification of those traits by natural selection. The microbiologist Lynn Margulis (1938–2011), who devoted her career to persuading her colleagues of the evolutionary importance of symbiosis, considered that it was one of the principal engines of biological novelty, adding another possible answer to Darwin's persistent question about the origins of innovation in nature. Symbiosis, Margulis argued, helps to explain the appearance at least three billion years ago of game-changing novelties of nature like the enclosed nuclei and other internal cell structures of multi-celled organisms (the *eukaryotes)*, a category of organisms which includes most of the organisms visible to the naked eye, including *homo sapiens.*

For Margulis, symbiosis was a kind of "evolutionary lightning." In a mere moment of evolutionary time, symbiosis could create new forms which would, otherwise, have taken tens of millions of years of random mutation followed by natural selection. This is important for any musician whose professional curiosity is piqued by biological discovery. In nineteenth-century biology, gradual evolution measured in millennia was the only kind of variation which nature appeared to offer as a model for musicians. If the glacial pace of species change was still the current state of knowledge, then there would now be another reason to disregard symbiosis (and, by implication, any other biological phenomenon) as models for music-making. But it was the clear evidence of symbiotic lightning in nature provided by Margulis and others which triggered my own musical curiosity. It was the rapid variability visible in nature which put me in a state of mind which I recognized as musical: it felt like when I witnessed the nano-variations in the performance of jazz and chamber music and when I recognized the behavior of folk songs in the subtle flickering of movement in starling flight.

For decades, there was stiff scientific opposition to Margulis's argument that some of the essential cell structures of organisms with a walled nucleus were originally single-celled organisms with no nucleus, members of the immense domains of bacteria and archaea. She argued that the *mitochondria* of all cells (specialized internal structures essential to energy production) and the *chloroplasts* of plant cells (structures essential to photosynthesis) were originally microbes without a walled nucleus which had been engulfed – "eaten," as she often put it – by a host cell and put to work as enhancements to the host's metabolism.[9] As symbionts, the engulfed cells, even as they were now supplying services to the host cell, benefitted directly or indirectly from what the host cell had to offer, such as, its more elaborate energy and reproductive systems, its protective cell wall, and its genetic code which interacted with and supplemented its own.

The lichen was one of the principal exemplars of symbiosis for Margulis, but her examples of the mitochondria and chloroplasts represented a stage of evolution which was a considerable step beyond that of the lichen. What might have begun as symbiosis – the mutualism of two distinct species – could develop naturally into *endosymbiosis,* where one organism lives symbiotically, not just adjacent to another, as in the case of the lichen, but inside the body of another. Nevertheless, an endosymbiont is still not to be confused with a hybrid. Looking beyond the examples of the mitochondria and chloroplasts, colonies of *Wolbachia* bacteria, for example, are integrated into the digestive systems of sixty percent of invertebrates such as insects and crustaceans and play an active role in signaling crucial changes in the development of the invertebrate embryo.[10] But like other endosymbionts, the *Wolbachia* continues to add its store of genes to that of its host without merging with the host's genome in the way that the genomes of the horse and the donkey do in that famous hybrid, the mule. The same is true of the stable populations of bacteria in the human alimentary canal which for millennia have made possible the digestion of specific nutrient types.[11]

9 Margulis's argument for the origin of mitochondria and chloroplasts was originally proposed in 1967 in "On the Origin of Mitosing Cells," in the *Journal of Theoretical Biology* 14:225–274. By the 1990s the theory was being confirmed by other researchers and on the way to becoming scientific consensus. She also argued that the tails *(flagella)* of many microbes were originally thread-like microbes known as *spirochetes* which were engulfed and used for motility in other microbes. This part of her thesis never received the same level of confirmation in other studies.

10 Pietri, J.E. et al. (2016), 923–936.

11 These colonies are different in each individual human, and different locations in an individual's intestine host different bacteria families.

Margulis also saw the prominence of symbiosis in nature as an argument for collaboration and mutualistic interaction as a foundation of evolution. She considered symbiosis on a par with the staple concept of "the survival of the fittest," a phrase coined, we should remember, not by Darwin but by the man who became Darwin's nemesis, the social scientist Herbert Spencer (about whom more will be said in Part Four). For Margulis, the collaborative complexity of the lichen, the Wolbachia, and the human gut microbes were not just occasional and exceptional but were signs that symbiosis was the cornerstone of a "symbiotic planet." She argued that symbiosis offers a view of nature that contrasts with the famous "red of tooth and claw" view first articulated, not by a scientist, but by a poet who was a contemporary of Darwin. In a famous poem published in 1850, nine years before the publication of *The Origin of Species,* the poet laureate of Victorian England, Alfred, Lord Tennyson, expressed a widely held view of the supposed clash between Christian love and the ruthlessness of nature. In a poem in memory of a friend he asked, "Are God and Nature then at strife?" His answer was that, although Man – "her last work, who seem'd so fair…Who trusted God was love indeed, and love Creation's final law" – is locked in a struggle with nature which, "red in tooth and claw…shriek'd against his creed."[12] Margulis argued that cooperation and mutuality were introduced into biological thought by way of symbiosis, and it can be said that, for now at least, these qualities, and the scientific acceptance of them in natural processes, are here to stay. No longer marginal, symbiosis and endosymbiosis have become central to the debates among the majority of biologists about living systems alongside natural selection and survival of the fittest.[13]

Identities biological and musical

In the lichen, what you see is not what you get. This is an easy lesson for scientists, and a potentially disruptive lesson for musicians. The lichen presents itself and even behaves for the most part like any other organism. It is a free-standing organismal type like a stickleback, starling or finch. The lichen also persists over time, grows, replicates its form in future generations, and changes through adaptation. But unlike the stickleback, starling or finch, it is also something more than unitary in the way that we commonly understand an organism to be. The lichen is chimeric in the

12 Alfred Lord Tennyson, "In Memoriam, A.H.H." (1850).

13 For a recent summary of the current scientific consensus on symbiosis, see: Raina, J.-P. et al. (2018).

general sense that it is composite like the mythic Greek animal. But more importantly for our purposes here, it alerts us to the potentially composite nature of many other forms of life. What if this chimeric quality turns out to be a necessary ingredient of living systems of many kinds?

Once again, this question poses a challenge that extends beyond the boundaries of the sciences by suggesting a greater complexity of identity itself – any identity, in any discipline. The identities categorized by Linnaeus and two further centuries of scientists, expressed in hierarchical Latinate categories like *phylum, class, genus* and *species,* were overwhelmingly based on observations of *morphology,* on features like appendages (as in the jointed limbs of insects in *Arthropoda*), body shapes (as in the hollow and radial bodies of corals and jellyfish in *Cnidaria)* and reproductive structures (as in the flowering and fruiting plants of *Angiospermae*).

There had always been a tension in the Linnaean categories between the grouping of organisms according to shared features – *taxonomy* – and the grouping of organisms according to a common ancestor – *cladistics.* Musicians might recognize similar tensions in musical categories. Names for musical types such as *prelude, sonata* and *reggae* stand both as markers of similar structures or similar functions and as markers of how musical practices persist and modify over time. For example, for eighteenth-century composers a prelude was a short introductory piece (most commonly, for keyboard) which was generated as often by a performer in performance as by a composer at a writing desk. For nineteenth-century composers, a prelude had become a composed piece which stood at the opening of an opera (as in Wagner's *Prelude to Tristan and Isolde*) or a free-standing orchestral piece based on a work of literature like Debussy's *Prelude to an Afternoon of a Faun,* based on a poem by Mallarmé.

Musicians, scholars and teachers normally assume that the two usages are connected by an evolutionary story which explains how the eighteenth- and nineteenth-century preludes are related to each other in a family tree of musical development. The terms prelude, sonata and reggae each stand for distinct ways of being musical, of making music, ways which retain certain characteristics even as they transform over time. The symbiotic behavior of the lichen, like the adaptive, collective and ecological behaviors of the finch, starling and stickleback, offer us models of how we might think about music because they are ubiquitous templates of natural events. We might even consider them benchmarks for how things work here on this planet.

In biology in the late nineteenth century, attention had begun to shift to the cellular and eventually molecular levels of comparison and classification, and the classic tension between taxonomy and cladistics was carried into these new discussions. In biological conversations today, it can be said that what establishes relatedness among life forms are the largely invisible signatures in an organism's genetic code consisting of recipes for specific molecular structures in the organism. Fortunately for non-scientists, despite the increasing complexity of the familial relationship described in taxonomy and in cladistics, the everyday concept of "family" still retains its usefulness in the conversations among geneticists. In a single term, family embodies the tension between these two ways of looking at living forms because it encapsulates in one commonplace term the legitimate groupings of organisms by how they look *and* their equally legitimate groupings by inheritance.

For these reasons, in the huge family of lichens, symbiosis demands our musical attention. If our musical curiosity were triggered by anything observable in lichen behavior, what might the lichen's symbiotic qualities look like when placed into a musical context? What musical creations come to mind which could be considered symbiotic – a mashup of pre-existing creations? Starting from symbiosis as it is defined in the life sciences, what musical practices would usefully be considered a partnership of qualities or practices from distinct origins which continue to develop as distinct genres on their own, even as the mashup itself develops as a wholly new line of creation?

Immunity as immunitas

Before pursuing these musical questions directly into actual repertoires, I am obliged by current scientific trends to first follow the thread of biological thought about symbiosis to where it has now led beyond the place in nature first marked by the mythic chimera. As the Neo-Darwinian resistance to the idea of epigenesis dwindled in the late twentieth century and as symbiosis has gained acceptance, even the function of the cell wall, that most elemental of identity boundaries, has come under review, with biological consequences in the field of immunology and with philosophical consequences for the very notion of identity itself. The cells of all living organisms, those "lipidic [oily] bags,"[14] as Margulis referred to them, are now no longer being seen as simply a battlement fortified with antigenic weapons. Instead, these membranes have become regions of exchange. Thomas Pradeu calls the older view of immunity "the science of self and non-self," which was first articulated in 1949 by the virologist

14 See a review of Margulis on the origin of cells in Gray, M.W. (2017).

Frank Macfarlane Burnet.[15] "It is as if the body can recognize its own individuality," Burnet declared, "and will accept nothing that is inconsistent with that individuality."[16]

Research in tissue transplants provided the early support for Burnet's theory[17] and – as happens repeatedly in science – developments in transplant research later on supported the theory's rebuttal.[18] In many species, "tissues implanted early enough in the embryo or the newborn... can be tolerated indefinitely," says Pradeu. Immunity is no longer the process of distinguishing between what is, so to speak, foreign and what is domestic. What is accepted as self at an early stage might be rejected as non-self at a later stage, and vice versa. Thus, an organism's intolerance and tolerance of what impinges on it, from within its own system and from without, are fundamentally historical phenomena, emerging over time through continual negotiation with environment. Immunology, in this way, is a theater of epigenetic interaction, as in the stickleback's combination of sensitivity to interactions with its surroundings and its deep memory of those interactions. For Scott F. Gilbert and Alfred I. Tauber, "Immunity is more than immunity, it is *immunitas*,"[19] an expression of *communitas* among living things, a special association of organisms which tailor-makes a niche benefitting both host and symbionts. Immunitas is as much an expression of context as of self. This entanglement of immunity with symbiosis and definitions of self adds up to what looks like a surprisingly "chimeric" identity for many, if not all, organisms, including homo sapiens. Contrary to August Weisman's famous "one genome one organism"[20] summation of the biological view of individual identity in 1893 – one of the pillars of Neo-Darwinism – there is now little support for the idea of monogenetic individuals.

The evolution of the cow's four stomachs – and those of the buffalo, reindeer, caribou, zebra and all other hoofed ruminant animals – has been a co-evolution with the bacteria in its gut which allows it to digest the grasses on which it learned to feed over millions of years. Thus, as Margulis put it, "there is no understanding a cow without its symbionts."[21]

15 Pradeu, T. (2012), 41.

16 Burnet, F.M. (1969), 14, quoted in Pradeu, T. (2012), 42.

17 Billingham, R.E. et al. (1953), 603–606.

18 Pradeu, T. (2012), 136.

19 Gilbert, S.F. and Tauber, A.I. (2016), 849.

20 See discussion of the public debate between Herbert Spencer and August Weisman after the publication of Weisman's one genome one organism theory in 1893 in Mitchell, P. (1894).

21 Margulis, L. (2000).

Following Margulis, many biologists have taken a further step into epigenetics by insisting that the cow's co-symbionts would even have to include the vegetation and microbial soils which coevolved with the cow's gut. In the vast tracts of grasslands of North America, Asia and Africa, the ruminant's digestive tract, the chemistry of the soil, the microbes in the soil and the species of vegetation growing in the soil have all adapted to each other, co-creating a niche which is uniquely tailored to support them all. "All evolution is coevolution," is the oft-quoted expression of biologist and systems theorist Stuart A. Kauffman.[22] To this, Gilbert has added, "almost all development is *codevelopment*," a term which he defines as "the ability of the cells of one species to assist the normal construction of the body of another species."[23] Margulis points out that in the human gut there are some 150 essential partnerships with bacterial species. For Gilbert, this means that "the gene set contained by this symbiotic *metagenome* is about 150 times larger than that of the human."[24] The human genetic identity is dwarfed by the collective genetic identity of the single-celled symbionts which enables human functions like embryonic development and digestion.

The century of the holobiont

The term metagenome implies a meta-identity for an organism, a concept now commonly represented by the term *holobiont* ("whole" + "living entity"). The term holobiont and the related terms *holobiosis* and *holobiontism* have taken hold in biology only in the past generation, but the idea has its origins in, once again, the nineteenth century. During Darwin's lifetime, a few scientists were already giving credit to symbiosis and interspecies cooperation as a general principle in evolution. In 1872, the biologist Johannes Reinke used the word "Konsortium" to describe the mutualistic relationship of algae and fungi in the lichen, and in 1877, Albert Frank, a botanist and mycologist, used the term "homobium" to describe the creation of a new organism from cooperating partner organisms.[25] But it is the biologist Adolf Meyer-Abich in 1943 and Lynn Margulis in 1990 who are credited with bringing the concepts of holobiont and holobiosis into the center of biological discourse. Margulis, in her decades-long argument for the importance of symbiosis in evolution, did not use the

22 Kauffman, S. (1995).
23 Gilbert, S.F. et al. (2012), 286.
24 Ibid., 327.
25 Baedke, J. et al. (2020), 150.

term holobiont, but she referred to mashups like the lichen as Meyer-Abich had, as a "symbiotic complex."[26] In 2023, the holobiont idea was defined by Jan Baedke and colleagues as "a close association between different individuals…that together form anatomical, physiological, immunological or evolutionary units."[27] A few years later, Gilbert and Tauber defined the holobiont as "an integrated organism comprised of both host elements and persistent populations of symbionts." Gilbert summarized the development of the concept in this way: "If the twentieth century has been characterized as 'the century of the gene,' the twenty-first century may become characterized as the century of the holobiont relationship."[28]

The symbiosis factor in evolution is carried a step further in the composite makeup of homo sapiens when viewed in terms of ancestry. In the context of all the other complete genomes now known, only 6% of human genes are shared with our nearest evolutionary relatives, the primates. We share far more of our genes – 37% – with bacteria, our forebears from the first billion years following the cooling of the forbidding volcanic surface of planet earth.

Symbiosis, in other words, is no longer a special category of creatures which are monstrous exceptions to the rule of the pure species. If, in the current era of global pandemics, the assumption of self vs non-self becomes less persuasive to immunologists – the very professionals devoted to protecting us from microbial threats like the CoV-2 or Ebola viruses – then we should expect that a similar blurring of identities might appear in other fields. We can expect that the confidence in discrete, pure and uncomplicated identities of any kind will erode, if they haven't already.

Within a few generations in the late twentieth century, organisms like the lichen and coral reef that were looked upon as exceptional and chimerically patched together were now being considered "integrated." We should take this as another of the lichen's bio-musical lessons: *Sometimes it is not new observed facts which alter understanding of a phenomenon, but it is changed understanding which alters what we can see in the facts we observe.* Putting this another way, observable phenomena may remain the same while all that changes is our point of view. In the case of symbiosis, the change was a turning away from an earlier consensus emphasizing standalone individuals and a turning toward a consensus which emphasizes systemic interactivity.

26 Ibid., 150.
27 Ibid., 151.
28 Gilbert, S.F. (2023).

Musical hybrids and holobionts

Hybridization in nature is much less common than symbiosis. It could be argued that in music the reverse is true. This statement should not be taken as an attempt to score a point in a contest between competing bio-musical terms, but is meant as an invitation to readers to stay open to considering the full spectrum of implications of biological thought in music. The chemical and physical constraints of cell walls and the protein chains of genomes make mutualistic collaboration more likely than hybridization, where traits are blended. In other words, in nature, physically distinct entities whose bodies remain physically separate as they act cooperatively tend to be the rule. In music, the combining, merging and exchanging of traits is experienced by musician and listener both as a physical phenomenon and as a mental phenomenon. Any musical experience registers in the human body, in its interconnected systems of nerves and muscles. However, collaborating or merging between two different musical phenomena is common, as when a tune in one genre and the instrumental colors of another genre combine as physical sound patterns in the air. An example of this might be when a European pianist playing an instrument designed for an urban concert hall performs one of Béla Bartók's piano arrangements of folk songs and dances from an Eastern European peasant village.

The sound of the pieces from Bartók's six volume *Mikrokosmos* (1926–1939) strike the eardrum and registers on the nervous system as an integrated event. The sounds are novel in the unfamiliarity of the new combination, made more novel by the dissonances and harmonies which are native to neither a village nor urban habitat. Regardless of whether we would choose to call these pieces a hybrid or a symbiosis, the two separate sources reflecting distinct habitats and histories do not need to jump physical barriers in order to achieve this merger. The merger takes place in the sound waves impacting a musically sensitive human ear.

Exploring the model of mutualism in nature a bit further, what if we were to step outside of the conventional category of composition and simply juxtapose two musical phenomena the way a fungus and blue-green algae are physically positioned next to each other to create a lichen? For example, what if two recordings of two pieces of music were played – side by side, as it were – in the same room simultaneously? Though the two recorded versions of the song "Stop Breakin' Down" by Robert Johnson (1937) and The Rolling Stones (1972) exhibit traits which are objectively similar, when played at the same time, they will most likely not be experienced as mingled, but will strike the listener as a cacophonous jangle.

In this example, I have chosen to focus on The Stones' version, which is itself a recombination. The band openly acknowledged that their song was a conscious reworking of Johnson's 35-year-old recording made a year before Johnson's death in 1938. The Stones' song was, in other words, the product of human brains that had reconceived the sounds of the original as a chimera of sorts, as a mashup of a traditional African-American blues idiom and the rock 'n' roll idiom which was dominating popular music at the time. The actual sounds of the two recordings played simultaneously do not themselves merge. It takes the interaction of brain, muscle and ear in human musicians to accomplish the new combination.

Both versions now strike me as mutualistic in the way that the two strands of music-making combined but also stayed separate. With more listening, the Stones' song impresses us as the result of a collaboration, that is, as separate organisms with separate histories and habitats which persist independently of the symbiotic union. In the standard definition of symbiosis, the relationship between symbionts of different species is *beneficial to one, neutral or beneficial to the other.*

When speaking of benefits in music, it is customary to consider benefits primarily with respect to the artists – the furtherance of the reputations, influence and bottom lines of both Johnson and The Stones. But following where the idea of mutualism and symbiosis lead us, we can see what I would call the systemic benefits of the song. When "Stop breakin' down" appeared in the Stones' *Exile on Main Street* album in 1972, it was well received, and since then it has been replayed countless times globally. In the decades ahead, it triggered other versions, such as Eric Clapton's famous 2003 rendition. Johnson's song, for its part, had been recorded for a highly restricted, niche African American audience. This was the category of the music industry known as "race records," the only platform from which songs by Black artists were permitted to be sold in the United States in the segregated music market of the 1930s.[29] In the Stones' version, Johnson's song was immediately catapulted into an international habitat where, like the separate symbionts of the lichen, it persisted, thrived, spread to other environments and replicated itself through versions of it by solitary guitarists and garage bands. As a symbiotic union, The Stones' song could be said to have benefitted from its harnessing of the qualities embedded in Johnson's lyrics, melody, rhythm and performance, and Johnson's song benefitted from a new and much larger habitat composed mostly of young white people in the United Kingdom and the United States.

29 Oliver, P. (2001).

Keeping this lichens-eye-view of a hypothetical symbiotic linking of the two songs, I believe that exactly which term we choose to apply to this song – is it a product of symbiosis or is it a hybrid? – is less important than the fact that each of these terms describing natural processes asks us to examine the kind of mutualism at work in the creation of The Stones' 1972 hit. My own preference is to view the later rock 'n' roll version of "Stop breakin' down" as a holobiont. Taking this view, the song is, using Gilbert and Tauber's words, "an integrated organism," even as it retains both distinctive qualities of African American music in the rural south in the 1930s and the qualities of popular music rooted in North America and evolving in 1970s England under the label rock 'n' roll. Each of these two symbiont qualities existed before the Rolling Stones' song came into being and have persisted independently apart from it, right up to the present day. The strains of algae and fungi in the lichen also continue on their own, even as each individual lichen growth on a wall or tree stump looks and acts as any other integrated organism worthy of a place in Linnaeus' catalogue. Each of the 18,000 catalogued lichen species has a name in Latin and a scientific description. Each one is as distinct from other lichens as The Stones' "Stop Breakin' Down" is distinct from Johnson's song by the same name and from the other Johnson song (recorded in 1938) which The Stones recorded in 1969, "Love in vain."[30] New generations of musicians listen to the scratchy recording of Johnson's voice and guitar because they knew the Stones' supercharged version, and the two versions continue to develop audiences on their own, each version sinking independently into the memories and voices and fingers of new listeners.

We are all lichens

As a holobiont, the lichen's most disruptive lesson now may be in its implicit challenge to individualism itself. The cluster of terms describing the collaborative and mutualistic activity of living systems at times seem to be competing for the truth. But in keeping with the view of many life scientists, I am preferring to view these terms as layered, at times seeming more like Russian doll concepts, and at other times as stages in a developmental continuum representing different levels of systemic complexity: from symbiosis (the cooperation of different species side-by-side) to endosymbiosis (the cooperation of different species by absorption of one into the other) and holobiosis (the recognition of the composite nature of living things on a cellular and even molecular level). At the very

30 The Rolling Stones (1969).

least, the terms may simply represent different perspectives on a mutualistic dynamic which is fundamental to all living processes.

In the 1940s, the older, non-symbiotic (or pre-symbiotic) view of immunity – Burnet's self vs. non-self – had been confidently proclaimed as the way any organism systemically rejects any influence which is inconsistent with its own individuality. Today, against a growing awareness of organisms in a system characterized by complexity and adaptation, if individualism is embraced at all by life scientists, it is likely to be qualified by words like "promiscuous."[31] This is a word consistent with Margulis' earthy descriptions of the evolution of ingestion, cannibalism and sex, descriptions in which the consuming and reproducing processes overlap. As Margulis put it, "once upon a time, we think, eating and mating were the same."[32] Both eating and sex are cases where in nature the essential boundaries of organism and even cell wall are systemically – lawfully – breached. Promiscuous individualism is also an idea consistent with James Lovelock's characterization of life as something edible, lovable, or lethal, and it echoes what physicist Nick Lane described as the attractions, repulsions, wants and discharges among atoms and molecules. The symbiotic idea, with its implications of porous boundaries and categories may initially attract the attention of nerds in laboratories. For twits in the humanities, the centrality of symbiosis demonstrated in the laboratory and in the field must suggest a rethinking of assumptions about the place of the individual in all creative processes. More will need to be said about this, especially in Part 3, but it is worth offering, once again, what Gilbert and Tauber, the developmental biologist and the immunologist, had to say about the view of individualism itself in recent science, this time within the context of symbiosis and holobiosis: the human community as a whole, they say, is forced to recognize that "for animals, as well as plants, there have never been individuals…We are all lichens."[33]

31 Dupré, J. (2012).
32 Margulis, L. and Sagan, D. (1995), 115.
33 Gilbert, S.F. et al. (2012), quoted in Baedke et al. (2020), 153.

7

The music of lichens, fish and birds

Looking back over the past 200 years, we can now appreciate that ideas like adaptive radiation, epigenesis, symbiosis and co-evolution percolating through the life sciences – each idea accompanied by its own unique cluster of controversies – were pointing in a similar direction. With the benefit of hindsight, we can see that together they were arguing for a biology which was less visible at the time, but which resembles the biology of today in at least two respects.

First, the way was being pointed to a biology which has set aside the presumption of stable and unchanging life forms, determined either by God (in the eighteenth and nineteenth centuries) or by the genome (in the twentieth century), in favor of life forms characterized by ceaseless adaptation.

Second, the gradually emerging biological consensus of today is a science which emphasizes, as never before, the importance of environmental factors in species change. Gilbert and Tauber described this trend in 2016 as a general shift "from a molecular-centered science towards ecology,"[1] an idea already articulated in 2004 by the biophysicist Harold J. Morowitz, who declared that "life is a property of planets rather than of individual organisms." (See the epigraph to Part 2) My own way of putting this complexity has been more home-spun: *in order for there to be an Adam, there had to be an Eden.*

The understanding of nature which in current biology shines through Linnaeus' well-worn categories of life forms is looking more like Denis Noble's theory of biological relativity, where any concept is inadequate which does not at least imply a hyphenated definition – not simply "creation" or "evolution" or "development," but "co-creation," "co-evolution,"

1 Gilbert, S.F. and Tauber, A.I. (2016), 849.

and "co-development." Therefore, the question arises: how is this relational approach embodied in biology applicable to music? How might "co-" concepts like co-evolution and co-creation be usefully applied to musical practices? If this is the direction toward which the science of life on this planet has moved in this century, why wouldn't the science of the planet's human-made sound art not move with it?

Darwin's finches and Lamarck's giraffe

Exactly what these general trends mean – and what they mean for the special case of music – is clearer when they are seen in the context of the changing status of yet another nineteenth-century idea. The so-called "theory of the inheritance of acquired characteristics" is still generally associated with the writings of Jean-Baptiste Lamarck (1744–1829), the most prominent of French naturalists in the generation before Darwin.[2] In much twentieth-century popular science writing, Lamarck's theory of inheritance of acquired characteristics has been misleadingly presented as competitive with Darwin's theory of natural selection. Lamarck famously argued that environmental influences can bring about changes in the morphology of an individual organism, but this is a view which Darwin himself shared – the beak of the finch, after all, changed as it adapted to new habitats. More controversially, Lamarck also claimed that those changes brought about during a given individual's lifetime could be passed on to succeeding generations. About this claim, Darwin was much less supportive.[3]

The most famous of Lamarck's own examples, which has served as a lightning-rod for anti-Lamarckist sentiment for at least a century, was the giraffe.

> The giraffe is known to live in the interior of Africa in places where the earth is nearly always arid and without pasturage, obliging it to browse on the leaves of the trees and to continually strive to reach up to them. It has resulted from this habit, maintained for a long time by all the individuals of the race, that the forelegs have become longer than the hind legs and its neck so lengthened itself that the giraffe... reaches a height of six meters.[4]

2 Gissis, S.B. and Jablonka, E. (2011).

3 Both Darwin and August Weisman shared Lamarck's view of acquired characteristics, though Darwin hesitated to affirm that those characteristics could be inherited. Weisman was more outspoken in opposition to Lamarck's inheritance claims in his theory and led the chorus mocking Lamarck's giraffe example. See: Mayr, E. (1988), 508–509.

4 Lamarck, J.-B. (1809), 256–257, quoted in Burkhardt, R.W. Jr. (1977), 32.

This is not the only evidence of his theory Lamarck offered, but it is the one which has attracted the most attention and has been endlessly ridiculed ever since it first appeared in print, continuing into the twentieth century. But the science historian Richard W. Burkhardt has seen in the anecdote a larger, more enduring insight. Behind the image of the hungry giraffe stretching its neck to feed on treetops was the profoundly disruptive idea that *behavior could be an agent of organic change.* The flimsiness of Lamarck's giraffe example has been demonstrated repeatedly over the years, but Lamarck's intuition which admitted behavior to the understanding of material changes in organisms – that is, *what an organism does, how it interacts with context,* and *with what it interacts* – has borne fruit in recent science. Twentieth-century studies of changes in finch, starling, stickleback and lichen bodies contribute to a (giraffe-free) *neo-Lamarckism,* which now stands with ecology, epigenesis and symbiosis as a key topic of serious discussion and controversy.[5] This is a legitimacy which in the nineteenth-century Lamarckism lacked and which twentieth- and twenty-first-century tools would ultimately give it.

Like epigenesis and symbiosis, Lamarckism had to get through a prolonged period of marginalization in the neo-Darwinian, DNA-centered world of the life sciences in the twentieth century before being accepted into the mainstream. By 2011, in a volume of studies by an international panel of biologists and science historians re-examining the legacy of Lamarck, the supposed competition between Lamarckism and Darwinism was being replaced by arguments for their "complementarity."[6] It is worth noting that the implications of Lamarckism's recent advance to a position of greater respectability are perhaps greater for science than for music. The musicologist Ruth Solie may have been correct when she pointed out in 1982 that the idea of the inheritance of acquired characteristics, however controversial it may have seemed to twentieth-century biologists from the point of view of genetics, has always been "plausible in the history of an art form" and "it may be this very plausibility which has kept the [Lamarckian] metaphor alive in music historiography."[7]

5 Burkhardt, R.W. Jr. (1977), 32.

6 Gissis, S.B. and Jablonka, E. (2011), xii. See also Newman, S.A. and Bhat, R. (2011), 157–169.

7 Solie, R.A. (1982), 300.

Species and songs in an interactionist consensus

In 2001, the philosopher of science Philip Kitcher looked back on the previous generation of scientific debate in fields as diverse as the old disciplines of biology, sociology, economics and philosophy and the young fields of cybernetics and systems theory, and seeing a trend. He summarized this multi-disciplinary conversation as an emerging "interactionist consensus."[8] In all the interdisciplinary turbulence, Kitcher saw expressions of a growing opposition to "genetic determinism," the idea that genes were a conceptual one-stop-shop, the one real answer to any question about causes in human development,[9] and therefore, in nature in general. Of course, putting it this way did not end the debates. His attempt at defining what was a potentially unifying feature of the trend was itself just the tip of the iceberg of controversy and the iceberg was continually growing.[10]

The issue of genetic determinism is not likely to suggest its relevance to music. But for a conjectural bio-musical inquiry like this one, there is at least one potentially useful take-home from the controversy over interactionism. Beyond the focus on genes, Kitcher saw in all these disciplines a coalescing of approaches to causation which stresses the reciprocal effects of individual agents, whether human or non-human, living or non-living. The phrase "inter-identities" in one editorial in a psychology journal[11] hints at how the "inter'" prefix could be meaningfully attached to concepts in other fields. Within interactionism, hyphenated "co-concepts" – as in Scott Gilbert's "all development is co-development" – find a comfortable home in which the scare quotes around them begin to become unnecessary. To be credible in an interactionist context, any account of how life forms originate, vary, adapt, thrive and die will need to address the ecological principle – the obligatory back-and-forth of living systems with their surroundings. In this way, the back-and-forth in the accounts of starlings in flight, of the adaptations of the sticklebacks and finches, and of the lichen's co-symbionts would all qualify as instances of interactionism. No description of how all the life forms of a forest wetland, an arctic tundra, an undersea volcanic vent, or a tropical rainforest behave ecologically can avoid language which is interactionist.

8 Kitcher, P. (2001), 396–414.

9 Ibid., 397.

10 See Oyama, S. (2000) and Kitcher, P. (2001). For a summary and critique of the debates over interactionism and the resistance to the concept as a unifying theory of complex systems, see Griffiths, P. (2006) and Ruiz, M.F. and Umerez, J. (2021).

11 Etxeberria, A. et al. (2021).

The breadth of the interactionism debate suggests a potential place for music. In the way that Ruth Solie suggested with regard to Lamarckism, interactionism may also have a simpler, more commonsense usefulness in music, one more or less interchangeable with ecology. Consider the many adaptations of basso continuo practice over centuries, the infinitely variable transmission of songs in oral tradition, the micro- and macro-variations in the performance of a Schubert quartet or a jazz standard, and the symbiotic absorption of blues songs into rock 'n' roll. These phenomena, from a musical point of view, are reasonably understood as interactionist. To be clear, an openness to the interactionism in any of these musical practices does nothing to change their musical details. But an interactionist perspective does shift our focus on musical events from the individual artist as the primary agent to the artist's role as an actor in a web of transmission, memory and pedagogy. Artists are, from this point of view, the active creators of moment-by-moment events and they are also recipients – the nodes of intersection among the inputs of countless others in a multi-generational web.

As actors within a received web of constraints, inputs and options, musicians are, so to speak, *inter-actors*. They are, at the same time, the entirely necessary agents of each musical experience before us. Musical practices may be usefully called interactionist, in the same way that they may also be understood as examples of epigenesis, symbiosis or of "Lamarckian" exchanges with environment. When we lift terms from biology and apply them to music, we test their efficacy as musical terms, but we are also demonstrating to ourselves how work done by life scientists and musicians may comfortably coexist under the same conceptual tent. What Scott Gilbert saw in the sciences in 2014 might also be applicable to music: an increasing comfort with the idea that "information [is] not a matter of *essence* but of *relation*" (emphasis added).[12] Speaking from the point of view of the humanities in 2018, Donna Haraway said the same thing in *Staying with the Trouble*, but in a way which is perhaps more suggestive for music: "relationships are the smallest possible pattern for analysis."[13]

12 Gilbert, S.F. and Bard, J. (2014), 132

13 Haraway, D.J. (2008), 25–26. Other examples of this trend include: in systems theory, Deacon, T.W. (2012); Holland, J. (2012); and West, G. (2017). In the social sciences, the Agent Network Theory (ANT) of DeNora, T. (2000) and Latour, B. (2005). In the humanities, the post-humanist and interspecies perspectives of Wolfe, C. (2003) and a smaller number of writers on music, including Born, G. (2010) and Hennion, A. (2015). In cybernetics, Von Foerster, H. (2003) and Clarke, B. (2020). The field of ecomusicology, with its roots sunk deep in the societal orientation of ethnomusicology, is

Frozen development in nature and in music

The evolving beaks of finches, the starling's murmurations, the stickleback's nimble adaptations to salinity, and the intimate chemical collaborations of algae and fungi can each be conceptually frozen into momentary snapshots. Music can also be frozen in time and for the same reasons as living processes are commonly frozen by scientists – to study them or teach them. Nevertheless, the enduring attraction to humans of both nature and music is not in captured images, but in real time behavior. The aesthetic impact of the starling ballet,[14] as well as its scientific significance, is in its motion and in the individual movements of many avian actors. When contemplated as a studiable object, a musical event is a butterfly pinned to a tray, but so is a snapshot of a murmuration in an evening sky. A lichen on a garden wall and a stickleback thriving in a new habitat also offer opportunities for snapshots, moments in a developmental continuum. Any of these freeze-frames is meaningful and useful as knowledge about nature and music. What is more, beyond their importance as knowledge in this sense, these snapshots are also enduring images of wholeness. They are icons of the simultaneous enactment of individual agency *and* system-wide self-organization. They serve as examples of both chaotic variability *and* compositional coherence, of stand-alone beauty *and* the beauty of interactivity itself.

In 1953, the developmental biologist Paul Weiss spoke on the continuity of natural and artistic processes to assembled colleagues at an international conference. "Static form is only a *precipitate* [emphasis added] of underlying and antecedent formative dynamics," Weiss said.[15] The examples he offered were cases where the development of a living thing is recorded in successive layers of hard deposits, as in the annular growth rings of a tree (Figure 7.1a), and the stripes of calcium in the shell of a snail (Figure 7.1b). Both examples strike the human eye as harmonious, even beautiful.

the latest branch of this movement and the one most directly addressing music and nature, as in Allen, A.S. and Titon, J.T. (2023).

14 See, for example, the thousands of images and videos of Starling swarming which can be found on the internet, offered by photographers and the daily press. https://www.youtube.com/watch?v=TLDpHmwRGHU; https://www.youtube.com/watch?v=uV54oa0SyMc (accessed 1/3/26).

15 Weiss, P. (1955). This article was based on Weiss's address as vice president of the Rockefeller Institute for Medical Research to the Boston meeting of the American Association for the Advancement of Science in December, 1953.

Figure 7.1a The development of a tree recorded in its annual growth rings (*Popular Science Monthly*. Volume 3, 58. [Public Domain. Author unknown, ca. 1871])

Figure 7.1b The development of a snail recorded in the layers of calcium deposits in its shell (H. Zell. *Cepaea hortensis* var. *lutea lutescens fasciata* 103:5, White-lipped snail; diameter 1.9 cm. Naturkundemuseum Karlsruhe, Nr. SMNK-CONCH 4738; Without indication of provenance. Wikipedia Commons).

We should remember that the dynamics which Weiss was referring to – the chemical processes of an organism's metabolism – are ones which scientists for centuries had been able to theorize but which were not directly observable to them in vivo. Weiss belonged to the generation of biologists who, for the first time in history, were given the tools which promised that within their own lifetimes it would be possible to observe in real time new cells being added to plant and animal structures.[16] Speaking from this experience, Weiss said that what humans admire "as order and beauty in the final form is only a product and an index of the measured orderliness of the developmental actions and interactions by which it has come about." He then cited a historical figure most famous as a poet, but who was also a leading naturalist in the generation before Darwin. Johann Wolfgang von Goethe, he reminded his audience, called architecture "frozen music." In the same sense, Weiss said, "organic form is frozen development; and formal beauty reflects developmental order."[17]

In music, too, we are accustomed to focusing on precipitates. The form of a minuet has none of the material solidity of a tree trunk or snail shell, but the forms are as quickly identified in the ear as a solid object is identified by the eye because memories of other minuets stored in the brain, voice and muscles are triggered by even a few seconds of a musical performance. In the historical periods and regions in which the minuet has flourished, the solidity of the minuet phenomenon is the result of its re-creation by many generations of musicians. The re-creation is never identical, and always happens with variation and adaptation to local circumstances. It is a musical truism that only humans whose experience falls entirely outside of the habitats where minuets flourish can remain neutral to them. Non-neutrality is a characteristic of any ecology, if we understand ecology to be a network of interacting forces and agents which sustain each other and which sustain the whole. An ecology always has a point of view, a tendency to favor some life forms over others.

I am tempted to call musical forms such as the minuet *audible precipitates*, the residues of behavior, as Burkhardt understood behavior in Lamarck's giraffe story. Through the countless individual acts of creation, re-creation, remembering and teaching by generations of musicians and listeners, each occasion when a minuet vibrates the air leaves an imprint

16 Electron microscope technology was already able to provide still photography of cell activity in the 1940s, but the technology for viewing and capturing the real time activity of live specimens was not generally available as a research tool until the 1960s. See: Jensen, E.C. (2013), 1–8.

17 Weiss, P. (1955), 288.

on the consciousness of humans. We know that, as Weiss put it, "trees record their spurts of growth as tree rings...[and] animals preserve a record of their periodic growth in hard deposits, such as shells or horns or nails."[18] These "hard deposits" are the "transcription" of the organism's "formative dynamics" into solid materials.

Musical forms, like natural forms, do not immediately reveal the processes which created them. In both the musical and natural cases, the hardened shapes of the tree ring and the separate lines of a minuet are the residue of repeated processes which are not easy to catch as they fly by. Only after a phrase is executed is its shape apparent, just as the process of tree growth, operating on a much longer time scale, is only perceptible after a full cycle of seasons. The mental processes shaping the musical phrase are hidden from the listener and even the performer at the moment those processes are being embodied through performance. The performance is the tree ring or, to put it more precisely, the tree ring is the lasting impact of the performance on human minds and muscles. The experience is transcribed, as Weiss put it, in the only way that musical experience registers on human beings – as sound waves which excite the nervous system and muscles by way of the ear. Musicians in the act of performing capture the occasion more directly and perhaps more deeply than passive listeners or other kinds of observers (for example, scholars). But for all those participating in sound art, the complex experiences are planted deep in the memory as neural and muscular phenomena.

A minuet is a series of body movements, so we might ask if these movements are the "formative dynamics" of the minuet's music. If the perceptible and memorable forms of the minuet are "frozen development," how would we identify in musical terms the developmental forces which we see embodied in the dancers' movements and which we hear transcribed in patterns of sound? When we talk about music, the reflexive response to any question about the fashioning of a particular musical event is to point to the artist, as if to say: "she or he did it. By sheer force of personal discipline or genius, the piece emerged." The presence of the artist irresistibly draws any questions about music into a vortex, as when all the water in a bathtub is drawn to a single drain in a tight spiral. The full scope of knowledge relevant to a particular musical work is shaped by its ultimate destination, the artist, whose life and personality answers all questions.

18 Ibid.

Context, historical and biological

But there is another thread of musical thought which runs parallel to this individualistic focus: the tendency in both scholarly and popular speech about music to take into consideration the "context" surrounding the artist's work. The artist's body, brain and accumulated skill, his or her biography and compositional models, the traditional practices inherited from teachers and the specifics of human culture from which musical practices have sprung – all these aspects of "context" in musical thought began to make their appearance here in this book in Chapter 3 as part of the discussion of the adaptive radiation of finches. As a contextual way of thinking surfaces in conversation about music, musical thought begins to resemble the contextualizing tendencies of science. of which adaptive radiation, epigenesis and symbiosis are examples.

If we choose, as musicians, to dive deeper into the questions raised by Weiss, we must, once again, acknowledge the differences between the medium of living tissue and the medium of sound. The tree and the snail shell register development in fibers of cellulose and in calcium deposits, while musical form registers invisibly in the ear, in muscle and in memory. Moreover, the significance we give to those sound events – their function as love song and as dance, for example – is mental and social. At the same time, the sounds also leave their imprint in our bodies' responses to them – in the way our heart rate and breath speed up or slow down or the way we involuntarily move or sing with what we're hearing, or the way the composer or improviser's hands, voice and body movements participate in the fashioning of a musical moment.

And yet, if we are to continue playing along with Weiss's model, we need to probe still further and ask what is the developmental order of which the minuet form is a precipitate in sound? The minuet form, shaped in part by the social phenomenon of millions of human bodies moving to melodies in a 1-2-3 rhythm over centuries, is as persistent as the infinitely replicated and varied form of a leaf. If this is so, then how might we think of frozen development in music? How can we put our finger on that development, the phenomenon as distinct from its transcription? These are questions which I am unable to answer now, but they arise when we choose to cross the boundary between what nature makes and what humans make. Individual will and action alone do not get us closer to these answers. A bio-musical perspective can compel us to consider a role for the human artist as an actor who is, at the very least, more integrated with the workings of nature.

Part Two
Finding biology in music

Human creativity and innovation can be understood as the amplification of laws of nature already present in physics and chemistry.

Prigogine, I. (1996), 71

The music of biology and the biology of music are alternate views of the same planetary maze where, as I am proposing, life forms and musical practices are co-created and co-developed. In Part 2, the entry point into the maze is not selected species as it was in Part 1, but selected musical practices. Whether we look for the signs of music in biology or the signs of biology in music, we can expect to pass through the same bio-musical terrain.

Variation is a commonplace concept in biology and music for centuries. In the following four chapters, music will be viewed with a biologist's eyes, so it is variability, biology's view of the systemic changes that ceaselessly churn all living matter, which will shape the examination of variation in music.

8

Making sense of musical variability

Consider one approach to variation in music voiced by an eighteenth-century Boston clergyman, the Reverend Thomas Walter. The subject of his sermon is congregational hymn-singing:

> Our Tunes are, for want of a Standard to appeal to in all our Singing, left to the Mercy of every unskillful Throat to chop and alter, twist and change, according to their infinitely divers and no less odd Humours and Fancies…there are no two Churches that sing alike. Yea, I have myself heard *Oxford* Tune sung in three Churches with as much difference as there can possibly be between *York* and *Oxford,* or any two other different Tunes.[1]

Here the concern for variability takes the form of an anxiety of authority, an appeal to a standard which could somehow bring order to all the unsettling promiscuity of musical practice. The reverend's particular concern for uniformity of performance expresses a larger concern for uniformity of belief. Here, variation is the problem and fixity is the solution, with the notated hymnbook, like scripture, as the ultimate disciplinarian.

On the other hand, we have already learned that for Johann Gottfried Herder, Reverend Walter's contemporary in Weimar, variability seems to cause none of the Reverend's discomforts. "All folksongs on this earth are altered," says Herder, without flinching. This is an attitude toward variability which has tended to cling to folk music over the centuries far more than to art music. In the 1970s, the Irish *sean-nós* (old style) singer John MacGuire voiced a similar embrace of variability without even mentioning it:

1 Walter, T. (1721). Quoted in Barry, P. (1937), 2.

> I'd remember the story and if I remembered the story you see it wasn't too hard to put in a word, if you wanted, into it. Once you have owned the story of a song and the air of it you'll have very little trouble in carrying on...if you remember the main parts of it you'll remember the rest – you can make it... *Everyone had a bit of it and no one had it all*."[2]

For MacGuire, there is a wholeness at the heart of variability, a belief that variants occur in the process of recreating an entity – both "the story of a song" and "the air of it" – which is in some way public property and ubiquitous. Within this collectivity, individual variations – the particular bits of it we each have (though no one has it all) – are secondary in importance to the act of remembering and carrying on.

Such examples in no way exhaust the roles of fixity and variability in music. But the observations of Walter, Herder and McGuire remind us that the questions raised by variability are not simply academic, or relevant only to art music, but have wider implications for music in general. The speakers' contrasting tolerances for variability confirm that in the mind of those making and hearing music, a song's stability is not pre-ordained but very much an artifact of the task at hand. If that task is to conjure up a well-known popular song out of collective memory, its fixity in the minds of singer and hearer will likely be different when the task is to confirm the unity of a group's religious beliefs through song. Speaking of medieval sacred song and its patterns of transmission, Leo Treitler has put it this way: "Stability is not inherently valuable, but becomes more or less important according to the function of music and the degree of centralization in its transmission."[3]

Bungling pianists

As a teenager, the versions of Frederic Chopin's Preludes played by Alfred Cortot in my parents' collection of 78 RPM discs obviously differed from the performance of those same pieces by Artur Rubenstein and Vladimir Horowitz, even though I accepted them all as the same pieces notated on

2 Morton, R. (1973), 10–11, italics added.

3 Treitler continues: "the more that a genre entails the integration of musical parameters such as rhythm, modal structure, consonance, etc., i.e. the more complex it becomes, the more options become limited in its reproduction, and pieces become set – that is, the more the transmission becomes stabilized. Such integration, of course, requires increasing degrees of skill, and favors more centralized patterns of transmission." (1977, 211).

the pages of the C.F. Peters Editions I was following. My uncle, a Curtis-trained classical pianist, would harumpf over Cortot's renditions, calling him a "bungler," much like an organ teacher of mine who would speak condescendingly of Albert Schweitzer's a-bit-too-free interpretations of the Bach Preludes and Fugues compared to those of E. Power Biggs and Marie-Claire Alain. Looking past the screen of moral condemnation in these comments which implied that these variants were the result of error or incompetence, it becomes clear that in the hands of musicians, actual musical events tend inevitably toward difference, even when the musical example is staring at us from the pages of an authoritative text. The Cortot-Rubenstein-Horowitz differences were all small, of course, consisting of subtle shadings of rubato, accent, touch, pedaling and maybe a few ticks of the metronome in tempo. But it was these very details which were the focus of intense scrutiny in the rehearsals of the Juilliard and Guarneri quartets. The changes in the performance of compositions by Schubert and Beethoven which result from the intense focus of the two professional string quartets may be too subtle for the average listener to identify at the moment they are happening. And yet with repeated exposure, experienced listeners come to prize these differences as the signs of the unique voice of a particular musician or ensemble. In the world of the western European art music canon, musical differences, regardless of how small, and regardless of whether they were legitimized by a canonic standard, are actually considered indispensable.

What was less known in the 1950s and 60s was that recorded performances of canonic pieces by nineteenth-century pianists who lived into the recording age often showed even more dramatic differences from each other and from the meticulously edited *urtext* scores on the music stand. By the 1990s, studies by musicologists like Leo Treitler, Jeffery Kallberg and Jonathan Bellman of early 78 RPM discs of Frederic Chopin's works and of contemporary accounts of his teaching and performing would show just how much the tolerance for variability in performance had diminished during the twentieth century[4] (Music Example 8.1). The kinds of flourishes, cadenzas, extensions, harmonic alterations and tempo changes inserted into a Chopin waltz by Vladimir de Pachmann (1848–1933) or Joseph

4 Treitler, L. (2002), 57–77; Kallberg, J. (1996), 215–228. See further discussion of these and other examples in Labaree, R. (2013). Treitler 's list of performers whose early cylinder and disc recordings have been digitally remastered includes: Camille Saint-Saens (1835–1921), Raoul Pugno (1852–1914), Ignaz Paderewski (1860–1941), Ferruccio Busoni (1866–1924), Sergei Rachmaninoff (1873–1943), Alfred Cortot (1877–1962), and Mischa Levitski (1898–1941).

Music Example 8.1 Josef Hofmann's variants in the left hand of Frederic Chopin, Op. 64, no. 2 in 1922, compared with the C.F. Peters edition. Treitler, L. (2002), 64.

Hofman (1876–1957) – not to mention the few surviving recorded examples of how pianists of an earlier age inserted their own spontaneous preludings between pieces on a live program[5] – were entirely absent by the late twentieth century in the performances of those same pieces at the annual Chopin competition in Warsaw.

Clues to the late nineteenth-century sound world of Hofman's youth can be found in what we know of Pierre-Joseph-Guillaume Zimmerman (1785–1853), Frederic Chopin's next-door neighbor and the head of the piano department at the Paris Conservatory. In his *L'Encyclopèdie du pianiste compositeur* (Encyclopedia of the pianist-composer) (1840), he assumes the role of teacher. His pedagogical examples were an "Homage à Chopin;" where he supplied an extra line of music to demonstrate how, in the generations before Josef Hofman, a pianist in the Chopin style would be expected to elaborate on a lyrical line over a steady accompaniment. However, by the mid-twentieth century, such approaches to the performance of Chopin's compositions from the nineteenth and early twentieth century were being disparaged by piano teachers as eccentric deviations from what had become a universal norm, the printed page. In the 1930s and 40s, Hofman himself argued publicly for greater "objectivity" and "exactness" in the treatment of scores. And yet, as Treitler's account makes clear, Hofman's own recorded performances of Chopin would have never met the standards of objectivity displayed a few decades later by winners of the Warsaw competition.[6] The take-home from these changing

5 See Goertzen, V.W. (1996).
6 Treitler, L. (2002), 65–66.

benchmarks of musical variability is a sense that variation in the performance of notated music, while inevitable, was itself variable, a quality acknowledged, embraced and rejected in different degrees according to historical context, local cultural preference and personal taste. In any case, within the confines of western art music, such examples of variability open a door between the invisible web of sound and thought in music and the invisible web of energy and communication in nature. In both webs, replication of existing forms and behaviors is crucial. In nature, the finch remains a finch even as its body parts adapt to different circumstances, and in music, forms like the sonata, mass and blues remain recognizable through generations as revered pieces are played and replayed and new ones are created in their image. In the natural and musical cases, each repetition is accomplished with variation and in each case, variation is not only acknowledged but considered indispensable.

Comparing the incomparable

The forms, behaviors and histories of birds, fish and microscopic life, when viewed through a musical lens, provide doorways between biology and music. But judging from my own experience, those doors can take time to open and the details of those life forms can persistently look like anecdotal curiosities. The symbiotic relationships of the chimeric lichen can strike us as so different from the finch beak adaptations, the starling's in-flight interactions with its flock and the stickleback's responsiveness to its changing environment that we can fail to see what they have in common: *they are all cases where order is being created through systemic interactions among large numbers of forces and agents.*

The time scales of variability in each of these species and in music are registered in fractions of a second, in years, and in millennia, and this can be a further obstacle to our habits of comparison. The geological rate of changes in species (measured in millions of years) can look so different from the adaptations and variations in species and habitats (measured in years), and in real time musical events (measured in years and milliseconds) that they strike us as unarguably *other.* They are as different from each other as the two historical figures in the well-known drawing by the cartoonist Roz Chast.

The drawing poses the question: what do Davy Crockett and Franz Kafka have in common? How does "the Coonskin Congressman" and "American folk hero" who "could whip his weight in wildcats" intersect with the "depressed guy" who "lived in Europe" and who "thought the world had a lot of problems"? The answer the artist provides – "they both liked to eat raw dough" – tests the most absurd edge of my claim that music and nature, different as they are, actually have anything useful to do with each other. As icons of difference, the two cartoon figures invite us to consider that some things in the world might not be comparable. But more importantly, they also illustrate how comparison itself depends on the details selected. Davy Crockett and Franz Kafka are, after all, both human beings, revered icons of cultural values, and each one reflects the cultural values of his own time and place. Taking an epigenetic view of the two men, we can also see that their individual interactions with culture, time and place make them more similar than different. The invitation to absurd comparison eases the way into challenging comparison.

Looking back over the decades, it is easy to wonder at the many continuities between biology and music which I had missed. But I believe that the doorway most accessible at that time more or less opened itself. This was an opening which was a direct outcome of comparison itself, or more precisely, of the disciplinary habit of comparing just about everything to everything else and then asking questions about what the comparison revealed. The opening expressed itself in the commonplace, even humdrum, pair of notions – *sameness and difference* in music.

From the perspective of current science, this will seem like a crude key to open the door connecting music and a twenty-first-century scientific thought community described by the mathematician Ian Stewart as "relentlessly non-linear."[7] This is a community, after all, that has been looking, for the better part of a century, at the subatomic world through the *quantum* glasses of Albert Einstein and Nils Bohr, by which an electron, for example, is perceived as *both* a particle *and* a wave, depending on what it is interacting with. It is a path blazed by physicists, but then joined and broadened by biologists. Fritjof Capra, one of a growing number of physicists who have directed Isaac Newton's classical laws of inert matter and the quantum laws of Einstein and Bohr to the world of living matter, has even labelled this new perspective on physics and biology as *the principle of biological relativity*.[8] In such a world view, made up of identities that are contextually dependent, a musician's problems with sameness and difference might be deserving of no more than a condescending scientific shrug. Once we are in a comparative frame of mind, the experience of change – the observation of variation occurring in any object or phenomenon – remains an

7 Stewart, I. (2002), 83.

8 Capra, F. and Luisi, P.L. (2014). See also biological relativity discussed by the physiologist and biologist Denis Noble in Noble, D. (2017), 168.

essential step in experiencing it as *alive*. The act of comparison and the perception of variability together shape a platform from which music and nature can be viewed at a single glance, joined not by their superficial appearances or their emotional qualities, but by their ways of working, their processes of variation.

Twits and nerds contemplating variability

Moving objects and dynamic states are harder to define, measure and analyze than those that stay still. This statement requires no specialist knowledge to understand. But the biologists' relative comfort with variability in the living systems they study is one reason why musicians open to harnessing the tools of biological thought to musical tasks might feel like biological language is mapping an alien territory. Expressions like "non-linear causality," "emergent dynamics," "genomic metastability" and "far from equilibrium state" (all of which will find their way onto these pages) emphasize phenomena which are indeterminate or in motion. What is more, they also suggest a creativity in which the very agents of variation are murkily collective or at least hard to isolate, especially in terms like "co-creation," "co-development" and "co-evolution," not to mention the even more uncomfortable "self-organization" and "autopoiesis" (self-making). As will be asked repeatedly on these pages, where in all these expressions can we locate the revered creative agents of music – the composers, improvisers and performers prized for the distinctive qualities that make their accomplishments stand out as signs of individual achievement and style? And where in these obscure scientific phrases implying ceaseless movement and shifting identities can we find a place for the familiar piano lessons, globe-trotting virtuosos, garage bands, love songs, avant garde experiments and sacred rituals of human music-making? Or should we take the alienness of this scientific discourse as proof that music is, just as we thought, a thing apart, a discontinuity with nature?

There is certainly nothing new about these questions. They are what the ecological historian Alfred W. Crosby would have considered to be the perennial expressions of mutual discomfort between two professional tribes, the "twits" and the "nerds," his terms for the liberal arts and sciences. He does not address the difficulties of understanding phenomena in motion, the variability being tracked in these chapters. But any meaningful conversation between music and biology, in Crosby's view, would have to overcome the habitual wariness on the liberal arts side of those disciplines which are "fixated on physical minutiae" and the disdain on the science side for the "methodological sloppiness" of disciplines which are "concerned with matters that defy measurement."[9] Simply identifying these tribal differences does not resolve them or minimize them. But recognizing these differences might be a useful tool for understanding some of the problems we experience in our attempts to move more fluently between nature and art.

Developing an enthusiasm for variants

For the small branch of scholars in the twit camp working with the written sources of medieval music up until the late twentieth century,[10] variant details in song notations posed a problem. By now, the problem should be familiar. Although for more than a century, within the liberal arts, oral tradition folk song and notated song dating from before the modern era were not commonly spoken of in the same conversation, both sets of twit scholars were experiencing Gavin Greig's despair over the "endless variants" in their "ruck of records." Different versions of what otherwise seemed to be the same song were obscuring the identity of those songs and inhibiting an analysis of them as works of art on a par with a Verdi aria or a Schubert Lied. Which version was the *real* song? With such a blurry legacy, how would it be possible to study and evaluate the quality of a musical object of the past?

Music Example 8.2 is an example of the range of variability that provoked such questions: three versions of just the first four lines of a single song, "No m'allegra chan ni critz," attributed to the twelfth-century singer-poet, Gaucelm Faidit. The versions appear in three different manuscripts (commonly referred to by the letters *W, G and R),* written by three different scribes in France and Italy between the late thirteenth and early fourteenth centuries.

Neither pronunciation nor spelling in medieval France were standardized, so we should not be surprised by the variability in text we see here. As for melody, the majority of troubadour poems were passed on without a melody that was its vehicle in performance, and the extant troubadour melodies, only about 340 in all, representing just 180 separate poems, are mostly one-of-a-kind. But the relatively few cases where two or more melodic versions of a poem have survived impose an obligation to compare.[11]

9 Crosby, A.W. (1994), ix.

10 See: Aubrey, E. (1982); Van der Werf, H. (1984); and Treitler, L. (2002).

11 About 2500 troubadour poems have survived in approximately 30 manuscript collections, but only four of these manuscripts contain any music at all, leaving some

Music Example 8.2 "No m'allegra chan ni critz" (lines 1–4 only), attributed to the twelfth century singer-poet, Gaucelm Faidit, as they are found in three manuscripts, *W*, *G* and *R*. Van der Werf, H. (1984), 137–140. *W* is the *Manuscrit du Roi*, perhaps written in northern France in the early fourteenth century; *G* is the so-called *Florence Manuscript*, written in northern Italy around 1300; *R is* the *Manuscrit le Valliere*, written in southern France around 1300.

The *R* ms. melody exhibits the classic pattern found in the first four melodic lines of many troubadour songs: A B A B. However, in the *W* ms. version, the internal melodic repetitions are somewhat more varied, A B A1 C, while the *G* ms. version shows even less internal repetition, A B C D. Comparing these three versions, they suggest different positions on the spectrum of this song's variability: *G* most approaches a through-composed form with no repetitions, while *R* is the most clearly marked by repeated material, with *W* falling between these two extremes. This diverse array of melodic behavior from repetition to through-composition allow us to entertain the possibility that all surviving troubadour poems and notations stood for songs which differed in each performance.

---

280 poems with tunes, or slightly more than 10 percent. See: Beck, J. (1908), 7ff.; Van der Werf, H. (1984), 3.

What's in a variant?

For scholars working in textual editing projects in fields outside of music, an intolerance of variants has always made intuitive sense. Beginning during Chopin's lifetime, the biblical scholar Karl Lachmann (1793–1851) led the way to creating a critical edition of the New Testament, but he was faced with the variant accounts of the life of Jesus and the disciples in his Greek sources. Did Jesus have brothers and sisters? Did he ever marry? What is to be done with the versions which make no mention of virgin birth? Faced with questions like these, the presence of mouvance in a text bearing the full weight of cultural significance could not have been seen as anything but an impediment. Lachmann's answer to these questions was the new science of *stemmatics,* based on his "method of common error": two sources were considered to derive from the same ancestor if they reproduce an identical error which could not have been arrived at independently by two different scribes. Stemmatics sought to provide objective criteria for the creation of a family tree *(stemma)* for each manuscript with the goal of reconstructing the single "purified" source from which a particular extant literary work was presumed to have derived.[12]

Scholars of folk tale in the twentieth century dealt with their own embarrassment of variants by collecting and meticulously cataloguing story versions – as many as possible from as many different sources as possible – and then categorizing the variants by theme, as in Richard Dorson's tale-type index created in 1948.[13] The emphasis on assembling variants and on gathering them into classes inverted the emphasis of earlier folklore work, when variants were viewed as a problem in a field where the logical end of collecting was a purified standard version of which Reverend Walter would have approved.

By the 1990s, around the time that Leo Treitler was writing about the distinctive details of recorded performances of Chopin Preludes, he was also proposing a revised view of musical notation in medieval sacred song.

12 See also the discussion of Lachmann's lasting effect on textual scholarship in many fields in Foulet, A. and Speer, M.B. (1979), 1–39 and Aubrey, E. (1982), 224ff. For Lachmann's influence on the early scholars of the folk ballad, especially in the United States, see Wilgus, D.K. (1959), 118.

13 See Thompson, S. (1950). For a good summary of the cataloguing efforts of folk-tune collectors, see Shapiro, A.D. 1975, Chapters I and II. See also the discussion of Lachmann in Foulet, A. and Speer, M.B. 1979, 1–39, and Aubrey, E. 1982, 224ff. For Lachmann's influence on the early scholars of the folk ballad, especially in the United States, see Wilgus, D.K. (1959), xiv, 118.

He argued that the familiar neumes on a four-line staff were, in medieval practice, not prescriptions of musical performances in the modern sense but rather a type of performance in themselves. For scribes in a monastic setting, music notation was, he said, less a coded recipe for a particular work than "a way of exemplifying pieces, just as performing it was a way of exemplifying it."[14] The neumes provide a model for the song – the song less as a fixed object than as a song type, as if to say, *here, the song goes like this.* As with the Chopin variants, the medieval notations are expressions of a tolerance for variability which are characteristic of a particular time and place.

Following Treitler's lead in sacred medieval song, a comparison of three medieval secular song melodies in Music Example 8.1 reveals obvious differences, but all three also to follow a similar contour. For much of these two centuries, the explanation readiest at hand for these differences was like the one offered by many folk song collectors for the variants they confronted: surely these discrepant versions were the result of the inevitable errors by the medieval equivalent of William Chappell's "untutored singers" in peasant villages. In this explanation, the "errors" in a courtly song in Provencal or a sacred chant in Latin were amplified by repetition and miscopying by scribes. The stemma itself sought to relate individual songs to each other in a familial structure but, like many folk song enthusiasts of the time, there was ultimately less interest among medievalists in the most recent generations of variants than in their presumed progenitor, the "original" tune. Although it was an imaginary construction, this original was still considered by many scholars to be the only tune of interest. All other versions were generally presumed by those writers to be faulty and watered-down redactions, exhibiting the signs of historical wear and tear. The passage of time was presumed to show – not the inevitable signs of adaptation and evolution as biology would see it – but deterioration.[15]

Another "evolutionary" theme was echoed in twentieth-century medieval study: written expression (a later development) was assumed to be more "sophisticated" than oral expression (an earlier development). In 1983, the musicologist Dennis Green argued that, in writing, the author, "unlike the oral poet, has the time and privacy in which to compose in advance on parchment, to iron out inconsistencies and to build in new

14 Treitler, L. (1981), 209.

15 See Joseph Bédier's critique of the Lachmann method summarized in Aubrey, E. (1982), 225ff.

levels of significance."[16] Alongside the drive to establish origins there was a drive to establish a timeline of progressive musical development. We now know both of these views suggesting improvement through evolution belonged less to Darwin himself than to certain Darwin enthusiasts in his own day, about which more will need to be said in Part 4. As biological thought is absorbed into reflections on musical repertoires, it becomes easier to see that attention to variability itself is as indispensable in music as in biology. However, it is also true that biologists' curiosity about the constantly changing state of the organisms they study was not matched to the same degree in these examples of the study of medieval music.[17]

Looking beyond the European middle ages, the list of discrepancies, versions and variants in the written records of European art music is crammed with examples of variants meticulously collected in the notated and published versions of Renaissance polyphony, Baroque instrumental music, Schubert Lieder, Beethoven symphonies, Verdi operas, and more. Such notated sets demonstrate a persistent scholarly interest in the artistic options, the mind-changing, and the practical adjustments that contribute to what is presumed to be the artist's final product. But by force of scholarly habit, these variations more often than not are measured against the default benchmark of the stable, replicable musical object. This tendency is strikingly similar to the way that eighteenth- and nineteenth-century natural scientists measured the repertoire of all living creatures against the benchmarks of the Garden of Eden, Noah's ark and Linnaeus's hierarchical catalogue of phylum, class, genus and species. Ernest Ferand's anthology *Nine Centuries of Improvisation in Western Music,* published in English in 1960,[18] stands out as an early sign that scholarship was beginning to recognize that the teaching of European art music prior to the twentieth century embraced variation, not as an artistic problem, but as an artistic norm. In Ferand's case, the term "improvisation" was used to fill the conceptual gap between the presumed final artistic product and its indeterminate beginnings in the composer's imagination. Carl Philip Emmanuel Bach's *Sonatas for Clavier with Varied Reprises* published in 1760 is one of many examples of a famous composer and teacher providing exemplars to students of how to vary the music written between repeat signs.

16 Green, D. (1983), 7.

17 Van der Werf's *The Extant Troubadour Melodies* in 1980 is one sign of a change in how variants were viewed in medieval studies. See further discussion in Part 4.

18 Ferand's research, originally published in German, was actually conducted in the previous two decades before 1960.

Bach writes in his introduction that "variation upon repetition is indispensable today. It is expected of every performer," and then goes on to criticize the tasteless displays of variation which he said, "squeezes the bravos out of most listeners,"[19] giving us a hint of how common it was. His pedagogy seemed to be intended, not to promote or deny performer-controlled variation, but to correct its excesses. For classical piano students in a twenty-first-century American conservatory, Bach's examples (if the students were lucky enough to come upon them at all in their education) could be the source of some discomfort, since they would see little space for this kind of variability in the current norms of concert performance which were bounded by meticulously edited scores. Were they supposed to memorize Bach's reprises for their coming recital? Certainly, for most of them, creating their own variations would feel out of reach. In general, students receive little support or even interest from their own teachers in finding a place for themselves in these practices. Their jazz colleagues, on the other hand, could be expected to take brief interest in Bach's examples, but then shrug them off as musically too conventional to hold their interest.

Finding patterns in the variations of melodies and organisms

The work of folk song scholars like Sharp, Bayard, Greig and Cowdery came to me through ethnomusicology. From there, the way led naturally to writers in closely allied fields focused on the study of folk tales and oral tradition, and from there, to classical and biblical scholarship, which joined in by way of medieval studies. While all these separate efforts tend to stay siloed within professional specialties, taken together, they also testify to the existence of a diverse community of researchers in the liberal arts camp who have at least one thing in common: they are all students of variants.

Biology is the study of life, and life is a concept which has always enfolded into it the commonplace understanding that living things are always changing. But looking closely at the methods and habits of life scientists, their work also shares with music a pattern of vacillation between an emphasis on stable forms and an emphasis on variation. I believe this vacillation reflects a deeper societal ambivalence about fixity and variability more generally. Darwin's new idea of ceaselessly evolving life forms was not a comfortable fit for Linnaeus's fixed Latinate taxonomies of organisms any more than it was a comfortable fit for the biblical consensus

19 Quoted in Ferand, E.T. (1961), 20.

of the time. To most Europeans and Americans, if they thought about it at all, the species visible in the present world were assumed to be the same as those recorded in Genesis, fixed for all time by their creator. For naturalists during Darwin's lifetime, the view was, in the end, more ambiguous. Linnaeus's categories were rigid, each Latinate species name a separate item in a catalogue. And yet, the growing interest in the concept of evolution also allowed for the possibility of species change. In public and professional discourse alike, it was still possible to maintain a belief, not fully acknowledged, that the species categories were themselves fixed. This seemed to be confirmed by the definition of evolution current at the time: change that is too gradual to witness within a human lifetime.

A fresh wrinkle in the complex and ambiguous consensus on fixity and variability was added during the first half of the twentieth century when, for many prominent scientists at the time like August Weisman and Ernst Mayr, the central pursuit of life in motion was temporarily subordinated to the gene-centered preferences of Neo-Darwinism. James Watson and Francis Crick's mid-century work on the structure and function of DNA gave this view additional fuel. One of the prominent symptoms of the twentieth century's gene infatuation described by Stuart Newman was the neo-Darwinian dogma that the unique identity of any organism is fixed in its genetic code. The fixity itself invited a reading which was superficially compatible with the largely unspoken societal consensus on the biblical creation story. For all that it is today seen by many members of the public as in direct conflict with Darwin, the Genesis paradigm can still lend support (for the most part unconsciously) to any aspects of biology which have the look and feel of fixity.

A seduction into life forms

I have already said that time-lapse photography and electron microscope videography of developing organisms have played an important role in my own seduction into life forms. But just as important as these images, it was the ceaseless rethinking of assumptions about living systems in the accounts of two centuries of writers in the life sciences which suggested openings for the musical imagination. The internal scientific controversies over rapidly shifting paradigms among Darwinists, Neo-Darwinists and post-neo-Darwinists resonated directly with more familiar controversies in music. At times, it even seemed that music and biology were singing the same song – sometimes even in the same decades. It was hard to miss, for example, the parallel resurgence in the late twentieth century of two marginalized theories of the Romantic era in biology and music. One of these has already been mentioned – Lamarck's late eighteenth-century evolutionary theory of the inheritance of acquired characteristics which had been temporarily superseded by Darwinian evolution. The second theory was the nature-centered philosophy of organicism in which music had participated during the century between Ludwig van Beethoven and Heinrich Schenker, then marginalized as a holdover from a Romantic movement overtaken by twentieth-century modernism. As we have already seen, interest in both Lamarckism and organicism has been reignited in the twenty-first century in response to recent biological discovery.[20]

Less directly parallel, but potentially more disruptive to prevailing thought in both disciplines, were the late twentieth-century challenges to the idea that genes were the exclusive command and control center of biological processes and, within western art music, the growing critique of the ideology of the unique, authored, fixed and replicable musical work. In both of the biological and musical trends, it was centralized control, the prioritizing of a creative executive function, which was being challenged. Behind this surge of questioning in both fields is the larger point that these developments, one of them inspired by living systems, have inevitably intersected and fueled each other. The findings of interdisciplinary life science research clustered around terms like "the Gaia hypothesis," "systems theory" and "complexity" were filtering through such diverse fields as climatology, sociology, philosophy, economics, psychology and, tentatively, music.[21]

An important outcome of these developments flowing out of biology since the late twentieth century is that scientists of many kinds have been affirming a picture of the natural world which encompasses the worlds of both twits and nerds – both the bulky materiality of nature and the immaterial, fleeting and internal human processes characteristic of language and the arts. The natural world, said Capra and Luisi in 2014, extends without break "from the cell to the realm of consciousness, while remaining in the realm of biology."[22] The philosopher of science Mark Bickhard (drawing on the physicist Norbert Wiener) offered the expression "flame state" as a description of all mental states. Mind and flame are both processes, he said, and in that sense neither can be said to exist, a statement which is as much metaphor as science.

20 For example, in biology, Gissis, S.B. and Jablonka, E. (2011); in music, Watkins, H. (2018).

21 See: Rehding, A. (2021).

22 Capra, F. and Luisi, P.L. (2014), 143.

All the teeming life in the biologist's world reveals a combination of ceaseless shapeshifting, profligate variation and meticulous replication which also suggests the global spectrum of musical practices, past and current, oral and written, which over the past century has become increasingly accessible to anyone anywhere on a widening array of media. Even at an introductory level, the vitality of biological variation conveys the same sense of possibility and surprise that brings most musicians and music lovers to music in the first place. The encounter with these natural phenomena can even trigger a familiar emotional impact. They are all *beautiful.*

9

From the I-word to mouvance

Professional music-makers and music scholars in the west have long been quicker to acknowledge variability and to value it in a jazz performance than in a concert of art music. This means that any musical event qualifying for the label of jazz receives a special status as an exception to the standard of the fixed musical object, an exception usually marked by the term *improvisation.* Do you detect variations occurring in the performance of a musical text? Then it may be said that the performer is improvising – that is, unless the music is not a proper example of jazz, in which case, the variations are possibly mistakes, hopefully small ones that can be tolerated, or as is the case since the late twentieth century in certain avant garde repertoires with labels like *stochastic, chance* or *game music,* they are considered legitimate composer-designed and performer-executed variations. But with just one legitimate term at our disposal – improvisation – how are we to understand the full spectrum of variability as it happens in music across all repertoires in European history? If we are considering a spectrum which exceeds the boundaries of Euro-American musical conventions, then the inadequacy of our single concept increases exponentially.

Even within the narrow confines of western art music, the term falls short. For example, imagine going backstage at the end of a solo violin concert and exclaiming to the performer, "I *loved* your Bach Chaconne! It was just like the score!" No musician in the violinist's tradition (or very likely in any tradition) would take this as a compliment since it presumes that *lack of variation* – that is, the *absence* of the very quality which registers the performer's personal conception of the piece at that time and in that context – was somehow the point. This is one kind of disservice which the term improvisation does to music: with only one legitimate term in the lexicon, the full spectrum of variability – minute and grand, calculated and spontaneous, west and the rest – is out of reach, and therefore of no interest. Variability in the performance of canonic western art music repertoire like the Bach Chaconne is prized, as the violinist in my anecdote demonstrates, no less than in an avant garde game piece by John Zorn or in a moment of Inuit mouth music. Variability in all these repertoires is prized on a continuum with the variability of a performance labelled jazz.

A thought experiment

Living for decades in the intersecting worlds of western art music, jazz, folk music, and world music (the convenient categories defining segments of the music market), I have constantly put the term improvisation to use. But the word's legacy has been complicated. In the quest for a fuller understanding of the endemic variability of music – one of the essential qualities of music which, as I am arguing, it shares with nature – improvisation has played the role of both conduit and impediment. As a conduit, improvisation enshrines a general recognition that music is variable, reserving for variability a limited, but legitimate place in musical life. By directing attention to how music is made in real time, improvisation provides a consistent alternative to the view of repertoires as collections of free-standing, replicable and usually unambiguously authored musical objects. But the wide acceptance of the term as it is has been used in Euro-American music circles for nearly two centuries, and the comfortable place into which it has settled in talk about music, has allowed it to harden into a narrowly defined category of musical behavior. It has become the exclusive province of specialists who are the exception to the rule that the normal way to think about music is that it is an object rather than an activity.

About fifteen years ago, I tried to coax the conflicting potentials of the term improvisation into a thought experiment. For a number of years, I had been placing the word in writing within scare quotes ("improvisation") implying "so-called." It was an attempt to invite questioning about the canonic term. As an approach to classes and in professional conversations, I began to inject an ironic treatment of the term "improvisation" by referring to it as the *I-word,* as if it were a socially unacceptable expression that you don't say in front of children. My hope was that irony would trigger a discussion of our uncritical dependence on the concept. It was becoming obvious that, inescapably, the I-word was invisibly pulling the strings of many musical conversations, creating an "I-discourse" which was the source of many "I-problems," including the schism between "I-repertoires" and "not-I-repertoires," "I-people" and "not-I-people."[1] The goal was to create a dialogue about the unquestioned dualisms of conservatory culture: "western" versus "non-western," "composed" versus "improvised," "written" versus "oral," "normal" vs. "exceptional," and finally, "us" versus "them." These were categories which I eventually realized had made their home within a still larger dualism: "all things human" versus "all things natural." In support of my growing doubts about the I-word as a useful concept, for a brief period in my classrooms I set up a kind of seminar in-joke: whenever a student (or I) reflexively resorted to the I-word as part of a description of a musical example we were studying, we would have to put an imaginary dollar in an imaginary coffee can in the middle of the room. The offender would then have to find a way to say what he or she intended about the variability of the example we were considering without using the I-word.[2]

Over two centuries, the concept of improvisation has become so tightly woven into thinking about music across the spectrum of repertoires and traditions that it has become a kind of avoidance reflex, like when our hand recoils without thinking from a hot surface. In the everyday culture of a professional music school, students and faculty tend to look respectfully (though suspiciously) across a divide which separates those who "improvise" from those who do not. Improvisation has become a matter of unquestioned identity, sabotaging any attempt to understand or even recognize the diverse ways individual musicians think about variability occurring in a full spectrum of global musical behaviors. For our purposes here, normalizing variability rather than improvisation has the advantage of positioning conversations about musical practice on a par with nature, where variability as much as fixity is not only omnipresent, but also a requirement for life. This has already made itself clear in the stickleback view of performance, where improvisation is positioned within epigenesis, biology's concept, not music's. This is not exactly a neutral place, but it is at least equally uncomfortable to both disciplines.

The I-word, we should remember, only came into common use in European languages during the second quarter of the nineteenth century, when it began to establish itself as a form of specialty musicianship, separate from composition proper, involving the creation of music

1 For further discussion of the "I-word" idea, see Labaree, R. (2013) and Treitler, L. (2015).

2 For the record, it must be said that the joke quickly lost its shock value and therefore its pedagogical value.

in performance. Since that time, it has increasingly provided western musicians and scholars a limited zone of legitimacy for any instance of variation, indeed any creation, which cannot be accounted for as the considered choice of an individual composer sitting, as it were, off-stage. Although music composition in performance – on-stage, so to speak – had been at one time a default skill set of both music professionals and popular music-makers,[3] by the middle of the nineteenth century it had dwindled to a small list of performer-composers who were regarded as exceptional, members of a special class of acrobatic virtuosi from whom variant versions could be expected. This is the ecological niche into which jazz, carrying with it the reflexes and idioms of Africa, was to fall in the early twentieth century. The term could also be recruited to a larger cultural distinction in which global musical phenomena are considered to be of two kinds, those which are stable and replicable and those which are not. The Iranian ethnomusicologist Laudan Nooshin's summary of this view gives this dualism a Eurocentric interpretation: "Europeans have art; others, such as Indians, have improvisation,"[4] effectively equating composition/improvisation with us/them.

The normalizing of variability that I am attempting here has the advantage of positioning conversations about musical practice on a par with nature, where variability, not fixity, is a requirement for life, giving us a vantage point far larger than the "normal" that simply stands for what *we do here in my community.* Nature is a platform big enough to enclose all human sound art. The tools of biology operate without concern for music which is old or new, east or west. Viewed from nature, there is no cultural center, only what Darwin gave us at the end *The Origin of Species* and which invites us now to apply to music: "from so simple a beginning, endless forms most beautiful and most wonderful have been, and are being, evolved."[5]

The mobility of medieval poetry

When it comes to discussing variant versions in any repertoire or tradition, certain terminologies from outside of formal music, and even from outside music altogether, have provided surprising glimpses of approaches to the variability shared by biology and music. *Mouvance* ("movement") is one such term, which I offer here as an example of how concepts rooted in

3 See: Cumming, J.E. (2013); Wegman, R.C. (1994); and Schubert, P. (2014).
4 Nooshin, L. (2003), 245.
5 Darwin, C. (1850/1998), 649.

non-musical experience can model the connectivity between conventional music-making and the world of organisms.

The Swiss scholar of medieval literature, Paul Zumthor, first observed in 1972 that versions of the same lyric poem in different manuscripts seldom coincided in all their details, and were fundamentally mobile.[6] With the possible exception of the first stanza, the rhyme words and the short final stanza or *tornada*, the precise composition of a given lyric on the page was often indeterminate. Versions might vary in the number and order of stanzas, the precise composition of individual lines, the number of lines per stanza, sentence structure, word choice, and so on, producing sometimes widely varying utterances.[7] Rather than dwelling on this phenomenon as a problem to be overcome by choosing the best or earliest or most popular variant (which might be the legitimate choices made by an editor of any historical document heading toward publication), Zumthor captioned the variants collectively with the neutral term *mouvance*. From there, he went on to focus on the variation process itself and especially on the interplay of written and oral transmission which surrounded it. Mouvance is an example of the advantage which can come from embracing the systemic normalcy of variation, freeing ourselves of the implications of incorrectness or incompleteness. Even more importantly, Zumthor went on to apply the mouvance he observed in the European tradition to a host of global poetic traditions which exhibited similarly restless polymorphous tendencies, thereby giving European literature a legitimate place in the far wider spectrum of world literatures which are not limited to the page.

He also left a legitimate place on the spectrum of variability for those occasions when the term *zero mouvance* might apply, though the fixity implied by "zero" was understood to be more aspirational or conceptual than absolute. In any case, the term mouvance usefully shifts the frame of thinking about human creations such as the endemic variability I was finding hard to ignore in music more generally, and converting it from a problematic exception to a benchmark.

Turning again to the courtly love songs of the medieval European troubadours and trouvères, we find in them a modest trove of melodies where both the limits of the I-word and the usefulness of Zumthor's mouvance can be demonstrated. In the 1980s, the scholar Hendrik van der Werf succeeded in explaining the variability of troubadour song notations by interpreting the variants in different manuscripts as the normal course

6 "L'oeuvre est fondamentalement mouvante," Zumthor, P. (1972), 73.
7 Ibid., 19.

of events in oral transmission. But this acceptance of variability was complicated by his interpretation of the contrast he saw between the relative sophistication of troubadour and trouvère poems on the page and theit rather plain tunes. His solution to this conundrum was to turn to the I-word: "Their poems[,]" he said, "were performed to melodies which sound to us like improvisations, while the poems themselves are sophisticated creations in carefully designed forms [...] For with notation, the musician was able to compose new, artful, and complex sentences." Here van der Werf is referencing the detailed rhythmic notations of the new polyphonic pieces which were appearing in the fourteenth century, about the time that scribes were assembling these troubadour manuscripts using the older notation system of sacred chant. Without these notational advances, he says, the troubadours and trouvères "made up beautifully flowing and charming but short and simple phrases in a traditional and conventional vein."[8] This one scholar's denigration of the complex practice he saw in the notated evidence by calling it an I-practice is an example of how the I-word can do a disservice to music. Because musical development without notation was unimaginable, the making of music in real time fell (or was pushed) off the scholarly worktable, a situation addressed in Treitler's suggestion that medieval notation be considered a form of performance.

Literature as performance

During the same decades that Sharp, Bayard and Bronson were doing their work on folk song in the British Isles and North America, two classicists from Harvard, Hubert Parry and his student Albert Lord, were tramping through peasant villages in the mountains of Yugoslavia in search of living performance traditions of epic poetry. They were following Parry's hunch that such traditions would provide clues to how the poetry of Homer's *Iliad* was created 800 years before the birth of Christ. The scholars had noticed how often clusters of words or whole lines of poetry in the Homeric text reappeared verbatim or near verbatim in different contexts and proposed that these "formulas" were the building blocks of a literature whose genesis was not the written page. Athena was often "grey-eyed Athena" and the sea, "the wine-dark sea." On the basis of these and many other passages (also commonly called "Homeric epithets"), they proposed that the Homeric text was in fact the product of a poet or singer delivering

8 Van der Werf, H. (1984), 115.

lines before an audience which was only much later, perhaps centuries later, written down.

In other words, one hundred fifty years after Herder's first pronouncements on the variability of folk song, the very literary standard of civilization itself – the benchmark of the stable artifact which was the product of a single human author – was being challenged by a view of poetry emphasizing creation in real time, a view which also offered a revised model of the iconic "artist," Homer himself. Viewed from the Balkans, literature was therefore imaginable as an event, subject to the conditions of the moment, a viewpoint which biologists would be inclined to call ecological. We have already seen how this view was taking root among scholars of oral traditions in general, and with all these efforts, scholarship was stepping out of the library into the field. In the scholarly toolkit, observation of human-generated actions was finding a place next to the decoding of documents. This in turn suggested a fresh view of documents themselves as the medium through which creative actions and processes of the past can be revealed. Homer's "works" were becoming transcriptions of events, the way the form of any organism was a record of the processes which created it, as in Paul Weiss's examples of tree rings and snail shells.

Lord, carrying on Parry's work, in the 1960s established the notions of "formula" and "theme" to analyze the way an unlettered farmer from the mountains of Bosnia-Herzegovina could construct with his voice, while accompanying himself on a two-string fiddle *(gusle)*, tightly disciplined and rhymed verses in Serbo-Croatian totaling the length of the *Iliad* before a coffee house audience on successive nights in the month of Ramadan. Lord's recollection of a moment in the 1930s about the epic singer *(guslar)* Avdo Međedović is revealing. Avdo was a butcher by trade who had never learned to read, but he soon rose in Lord's estimation to become the hero of the oral epic, a stand-out figure, a virtuosic Balkan Shakespeare. Avdo was waiting outside the hall where another singer, Mumin Vlahovljak, was recording an epic tale of several thousand words, an epic Avdo did not know. When the first singer had left, Avdo commented that "it was a good song and that Mumin had sung it well, but that he thought he might sing it better." The result was a version considerably longer than the first.[9]

A closer comparison of two recorded versions of a short segment from just one epic tale, "Alijaga Stočević," performed by the guslar Halil Bajgorić in 1935 and 1950, illustrate how the formulas – segments of poetic lines that are remembered and transmitted as units – are assembled, adapted

9 Lord, A.B. (1960), 78.

and transformed to create longer events. In Serbo-Croatian, the chanted lines flow out in consistent waves of eleven syllables, punctuated with frequent assonances and rhymes and shadowed by the rasping tone and ornamented lines of the bowed gusle in the poet's hands. The formulaic process is apparent even in translation (Figure 9.1). The later version keeps the frame of the earlier tale, as well as some of the actual lines. But adaptations small and large are detectable even in this brief example, leaving us two written stories with close family resemblances – the same replication with differences apparent in folk song melodies and troubadour songs. Lord's many recorded performances showed that frequent repetition of a tale, executed in real time before an audience, can shape the historical record left behind. It is no surprise that the most popular tales, through frequent repetition, tend to move toward zero mouvance over the years sooner than tales that were less popular or newly added to a singer's repertoire. In this sense, the tales were alive – changing and adapting as they settled into a niche in the human society from which they had arisen.

The Chopin problem

It is no accident that the I-word is rarely applied to the musical genres discussed so far in this book. Out of the limited set of world repertoires with which I am personally familiar, the epic song in the Balkans and ancient Greece, the piano music of Frederic Chopin, the songs of the medieval singer-poets, Gregorian chant, and the folk music of the British Isles were specifically chosen for the very reason that the I-word, if it occurs at all in writings about them, has played only a marginal role. Nor is the I-word resorted to by practitioners of these traditions. Neither troubadour, nor monastic singer, nor sean nós singer nor guslar spoke of what they do using any form of the I-concept in any language. However, in Frederic Chopin's artistic circle in the first half of the nineteenth century in Paris, the I-word does begin to show itself. Though we have no record of Chopin himself using the term or its equivalent, Chopin's close friend and fellow Pole Wojciech Grzymała, declared that "Chopin's improvisations were far bolder than his compositions." But Chopin's student Karol Mikuli also stressed that Chopin had "a memory as highly developed as it was reliable," and that he had at his fingertips a large repertoire of pieces by Bach, Mozart and Beethoven, even though his concert programs mostly consisted of his own pieces (how he played these now-canonic pieces is another matter).[10] It is also well-known that his instructions to students

10 Eigeldinger, J.-J. (1986), 282.

1935	**1950**
Stočević Alija fell ill	Stočević Alija fell ill
Midst Stolac, stony city.	Midst Stolac, white city.
He was ill for a full year.	
Ever the afgas of Stolac think	
That Alija has changed worlds.	
He was ill for two years,	
And he was ill also for a third year.	And he was ill for two years of days
Ever the aghas of Stolac think	
That Alija has long since died.	
	Dear God, thanks Thee
	for all things!
	But Alija has no one,
	Except his sister Fata.
	No one knows in the white city,
	Whether Alija is ill or not.
	Some think that he is not there,
	That he has lost his life.
	That marvel was heard afar.
A dark Arab heard of this	A black Arab heard of this
Across the dark blue sea, the deep	Across the sea, dark blue, deep,
Ane he mounted his Bedouin mare,	
Black as a balk hall she was.	
	That Stočević Alija had died,
	And he secured himself
	and his mare,
	And thus the hero spoke:
	"Come, I will go to stony Stolac
	To that Stolac in Herzegovina..."

Figure 9.1 An excerpt from the epic tale "Alijaga Stočević," performed by the singer Halil Bajgorić in 1935 and 1950 in Herzegovina, Yugoslavia. Lord, A.B. (1960), 115.

learning his pieces and his personal copies of them often differed from each other and from their published versions. Sometimes these differences were printed with the composer's approval by different publishers in the same year.[11]

Kallberg's comparison of variability of the left-hand accompaniments in Chopin's various drafts of the Waltz in C-sharp minor (Op. 64 no. 1)[12] resembles Treitler's comparison of Josef Hoffman's 1916 performance of the same waltz with the published score (see Music Example 8.1). This variability in Chopin's paper trail is well known, but in one version of the I-perspective it is interpreted narrowly – as a sign of his composerly perfectionism, his tireless search for the ultimate rendering of a musical idea. The I-perspective takes a different form when extempore composition – an exception to the paradigm of the fixed musical work – is justified by arguing for its place on a natural continuum between inspiration and museum-grade piece. Julian Fontana, the composer's lifelong friend, declared that "Chopin's most beautiful finished compositions are merely reflections and echoes of his improvisations," while Eugène Delacroix, another friend of the composer, likened his improvisations to the preliminary sketches for a painting. Kallberg has taken a more systemic view: the variability in Chopin's paper trail is a more general sign of how "composers, no less than poets, novelists, and painters in the nineteenth century, embraced a fluid conception of the work of art."[13] José Bowen takes this reasoning a step further, calling attention to the degree of tolerance for mouvance as characteristic of a particular genre and historical period: "Nineteenth-century musicians did not have a single aesthetic consciousness: composers like Rossini (from the opera-dominated South) continued to create unique musical events, while composers like Beethoven (from the symphonic-dominated North) began to create more permanent, fixed musical works, with composers like Liszt and Chopin usually somewhere in between."[14]

We have already seen how for folklorists, mouvance has been the cause of some discomfort.[15] However, in the twentieth century, scholars of folk tale have turned their energies to collecting and cataloguing story versions –

11 Kallberg, J. (1996), 215–228.

12 Ibid., 362.

13 Ibid., 267.

14 Bowen, J. (1999), 429.

15 See the discussion of Lachmann's lasting effect on textual scholarship in many fields in Foulet, A. and Speer, M.B. (1979), 1–39 and Aubrey, E. (1982), 224ff. For Lachmann's influence on the early scholars of the folk ballad, especially in the United States, see Wilgus, D.K. (1959), 118.

as many as possible from as many different sources as possible, and then categorizing them by theme, as in Richard Dorson's tale-type index created in 1948. The emphasis on assembling variants and gathering of them into classes in the manner of naturalists inverts the intent of folklore work a century ago, when variants were viewed as a literary embarrassment in a field where the logical end of collecting was a purified standard version of which Reverend Walter would have approved.[16]

I-music, I-specialization and I-studies

The Reverend Thomas's frustration with variants in the eighteenth century was directed at the singing of laymen, the non-musicians in the church pews. His complaints sprang from a view of authority tinged by theology (and perhaps tainted by class) which pushed his thinking about variability towards judgements of incompetence and illiteracy. The topic of variability tends to float to the surface in popular music studies more often than in art music, but the suggestion of incompetence in folk song was also reminiscent of the twentieth-century "bungler" epithet attached to Alfred Cortot's readings of Chopin compared to Rubenstein's. Carl Czerny (1791–1857) the student of Beethoven, a contemporary of Chopin and one of the premiere piano pedagogues of his day, may have shared a bit of Reverend Walter's disdain for lay music-making, but he had none of the Bostonian's discomfort with mouvance itself. This is an attitude he shared with other prominent pianists of the time. In his well-known 1836 manual on improvisation directed at a German-speaking audience of cultivated amateurs, Czerny devoted many pages to identifying several types of variation which he considered the legitimate stock and trade of professional concert improvisation.[17] At the time, improvisation (Czerny's German word was *Fantasieren,* or "fantasy-making," a reference to the familiar musical form *fantasia*) was a staple feature of public music-making for pianists and violinists. Most of his book, like similar publications by Johann Nepomuk Hummel (1778–1830), Franz Kalkenbrenner (1784–1849), Ferdinand Reis (1784–1848) and Ignaz Moscheles (1794–1870), is devoted to notations of his own graded and classified examples which

16 See Thompson, S. (1950), 1137–1138. For a good summary of the cataloguing efforts of folk-tune collectors, see Shapiro, A.D. (1975), Chapters I and II. See also the discussion of Lachmann in Foulet, A. and Speer, M.B. (1979), 1–39 and Aubrey, E. (1982), 224ff. For Lachmann's influence on the early scholars of the ballad, especially in the United States, see Wilgus, D.K. (1959), xiv.

17 Czerny, C. (1829/1983).

served as models for aspiring pianists. (For the less ambitious amateur, the pieces could also be memorized and spliced into a recital where a Hummel or a Moscheles would have put it.) Czerny's examples now look a lot like the direct heirs of the previous century's basso continuo and partimento exercises – elaborations of harmonic skeletons such as the rule of the octave model in Music Examples 3.1a and 3.1b.

For our purposes here, it is worth mentioning that in Czerny's day the musicianship required of improvisation was not thought of as different in kind from that of composition.[18] The eighteenth-century definition of musicianship, approximated in the hyphenated identity of the performer-composer, was actually in the process of changing in Europe, a fact which helps to explain the profusion of such manuals as Czerny's. The expectation of the previous generation of teachers like Carl Philip Emmanual Bach was that the well-trained keyboardist should be able to create a wide range of compositional types in the course of performance – including preluding before a composed piece, long and short cadenzas and lead-ins, the varying of reprises, variations on popular tunes, and personalized arrangements and adaptations of composed orchestral and opera repertoire. Czerny, perhaps in an effort to counter the larger trend toward a stricter division between creators and performers, went a step further than some of his contemporaries by even asserting that his own teacher Ludwig van Beethoven – the paradigm of what Peter Kivy in 1995 called the "composer worship"[19] in late twentieth-century music scholarship and education – was capable of improvising the form which had become the icon of the composer's singular craft, the sonata.

For us in the music world, more than in the world of biology, the gravitational pull to the solitary, heroic individual agent of variation seems to be irresistible. But looking into nature, what would the I-word, so inseparable from the unique spontaneous acts of individual humans, add to our understanding of the ceaseless variations in bacterial bodies in the human alimentary canal? Each of those bodies exhibits a tendency toward zero mouvance – that is, they are almost identical, but not exactly – and somewhat less frequently, they exhibit differences which are in some sense truly novel. (Some of these novelties might be experienced as threatening to the human whose bowel is the stage on which such novelties make their debut.) But no matter whether the variations are large or small,

18 For a recent account of important performers and pedagogues of improvisation in the early nineteenth century, see Gooley, D.A. (2018); Lohr, N.J. (1993); and Whitmore, P. (1991).

19 Kivy, P. (1995), 278. Quoted in Cook, N. (2001).

the tendency toward change is ceaseless. We are left to wonder how the performance of a bacteria in the human gut is different from the performance of a *canso* in the airwaves of a twelfth-century Provençal court or of a nocturne in the airwaves of a nineteenth-century Parisian salon or of a tale chanted in the smokey interior of a Balkan coffee house? I pose this kind of question as a thought problem which might lead to a variety of useful responses, mostly made up of further questions. Any entity that qualifies as living can be expected to exhibit variability. I suggest that if we try to understand the variability at work in the artifacts of both nature and music side by side, little room is left for a useful distinction between what is "improvised" and what is "composed."

10

Finding biology in melody

The archives of folk songs of the British Isles, of the songs of the troubadours and trouvères and of Gregorian chant resemble each other in one important way: they are all collections of melodies. They share this monophonic identity the way that fish and humans and rodents share an identity as vertebrates – they all have backbones – and their melodic identity sets them apart from the other predominantly multipart repertoires which have attracted the most attention in European music education and practice since the Renaissance. Different as these monophonic families are from each other historically and culturally, when taken together, they place before musicians an array which begins to approach the diversity and sheer numbers of the collections of lichens, bird beaks and fish feeding structures on the biologists' lab tables.

It is possible to see now that the outsized impact of folk songs and medieval songs on what was to become a biology-oriented approach to music was not due to any special musical qualities of these huge melodic families. Rather, the impact was due to the way that variability itself leaves its mark on the record of these repertoires. In a host of written and recorded sources, folk song and chant together present the student and scholar with a volume of variant versions which overwhelm musical curiosity. The sheer numbers of these variants are noticeably greater than that in the record of later art music repertoires. Their volume and diversity resemble the shifty, polymorphous patterning of nature I was seeing in, for example, the swarming of starlings by the thousands that Sharp saw in the sky over Somerset and the lichens appearing spontaneously on the scrub oaks outside my windows in New England. It was becoming apparent that the scale and variety of these musical and biological data sets taxed the ability of observers to grasp.

Melodic variants in a revival of Christian spirituality

It was Roman Catholic scholars of the nineteenth century who, undaunted by the immensity of the task, first leaped into the twelve-century accumulation of institutionalized and preserved European sacred song. In general, it can be said that the potential presence of Gavin Greig's vague "something" running through all the melodic detail of their sources was on the minds of the scholar-monks, but their interest in that something ended up being secondary. Beginning in the 1830s, as Darwin was starting his career as a naturalist by circumnavigating the globe aboard the H.M.S. *Beagle*, the monks' unprecedented scholarly project was taking shape under the direction of Dom Prosper Guéranger (1805–1875), the first abbot of the newly established abbey at Solesmes, in France.

This work had emerged from the shadow cast over the Roman Catholic Church by the secularizing trends of contemporary republicanism and Bonapartism. The scholars' mission was ultimately less about music than about faith: the restoration of Gregorian chant as a devotional practice for the revitalization of contemporary Roman Catholicism. Foremost in the scholars' minds was their maddening ruck of records – the thousands of the Church's handwritten sources housed in libraries, monasteries and private collections all over Europe. But the spiritual goals of the movement and the institutional goals of the Roman Catholic Church could not help but influence the way that all the variability in this huge manuscript archive of chant was ultimately viewed. The restoration movement, whose epicenter by mid-century was increasingly the atelier of scholar-monks at the Solesmes monastery in France, had its roots in a post-Enlightenment spiritual yearning for a return to an imagined Age of Faith. This is an orientation which required that, in David Hiley's words, "to the 'Gothic Revival' in architecture, a comparable revival in church music would be joined."[1] The reformers sought to reclaim the original practices of Roman Catholic sacred song after centuries of what they considered to be corruption. From a biologist's perspective, even in the reformers' own day, these corruptions would easily have been taken as signs of adaptation and evolution.

An account by none other than Felix Mendelssohn of chant practice at the Sistine Chapel in 1830 communicates something of the urgency which the Solesmes scholars felt, but its musical details also serve as testimony to how chant practice had evolved in a contemporary European habitat dominated by opera.

1 Hiley, D. (1995), 622.

> The intonation is entrusted to a soprano soloist, who hits the first note with some vigour, loads it with appoggiaturas, and rounds off the final syllable with an extended trill. The sopranos and tenors sing the melody as it stands in the book, more or less, while altos and basses sing in a third...the whole piece is sung in a sprightly rhythm.[2]

Faced with the changes in chant practice that time had allowed into the tradition, the monks chose a backward-looking reform program which they believed would lead forward to a re-invigorated liturgy conducted, according to Peter Wagner in 1901, "in the old venerable forms."[3] The monks' *Paléographie Musicale* editions of manuscripts from the ninth to seventeenth centuries were intended to stand behind the carefully edited and Vatican-approved publications of liturgical repertoire useful for daily devotions, the most widely circulated of which was the *Liber Usualis*.[4] These are the publications which Guéranger had hoped in 1843 would lead the way back to the chant "which most clearly reflects the reading of St. Gregory" – Pope Gregory I (540–604), the namesake of Gregorian chant.[5]

The study of musical variability was sharpened greatly by a new technology in the mid-nineteenth century, the photographic facsimile.[6] It should be remembered that, in a similar way, the study of variability in the forms and genomes of sticklebacks and lichens was advanced by the electron microscope in the mid-twentieth century. Both technologies enhance what we can see, collapsing into each other the common meanings of "see" and "understand." Facsimiles of multiple notations of the same chant, originally written down in the scriptoria of monasteries all over Europe in different centuries, could be positioned side-by-side for comparison like laboratory specimens. These activities of scrutinizing and comparing were shaping an emerging discipline considered by its practitioners at the time to be "scientific." These methods were not altogether new, but were borrowed from the methods of well-established traditions of classical and biblical philology and paleography. In all these disciplines, the goal was the production of authoritative texts, and to achieve that end, the variability of sources was a problem to be surmounted. By the end of the century, the French medievalist Pierre Aubry was the first to call this new music-centered discipline "musicology."[7]

2 Quoted in Cattin, G. (1984/2011), 96.
3 Wagner, P. (1901), 274–275.
4 *The Liber Usualis with introduction and rubrics in English* (1961).
5 Quoted in Combe, P. (1969/2003), 13.
6 See Immel, S.C. (2002).
7 Bergeron, K. (1998), 92, fn 1.

In the view of Dom André Mocquereau (1849–1930), for three decades the principle editor of *Paléographie Musicale*, the photographic facsimiles of those volumes were intended to be a "war machine" in the struggle against the corrupted chant editions in circulation and the practices of the kind documented in Mendelssohn's anecdote.[8] It is to be expected that the higher calling of the reformers would lead them to seize on any signs of what Paul Zumthor called zero mouvance as a symbol of what their movement longed to establish: a tradition which was authentically old, stable and enduring. An unusually invariant array featured in Volumes II and III of *Paléographie Musicale* – the antiphonal *Justus ut palma* – entered the polemic of the time as the model of an enduring tradition.

These exceptional examples of zero mouvance were offered in answer to those who felt that a restoration was impossible because of the bewildering diversity of versions. Looking back on the work of the Solesmes enterprise in 1920, Mocquereau declared that the 219 invariant versions of *Justus ut palma* "testified in favor of the Gregorian musical tradition of the Roman Church by contributing to our collection, with the exception of some insignificant variants of detail, *always the same melody – the very same melody as in the Solesmian Liber Gradualis*" (italics in the original).[9] In other words, despite all the variability of the multi-century tradition of ecclesiastical song, it is the tradition's remarkable stability which was emphasized in the official editions. Like Reverend Walter, the Solesmes reformers labored to create a scientific standard which would attest to the strength of the tradition shining through all those "insignificant variants." Once again, behind the concern for uniformity of performance stands a larger concern for uniformity of belief. If variability was the problem, then fixity was surely the solution.[10]

Order with tolerances

Laying bare the stunning variability of the tradition's melodies was not exactly the spiritual goal for which these early formers labored, but from our current position in the twenty-first century, all this work did invite

8 Hiley, D. (1995), 626. Hiley quotes Combe, P. (1969/2003), 126 for Mocquereau's war imagery, which continues: "...une sorte de 'tank' scientifique, puissant, invulnerable, capable d'enfoncer tous les raisonnements ennemis".

9 Mocquereau, A. Dom (1920), quoted in Bergeron, K. (1998), 174.

10 Leo Treitler's interpretation of this focus on stability calls attention to the ideological commitment behind the musical details: to the Solesmes reformers, according to whom, "Gregorian chant was a canon, authentic, uniform, inspired, as surely as were the books of the Bible." Treitler, L. (1975), 3.

questions about how this vast repertoire – and by implication, repertoires of all kinds – are best thought of less as collections of musical objects and more as exemplars of complex human activities. From a biological perspective, the process of boiling down the variability of the tradition into a spare and practical collection of fixed items also exposed musical behavior on a grand scale comparable to the lawful behaviors of organisms. All these behaviors, both natural and musical, were instances of collective actions which required individual actors possessed of highly developed capacities for adaptation to other actors and forces in their habitat. For a lichen, these adaptations are registered in biochemical exchanges. The great chant project suggests that in music the comparable exchanges are in social behavior and memory, transmitted aurally and registered in the human body.

In the case of both bird flocks and devotional song, inherent qualities in the bodies of birds and singers provide them with the basic capacities of cognition and response necessary for them to accomplish the tasks of flying and singing. These are the qualities which are built into natural and musical materials and which both limit and enable murmuration and song-making. Taken together, they suggest the outlines of an ecological view of natural and musical behavior. Just as important, there are also the constraints on the behavior of bird and singer which both sets of actors have acquired gradually over many generations These are the constraints which limit, and therefore shape, what is created in the sky and in the chapel. But within both of these types of constraints, the details of the birds' and singers' behaviors were determined locally in real time, suggesting the "plasticity of form" in nature which the biologist Paul Weiss had spoken of in the 1950s. In that same speech, Weiss went on to describe that plasticity of form in this way: "life is order but order with tolerances… true organic order sets only the general frame and pattern, leaving the precise ways of execution adjustable, and, to this extent, indeterminate." It is a principle which shows itself, as Weiss put it, in "laws of development which prescribe only the mode of procedure but leave the actual execution free to adapt itself to the exigencies of a world whose details are themselves unpredictable."[11]

In the case of the monks, the accumulated adaptations and experience which prescribed the "mode of procedure" were the product of years of practice, often beginning with a childhood spent in a monastery which was strictly governed hour-by-hour by the Rule of St. Benedict, introduced

11 Weiss, P. (1955), 297.

in the sixth century.[12] For any individual monk who aspired to the role of lead singer, his practice was further refined through imitation of a master practitioner. The exigencies of chant practice presented this singer at each of the eight prescribed devotional hours of the day with the need to render different biblical texts and execute ritual demands which were modified according to day and season. All of these requirements were executed according to the detailed requirements of melody-making established over generations of practice within the singer's religious community. This is where Weiss's plasticity of form applies: the many constraints on the singer were what he called the laws of development which prescribed a general procedure but left the actors free to adapt to the moment's demands. Ultimately, the actual form of a given musical item was indeterminate.

Looking through the lens of recent biology, the musical examples in this chapter may be viewed in a way that most biologists today would consider unexceptional in the workings of nature. In the vocabulary of biology, the examples exhibit the characteristics of living systems. The song-making practices hinted at in the documents, pedagogies and performative behaviors which were assembled and scrutinized by three generations of scholars in the Solesmes project add up to ways of working which are more complex than what can be conveyed to future generations by any one form of transmission. Neither notation, nor theory, nor reductive models, nor close observation of performance, nor direct pedagogy, nor imitation, nor (during the last one hundred years) audio recording, is alone able to fully represent the complex human behavior behind chant practice. The scholar-monks were perhaps more aware of this truth than most of their superiors in the Vatican, but these perceptions were pushed to the side as they pursued the Church's institutional mission.

The Solesmes project in particular and medieval repertoires in general demonstrate that comprehending living musical practices is like comprehending living organisms or ecosystems – we must exclude none of the available signs of life we find in them. In a tradition transmitted primarily in writing, the key to the living practice is the variability visible in those writings. Any species or musical tradition which generates variants is a candidate for the status of a complex system which is, in the language of current science, *living*. As the monks knew well, the signs of variability tend to be fragmentary, dispersed in their archives of notation collections,

12 *The Catholic Encyclopedia.* Chapters 9–19 are occupied with the regulation of the Divine Office, or Canonical Hours, seven of the day and one of the night. Detailed arrangements are made as to the number of Psalms, etc., to be recited in winter and summer, on Sundays, weekdays, Holy Days, and at other times.

pedagogical manuals and oral testimonies. It took special effort, and the help of new photographic technology, to bring them together and into the light. These signs, and the techniques for unearthing them, should all look familiar to students of musical traditions of the past. Nothing in them is alien to the practice of musicology today, nor to the liberal arts in general. They are the trademark methods of Alfred W. Crosby's whole tribe of twits.

Building with modules

After more than two generations of scholarly labor and controversy, in 1904 the Solesmes editions (some of which had begun to appear in print in 1889) were adopted as the church-wide standard by Pope Pius X, making available to Roman Catholic congregations examples of authentic ancient repertoire stripped of centuries of "insignificant" accretions.[13] The Solesmes publications, therefore, represent a different kind of document from the examples in Figure 10.1, which have a parallel but different emphasis: the exploration and preservation of the chant tradition as a performer-centered melodic practice.

Figure 10.1 is an excerpt from the edition of the Sarum (Salisbury) chant tradition of the thirteenth century prepared between 1901 and 1925 by William Howard Frere. Here, Frere's analysis of the repertoire preserved in the manuscripts reveals the melodic practice at work. In the introductory volume to Frere's edition he summarizes as either "typical" or "original" the melodic practice he observed in a pool of complete melodies consisting of different chant types.[14] Frere, like Mocquereau, was concerned with comparing and analyzing full examples of repertoire. He also shared with the rest of the nineteenth-century chant reformers the goal of establishing authoritative versions of repertoire. But where so much of the effort of the Solesmes workers was focused on uncovering the enduring melodies obscured by the manuscripts' "insignificant" variants of detail, for Frere the variants themselves also hinted at something significant in their own right:

13 For Dom Mocquereau, the examination of all the variant versions available would have rightly continued for at least another generation before a definitive edition of the repertoire could be completed. This position was decisively set aside by the Vatican when it adopted the Solesmes publications produced in 1904 as conclusive. From then on, in the mind of the Vatican hierarchy, the account was to be considered closed. Any further study of the manuscripts was to be in the interest of secular research only, with no further impact on the liturgy of the church. Bergeron, K. and Bohlman, P.V. (1992), 192–193.

14 Frere's study included some 800 antiphons, 100 invitatories, and 350 responsories.

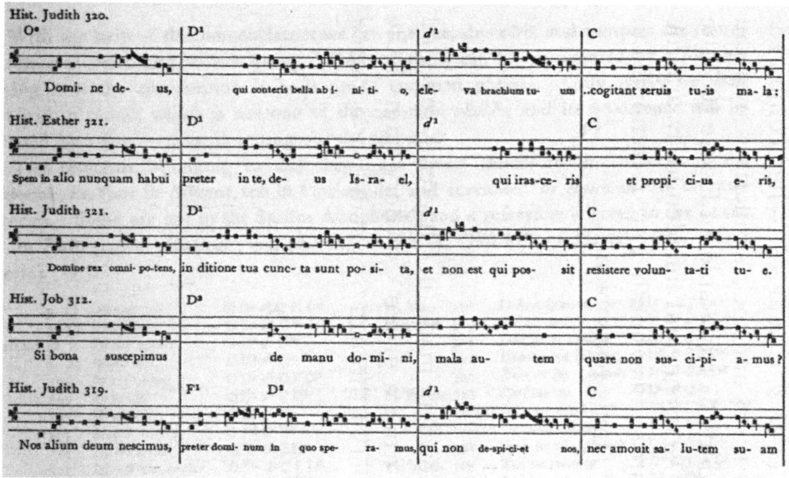

Figures 10.1a and 10.1b W.H. Frere, *Antiphonale Sarisburiense*, Introduction (1901), showing five examples of melodies on texts from the books of *Esther, Judith* and *Job*. (Note: the final phrase of each melody is omitted.) Catholic Church Diocese of Salisbury (England), et al. (1901), 6–7.

Adv.[1]	p. 13	*Letentur celi*	$O D^1 d^1 C k^1 D^3$	H. Job	312	*Si bona* (above)	$O D^3 - C f \Gamma D^3$
Adv.[2]	p. 25	*Docebit nos*	$O D^3 d^1 C F d D^2$	H. Judith	319	*Nos alium* (above)	$O F^1 D^2 d^1 C D^3$
Mens. Dec.	pl. l.	*Emitte agnum*	$O D^3 d^1 D^2$		321	*Dominator domine*	$O D^1 d^1 C A D^3$
	L. 26	*Rorate celi*	$O D^1 d^1 D^2$		321	*Domine rex* (above)	$O D^1 d^1 C F D^2$
XL[5]	194	*In te jactatus*	$O D^1 d^1 C F D^2$		320	*Domine deus* (above)	$O D^1 d^1 C A D^3$
	195	*Ne perdas*	$O D^1 d^1 C D^3$	H. Esther	322	*Conforta me*	$O D^1 d^1 C -$
	195	*Ne auertas*	$O D^1 d^1 D^3$		G. 412	*Miserere domine*	$O D^1 d^1 C A F \Gamma D^1$
	197	*Erue a framea*	$O D^1 d^1 D^3$		G. 413	*Da nobis domine*	$O D^3 d^1 D^3$
	197	*De ore leonis*	$O D^1 \delta^2 C D^3$		321	*Spem in alio* (above)	$O D^1 d^1 C \gamma D^3$
XL[6]	205	*Ingrediente domino*	$O D^1 d^1 C F D^2$	H. Mach.	325	*Tua est potentia*	$O D^1 \Delta^3 C A D^3$
	208	*Opprobrium*	$O D^1 \delta^2 C F D^1$		326	*Refulsit sol*	$O D^1 d^1 D^2$
	208	*Synagoga populorum*	$O D^2 d^1 C D^2$		324	*Exaudiat d.*	$O D^1 d^1 C D^3$
Cena	217	*Judas mercator*	$O D^1 d^1 C F D^2$		L. 299	*Tu d. universorum*	$O D^1 d^1 C D^2$
Parasc.	226	*Barabbas latro*	$O D^1 - C - K D^2$	H. Proph.	332	*Qui celorum*	$O f D^3 d^1 C A D^3$
Pasch.[4]	261	*Narrabo nomen*	$O F D^4 - k^2$ alleluias		L. 307	*Ciuitatem*	$O D^1 d^1 C F D^2$
H. Sap.	310	*Initium sapientie*	$O D^3 d^1 C A D^3$				

the presence of an enduring melodic *idiom*, his term for the "something" that Gavin Greig had sought in vain to confirm in Scottish folk song.

Significantly, Frere's synoptic array in Figure 10.1 compares different chants in Mode I (about modes more will be said below) with different texts rather than variants of chants with the same text. This is a type of comparison also taken up by Mocquereau in the commentaries to certain editions, a purely melodic interest sidelined in the Solesmes project by the overarching goal of creating fixed repertoire items for daily devotions. When Frere turns his attention from repertoire to practice, his emphasis is on the modular characteristics of a family of tunes, and therefore on the melodic grammar which enables the assembly of those tunes. Frere's analysis of the five melodies in Figure 10.1b lays out what he calls *themes* within a selection of responsories in Mode II which he considered bore a familial resemblance to each other through their use of shared *formulas.* According to Frere, these formulas recombined (to use Cowdery's description of certain folk song families) in a variety of forms.

Frere's theme IIa in Figure 10.2 consists of varying combinations of just five principal formulas labeled *O, D, d, C,* and *A.*[15] When compared, both the range of variability of the tunes and their similarities in detail and structure are immediately obvious. For example, in Figure 10.2 the antiphonals *Nos alium* and *Domine Rex* begin and end identically, and the component formulas of each song are similar, but they are recombined in a different sequence, obscuring the family resemblance.

Nos alium: O F1 D2 d1 C D3
Domine Rex: O D1 d1 C F D3

This was the kind of comparison Cowdery made (in Chapter 4, repeated here) of two recombining tunes from a group Cowdery named the "Blackbird Family" using the numbers 1 through 6 to stand for a shared repertoire of "melodic moves":

Lord Gregory: 156 123 345 45
Sail og rua: 345 561 1235 56

15 In Frere's method, the system of letters standing for particular melodic moves is consistent throughout all eight modes. Thus: *O* = a standard opening in a given mode; *C* (upper or lower case) = a phrase ending on C; *D* = a phrase ending on D, etc. He also uses Greek equivalents for pitches. For example, a Δ, a *D* and a *d* would all stand for different basic types of phrases ending on D. Further variations are indicated with superscripts. *X* = a penultimate phrase which tends to be less consistent than any of the others.

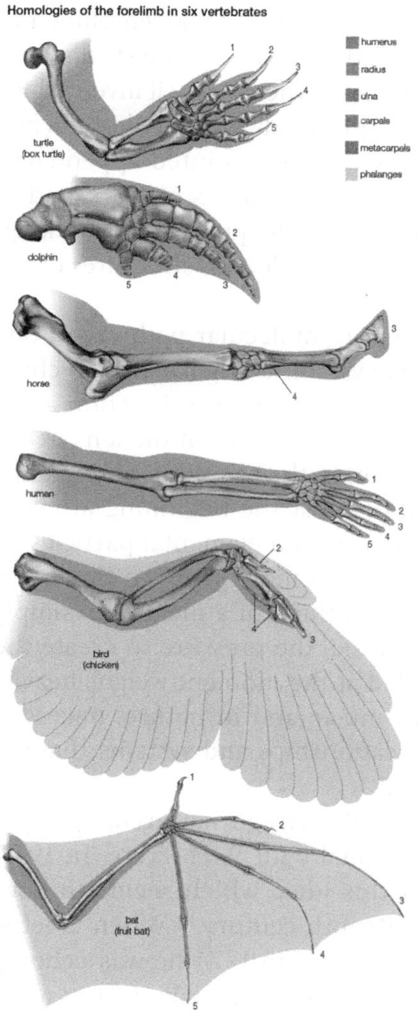

Figure 10.2 Homologous claw-flipper-leg-hand-wing structures. "homology." *Encyclopedia Britannica*, https://www.britannica.com/science/homology-evolution. (accessed 9/26/24).

Conventional explanations of difference – for example, explanations which contrasted sacred and secular, notated and unnotated, and high art and popular art – become less compelling in favor of concrete musical behavior. Gilbert and Barresi, in the twelfth edition of their widely read textbook on developmental biology, present modular construction as essential to evolution. They argue that modularity embodies a general principle of *parsimony* in nature which favors the simplest available solution to a developmental problem. Organs, tissues and adaptations of all kinds (they include the beaks of Darwin's finches as an example) are not made up of whole cloth each time, but are repurposed from preexisting structures through constant "tinkering." Complexity is created by assembling and modifying pre-existing structures. Life builds on life. No one structure is destined for any single purpose as, for example when a pencil invented for writing can also be repurposed "as a toothpick, a dagger, a hole-puncher, or a drumstick."[16] On the macroevolutionary level, jointed appendages have evolved repeatedly among vertebrates over millions of years, and in each case, existing structures were modified for the purposes of walking, grasping, swimming and flying, creating the homologous structures of the leg, hand (paw), flipper and wing (Figure 10.2).

On the microevolutionary level, an entire molecular pathway of one system is sometimes co-opted by another system, as when a protein chain that functions as an enzyme in the human liver can also function as an essential ingredient in creating the crystallin structure of the lens in the human eye.[17] In none of these examples, on both the micro- and macro-level, has development occurred by starting at the beginning at every moment and assembling a needed structure from elemental particles. A protein is made up of atoms, but it makes no more sense to think of a protein as simply a bunch of atoms than to think of a melody as simply a bunch of notes. In both nature and music, the pressure to create the structure or stage of development needed at the moment weighs heavily against starting from scratch. In both music and in nature, necessity requires the reuse and adaptation of relationships and systems that are already there and proven to be workable.

The idea that at least some songs in the Gregorian archive might be modular in construction was not original with W.H. Frere. In 1975, Leo Treitler recounted the history of this idea, which seems to have originated outside of music in the eighteenth century.[18] When Goethe spoke reverently of "sacred Homer" in a poem in 1774 he was echoing a sentiment at the time that the epic poems about the fall of Troy were, as Treitler put it, "paragons of the genius of European literature." No wonder then, that the philologist Friedrich August Wolf created controversy with his claim in 1795 that Homer's works "had in fact been handed down from diverse origins through a process of successive transmission and revision." The modular composition of the poems, says Treitler, was alarming to eighteenth-century thinkers because it suggested that "great Homer was a chimera."[19] Parry and Lord's descriptions of *The Iliad* as

16 Gilbert, S.F. and Barresi, M.J. (2016), 791.
17 Ibid., 791–794.
18 See: Wohlleben, J. (1967), 251–275.
19 Treitler, L. (1975), 1.

"formulaic" developed Wolf's theme still further and left a lasting impact on scholarship in folklore, anthropology and literature.[20] In music, similar claims of formulaic composition were first made a century after Goethe by François-August Gevaert (1828–1908), who declared that in the composition of liturgical chant, composers "worked in general with traditional themes, from which, through a process of amplification, they produced new chants." Gevaert compared this process with putting up a building: "in music, as in architecture, invention consists in constructing new works with the aid of material taken from the common domain."[21] This is a view which leaves open the possibility that a structure might be assembled from a mix of newly quarried stone and stone scavenged from an ancient ruin.

Order with tolerances

The most influential treatment of Gregorian modularity looked backward at classical antiquity as a model. The *cento* ("patchwork") was a late Roman poetic form assembled of lines from pre-existing poems. In 1921, Peter Wagner (1865–1931) reminded the music world of the classical model and went on to propose that *centonization* was an important general principle of Gregorian chant composition.[22] As a specific instance of the principle, Wagner cited a tendency of many Graduals of the Mass to employ the same standard melismas, that is, extended passages sung on a single syllable. Wagner compared these "wandering melismas," appearing in the notations of otherwise different songs, to the individual gems in a fine jewelry setting.[23] This is an image which captures a bit of the composite quality of chant melodies but it also misses the shifty and adaptative behavior of those melodies which large scale comparative scholarship has revealed in chant sources. The image of a necklace does not fully capture what Parry and Lord had struggled to define because the formulas they eventually identified took so many forms, adapted to so many different contexts, shaped by guslars as they sang. Individual diamonds do not modify themselves to fit their settings the way Frere's themes and Lord's formulas do. Viewing Frere's arrays of formulas in Figure 10.1a, biologists would probably note that the *D* formula *mutates* in a variety of ways (*D1, 2, 3, 4* and *d1,2*) in response to its surroundings. Gemstones do not show any of the adaptive behavior of the individual cyanobacteria and fungi

20 For an example of Lord's impact on music studies, see: Treitler, L. (1974).
21 Treitler, L. (1975), 6.
22 Wagner, P. (1921). Quoted in Treitler, L. (1975), 4.
23 Treitler, L. (1975), 7–8.

which co-evolve side-by-side in the lichen. The structures of the lichen's two symbionts remain distinct as independently evolving organisms with their own genetic structure. But the subtle bio-chemical adjustments each organism makes to create the lasting mutually beneficial link between them show that they are the farthest thing from a stone.

If, from a musical perspective, we are more comfortable thinking of fine jewelry and lichens simply as rough metaphors of melodic composition, no serious harm is done, but we have also learned nothing new. Comparing melody-making to natural processes now seems to me more useful, suggesting a framework for thinking about music which places front and center the dynamic qualities which music shares with the biome. The relationships and modularity so prominent in the huge collections of examples of melodic behavior in the Roman Catholic liturgy argue for placing it next to similar qualities of lichens, birds and fishes on our worktable. The fact that chant, preserved exclusively in writing, shares these same qualities of familial relationships and modularity with folk song (preserved orally as well as in writing) tempts us to expand the scope of our comparisons. The plasticity of form which Weiss saw in the biome, its order with tolerances, now looks increasingly like the more accurate and more musical description of the extremely diverse transmissions, social settings and histories of chant and folk song melodies. By comparing music with biology, rather than limiting ourselves to comparing music with other music (the default method of most musical analysis), we take a tentative first step outside of the enclosure which separates human sound art from nature.

In the final analysis, modular composition in monophonic music has been more often taken as a metaphor, rather than as a more direct description of musical phenomena. For more than two centuries – from Wolf to Gevaert to Wagner and Lord, and to Treitler's now-famous linking of Parry and Lord's formulaic composition to medieval music in his "Homer and Gregory" in 1974 – building with modules has remained a constant (and controversial) topic in literature and music. Together with the other signs of the music in biology and the biology of music in these chapters, these examples of modular thinking in music may be taken as symptoms of the gradual convergence of biological and musical thought. It is a development which leaves the two fields incrementally approaching a point where they are within reach of each other in a way that they were not over the past 150 years. Further demonstrations of this development lie ahead in future chapters.

But more importantly, it is worth noting that the many examples where modular construction has been taken seriously in the past, many of them either implied or declared that modularity was also a sign of a less developed (less evolved) music, appropriate to thinking about medieval or peasant monophonies, but not to modern art music. "Miserable patchwork" (the "übles Flickwerk" discussed by Treitler) is a contemptuous epithet proclaiming that scissors and paste construction made sense as a method for illiterate poets and musicians and simple monks constrained by orthodox faith. After all, how else could such people be expected to create complex works of art? But the biologist's view of the complex and dynamic structures of organisms and ecosystems casts modularity in a different light. Paul Weiss saw in the familiar patterns of nature, not a mechanistic reproduction of a rigid script, but as a sign of the orderliness-with-tolerances of life, an organic order which sets only the general frame and pattern. Order is bound by laws of development which reproduce past structures, but in execution, it is adaptation – variability – which is the rule.

11

Psalm-singing, an ecological perspective

In the medieval monastic communities where most of the chant tradition was developed and written down, psalm-singing was an expert soloistic practice based on learned conventions for the melodic rendering of biblical texts. In medieval documents beginning in the tenth century,[1] the practice is never fully notated, but illustrated (if it is illustrated at all) in the form of a notated skeletal *psalm tone* formula comprised of *intonation, recitation* (consisting of *tenor* and *mediant*) and final cadence or *termination*. This is an example of medieval melodic pedagogy which stakes out the more variable end of the chant spectrum for the singing of all manner of biblical texts.

Medieval writings abound in detailed discussions of how this structure is to be matched to the structure of the text at hand, with distinct approaches for psalms and for texts from the gospels and the epistles. Here we have a clear recognition of Weiss's biological principle as it might appear in music: the general melodic shape of the psalmodic repertoire was determined by tradition (represented by the skeletal notations in the modern *Liber Usualis*), while the rest of its melodic details were in the hands of the performer.

Music Example 11.1 is an excerpt from another example of medieval pedagogy, an eleventh-century Aquitanian *tonary*, a kind of anthology of materials necessary for teaching the singing of psalms.

The array here consists of ten fragments of different *antiphons*, the largest group of composed and fixed devotional songs in the Gregorian repertoire. On the right side of Music Example 11.1 the *incipit*, or opening

1 Bailey, T. (1978).

Music Example 11.1 From an eleventh-century Aquitanian Tonary: Terminations (*differentia*, or "differences") sung by a soloist, matched with sample antiphon incipits for Mode I, sung by the choir. Russell, C.T. (1966).

lines of ten different antiphons are shown, apparently all medieval scribes considered to be necessary for pedagogical purposes. Even though the notators of the melodies in all ten examples were seated in monastic scriptoria and were acting according to the spiritual and institutional expectations of the Holy Father in Rome (and no doubt sang in their daily devotions the repertoire eventually featured in the Solesmes editions), the purpose of the tonary's notator is not the same as that of the notators of the "Justus ut palma," or the fixed Salisbury repertoire in Frere's publications.

The array on the page in Music Example 11.1, made in the eleventh century, illustrates one facet of a particular musical skill which was as necessary for the conduct of daily services as the hundreds of separate pieces of repertoire collected in Mocquereau's *Liber Gradualis.* It provides a rare clue to the kind of connective tissue required to knit together bits of relatively fixed (gem-like) repertoire and the open-ended practice of singing biblical text (both prose and poetry) into an aesthetic whole which conformed with ritual standards. These connective adaptations were entirely the responsibility of the lead singer. The tonary lays out examples of how the largely unnotated art of singing psalm verses as outlined in Music Example 11.1 alternates with, and connects to, an antiphon sung by the choir.[2] In daily ritual performance, the particular psalm verse was different each day and at each of the eight devotional hours in the monastic routine. In these performances the resulting form was: *psalm v. 1 (solo) – antiphon v.1 (choir) – psalm v. 2 (solo) – antiphon v. 2 (choir),* and so on. In practice, this form required that the soloist create a graceful connection between the end of each psalm verse and the opening phrase of the particular antiphon being sung by the choir. The text the soloist was to use to create this connection was the Latin termination formula *saecula saeculorum amen* ("for ever and ever, amen").

The importance of exemplars

This example illustrates the moment in the transmission of sacred melodies when standard repertoire meets variable practice. But there is still much in these medieval examples which remains mysterious. The melodic differences among the various termination formulas (Music Example 11.1, left side) are clear enough, but the reasons for those differences remain obscure. For example, at nearly ten centuries' remove, it might strike us as logical if the last note of the termination formula on the left side of the example actually matched the first note of the antiphon to follow on the right side, thereby setting up the antiphon's beginning for the choir. But this is only the case in one instance. Instead, the terminations illustrate a different level of connection between psalm-singing and antiphon, a connection which in these examples is created by alluding to a similar kind of melodic movement in the antiphon to come. This allusion is comparable to the way an orator or storyteller ends his or her introduction by subtly alluding to the topic in the body of the speech which will follow. As notations, we must understand Music Example 11.1 as an *exemplar*, not a score in the modern sense. They suggest rather than prescribe: "here, in order to connect a Mode I psalm with an antiphon *like that,* terminate the psalm verse *like this*."

As exemplars rather than prescriptions, the notations were no doubt best understood through the live modelling of an expert practitioner. All we know from our vantage point in the twenty-first century is that the stereotypic, but flexible, melodic pathway of the soloist's psalm recitation is mutated by the presence of the particular melodies they must connect to.[3] In the delicate business of connecting a stem cell with the adjacent

2 For a full description of the medieval tonary, see Huglo, M. (1980).

3 As early as the sixteenth century, the intricacies of the psalm-antiphon practice was striking commentators adapted to polyphonic repertoires as mysterious.

cells in its location in an embryo – an action which must take place with unfailing accuracy at every moment during gestation – the very molecular structure of the stem cells' walls must alter in accordance with the structures emerging around them. A *dorsal* stem cell (positioned in the back of the developing embryo) becomes a neural cell for the spine, while the identical *ventral* stem cell (positioned on the embryo's front) alters its molecular structure to become epithelial tissue for the gut. As Weiss might have put it, the genetic structure provided the mode of procedure for the stem cell, but its position in the emerging body – the stem cell's immediate environment – provides the input which determines how the stem cell adapts itself to exigencies.

Viewing Gregorian practice from the perspective of recent biology, I feel comfortable viewing it *ecologically*, as an instance of how one living system (for example, one stem cell or one organism) co-evolves with other elements in a habitat. Stickleback epigenetics are a specific instance of the interactions of a fish's genetic make-up with its surroundings. The coded recipes in the genes for particular protein chains which shape specific features of the fish's organs are triggered or suppressed by the presence of conditions in its environment. I have likened these chemical recipes to the memory of practices which the musician acquires over a lifetime, as in jazz and string quartet ensembles. In both the biological and musical contexts, ecological behavior patterns are retained for re-use so that adaptations to neither fish nor melody needs to be made up each moment from elemental materials. Tinkering with preexisting structures is parsimonious, making it more likely that tinkering itself becomes a default way of working.

Following the biological model a step further, we could conclude that flexible melodic practice has been proven over time to enhance the repertoire's survival, the way the stickleback's sensitive adjustments of the chemistry of its cell walls to the salt in its watery habitat enhances its survival. In both cases, it is not just individual organisms or songs which benefit through sensitive adjustment, but all subsequent individuals of the same family – the successive generations of melodies and fish which are furthered by this sensitivity. In the case of the fish, the transmission of this information is by way of a plastic genome interacting with the stickleback's cell chemistry, enabling it to come up with nimble responses

---

Commenting uncomprehendingly on "the immoderate subtlety" of the practice that he found in the tonaries, the Swiss theorist Heinrich Glarean complained that they were surely the work of "meddlesome, not to say trivial men" who, "thinking about the easiness of...intoning antiphons, a thing clear itself, made it much more obscure." Glarean, H. (1547), quoted in Russell, C.T. (1966).

to conditions in its habitat. In the case of the chant, the necessary link in transmission is the modeling of a human being, a master singer with deep memory of melody-making over many years, without whom the notations in the tonary are nearly worthless.

Once again, the notations in medieval tonaries like those in Music Example 11.1 are more like pedagogical examples than they are repertoire and so they cannot be read as the same kind of musical artifact as those found in the Vatican's *Liber Gradualis* or *Liber Usualis.* When we consider biological examples alongside musical examples we can see in these tonaries instances of a process that is superpersonal. It is both uniquely individual and collectively systemic. To borrow the language of cell biology, we are observing how melodic homeostasis responds to disturbances by adaptation of an array of stored solutions. According to this way of thinking, the conventional, repetitive and predictable movement of the Mode I psalm formula consisting of initial, tenor, mediant and termination outlined in the *Liber Usualis* is the distillation of an evolved practice which is the product of many individuals over many centuries – in the biologist's terms, experience which has been "hardened" through generations of natural selection. This movement, though repetitive and predictable, is still adaptable. Its homeostasis is disturbed when it is confronted by the melodic requirements of one of the many hundreds of unique, relatively fixed, antiphons (as on the right side of Music Example 11.1). The ending of each psalm verse must be adapted by the singer, more or less instantly, to make the connection. The tonary provides not rigid solutions, but models of how that connection might be made.

Modes, families, species

Surrounding all these examples of melody-making in the monastic habitat and enclosing them in a larger system of inherited constraints, were the eight church modes. Like Linneaus's phylogenetic categories, the modes are heuristics, ideas about shared characteristics which are the residue, as Weiss would have said, of centuries of systemic behavior. These very large melodic families, grouped according to their typical behavior patterns, provided a way to think about Gregorian melodies comparable to the "order with tolerances" which Weiss saw in nature. The eight-mode system, probably created by Frankish musicians on Byzantine models during and after the reign of Charlemagne, was energetically developed by theorists over the centuries as a scheme for conveniently classifying, remembering and teaching the chant repertoire – a kind of Dewey Decimal system for practicing singers. For classification purposes, the scheme works imperfectly, but well enough, and remains the basic reference point for all discussions of chant performance in the same way that Linneaus's phylogenetic classifications in the eighteenth century continue to serve well enough today in a biological discipline. In both systems, there have been controversies over exceptions and needed changes, and both systems function equally as practical taxonomies and as implied theories of how the items in each system relate to one another. As early as the ninth century, European music theorists like John Cotton were calling attention to the relationship of the Gregorian eight mode system to that of "the Greeks" of previous centuries, the Byzantine *Octoechos*.[4] (More will be said about the reinterpretations of the Linnean system and their impact on musical thought in Part 4.)

Organisms and songs as exemplars

At bottom, the eight modes are an array of eight different sets of intervallic sequences or scales. This rational scheme of eight scales provides a fundamental palette of pitch relationships for the making of melodies native to the Gregorian sound world. Four diatonic scales based on four-note sets of whole-steps and half-steps beginning on D, E, F, and G (called *manneriae: Protus, Deuterus, Tritus, and Tetrardus,* remnants of Byzantine inheritance) provide a practical framework of interval sequences,[5] with an *authentic* and a *plagal* form for each (separated by a fourth), giving a total of eight.

Of special interest here is the way that the eight-mode system housed a wealth of melodic practice that far exceeded the rudimentary pitch description of a melody's scale, range and final pitch. A staple item in many tonaries up into the sixteenth century was a set of eight antiphon-like chants with mnemonic texts from biblical sources. (Music Example 11.2) (Note that the biblical text of each melody in Music Example 11.2 begins with the number of the mode: "*Primum* [1] querite," "*Secundum* [2] autem," "*Tertia* [3] die," "*Quarta* [4] vigilia," etc.)[6] These model antiphons first appeared in western European writings in the ninth century and seem to have been passed down mainly for teaching typical melodic behavior patterns to novices. The significance of these model tunes is that, in the

4 Afflighemensis, J. (11th c.). In Palisca, C. (1978).

5 The sequences of whole-steps and half-steps in each of the *manneria* are: *Protus,* DEFG (whole-half-whole); *Deuterus,* EFGA (half-whole-whole); *Tritus.* FGAB (whole-whole-whole); *Tetrardus,* GABC (whole-whole-half).

6 See: Bailey, T. (1978).

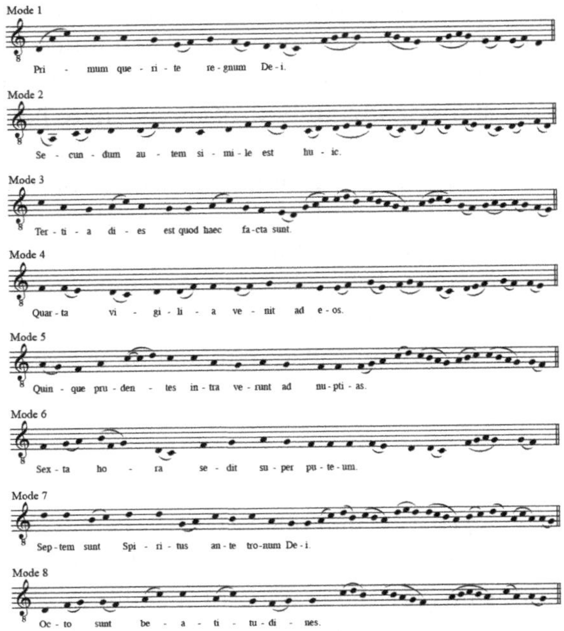

Music Example 11.2 Eight model antiphons (Johannes Afflighemensis, 11th c.). In Palisca, C. (1978).

minds of the singer-theorists who compiled the tonaries, the relatively technical matter of classifying chants according to scale, range and final into eight groups was inseparable from a more developed practice of melody-making. The wide distribution of these model tunes in medieval European sources (with the expected variant versions) suggests a widespread assumption that a singer's grasp of the vast liturgical repertoire depended on more than simply being able to sing a large number of songs, either from memory or from notation. The singer must understand any tune as a member of a family, that is, as a representative of a particular corner of a very large system of melodic behavior. Once again, as pedagogical devices, these eight little tunes are not scores, but exemplars, invitations to a novice singer to hear family relationships in song and even to make or modify melodies in real time: "here, a melody in Mode I (or II or any other mode) sounds *like this*."

In discussions of any of the families of organisms listed in the Linnaean phylogenetic system, it is common to let a single species within a particular family stand as a reminder of the whole family – a lobster for the phylum *Arthropoda,* for example, or a dog for the class *Mammalia.* The fact that a butterfly is also an arthropod and that a whale is also a mammal does not diminish the usefulness of the lobster and dog as icons of larger, more diverse groups. Likewise, the family concept borrowed from everyday observation of nature and applied to folk song and Roman Catholic sacred song – Cowdery's "Blackbird family" and Frere's "Theme IIA," for example – still serves a purpose as heuristic and as pedagogy. The separate worlds of melodies and organisms can be said to resemble each other in this respect: within each system, in describing individuals it is proper and useful to say that they behave *like this.*

The great chant project of Guéranger and Mocquereau, energized by a mission to revive western European Christianity, became an important launching platform for what was to become the secular science of musicology. The scholar-monks' hyper-focus on the resolving of ambiguities in transmission into stable and replicable musical objects became one of the trademarks of a century of musicological research and editing. And yet, despite its narrow focus, through the material the Solesmes scholars assembled, through the methods they used, and through successive generations of musicologists who followed their methods, they have left a legacy which exposes the variability of some ten centuries of melodic tradition for all to see. With recent biology's view of nature as the benchmark, rather than simply other human musics recorded in texts, it is now more possible to accept a musical world which is as fluid as the natural world without also abandoning the equal truth of music's moments of fixity. In nature, fixity is reflected in the replication and re-use of forms and processes that have sustained life and allowed it to evolve over more than three billion years. In western music, the timeline is far shorter, a few thousand years in all, but fixity registers in the functional stability of musical objects, in the tendency toward fixity in musical memory, and in social norms which treat achievements of musical individuals as enduring. And yet, music's variability – its dynamism – is never far from view and always explicitly acknowledged and celebrated by music practitioners, theorists and participants.

With his eyes fastened on the overarching social goal of confirming the stability of the Roman chant tradition (and perhaps of Rome itself), Dom Mocquereau looked on the slight variants among the 219 versions of *Justus ut palma* – and more importantly, on the highly divergent variants like those which preoccupied Frere – and judged them to be insignificant details. And yet, insignificant or not, we can feel grateful that the variants were duly noted and preserved in the Solesmes editions. Though the variants were generally marginalized in the scholar-monks' work, for the present generation they remain clues to mutations and adaptations generated in many different circumstances by singers and scribes of many nationalities over many centuries. We can imagine, for example, that a version of a chant in Burgundy might exhibit qualities which singers in Rome or Switzerland would have found distasteful, unmemorable or even incoherent, and therefore unworthy of preserving in writing. Yet the aesthetic and cultural tension caused by even the very minimal polymorphousness of songs like *Justus ut palma* was resolved in a stroke by editorially eliminating from posterity all but one of its manifestations for the publication of the *Gradual*. Under pressure which we can now understand was as social and institutional as musical, practice harden into repertoire. From a bio-musical point of view, under such influences the tangled bank is reduced to a topiary in a pot.

Musical inter-actors and the modern paradigm

The foregrounding of fixity as a solution to the problems posed by variability in the early years of musicology has proven to be formative. But it might be proper to view the discipline's leaning toward the fixing of musical objects as one more sign of the gravitational force which fixity exerts on human sound art more generally. When invariability is given pride of place near the center of gravity of musical thought, music tends to look less like a continuity with nature. As awareness of musical mouvance has sharpened, and as mouvance has been allowed to inch its way closer to the center of musical research and teaching, the inherent otherness of the biologist's and musician's respective materials diminishes. The possibility that Homer was a chimera was as threatening to literary thinkers as the idea of a chimeric lichen was to biologists and the idea of a chimeric Pope Gregory was to church scholars. These cases challenge any uncritical faith we may have in the ontological distinctness of creative agents. From all this variability observed across disciplines, a cascade of questions inevitably follows: If there is no Great Homer and no Gregory the Great, then perhaps there is no Great Progenitor Adam, either. If there is no Adam, but only a self-creating and unruly garden, as the current view of biology sees it, then what of the Great (divine) Creative Executive? Without the certainty of these creative executives, anthropocentric certainty begins to dissolve.

The Gregorian master singer, with all his or her experience and training, remains essential to melody-making. But if we allow biology to remain in our field of view as we approach musical creation, we are less tempted to see the singer as an independent actor than as something else – as an inter-actor, the necessary musical agent who mediates the systemic creation of his tradition. The stability of the psalm intonation, and of its place in the liturgy, with each hour of the monastic day a different psalm text and antiphon, is protected by its built-in capacity for creating variants which help it meet changing needs. A cell or organism is similarly protected by its ability to generate mutations which enable it to adapt to new circumstances. This is a situation described by Weiss's laws of development in biology *which prescribe only the mode of procedure but leave the actual execution free to adapt itself to the exigencies of a world whose details are themselves unpredictable.* In neither the biological nor the musical context is a successful adaptation guaranteed. No organism, no genome, no musical practice can be expected to take on all potential perturbations to the status quo. The transmission of life and of music is as vulnerable and fragile as they are robust and persistent.

Maintaining this level of readiness for new and challenging contexts requires something other than the fixity pictured in the kind of meticulously edited publications of the Solesmes workers, focused as they were on preserving distinct, free-standing items of repertoire. In Treitler's words, Dom Guéranger and his colleagues were engaged in "specifying the musical artifact," a goal which Treitler calls the *modern paradigm.* This is a model of musical transmission with the individual musician at the center: "the model of the composer producing finished works that circulate in stable form through closed written channels in scores that are blueprints for the performers, who must be able to read them in order to produce performances that are essentially identical from one time to the next."[7] In medieval music, Treitler says, "we are unable to look at four scores that resemble one another more or less and say with confidence, 'this is a score of the work, that one is a variant of it, the third is a copy of it with errors, and the fourth is a score of a different work.'"[8] The goal of specifying the musical artifact can go a long way to clarifying an important portion of the Gregorian tradition, but by no means the whole of it. Similarly, cell biologists must continually verify the separate and distinct objects before them – a molecule, a cell, a tissue, an organism – while at the same time they attempt to understand how these entities must mutate in order to survive under new stresses without simply relying on an all-wise god or its equivalent to answer all their questions.

It remains to be seen whether these continuities between the ways of monophonic practices in oral transmission and the ways of the biome can also be found in musical practices which are polyphonic and written. This question will return in Parts 3, 4 and 5.

7 Treitler, L. (1981), 473.
8 Treitler, L. (1977), 204.

Part Three
The self-organization of everyday life

Traveler, your footprints
are the only road, nothing else.
Traveler, there is no road;
you make your own path as you walk.
As you walk, you make your own road,
and when you look back
you see the path
you will never travel again.
Traveler, there is no road;
only a ship's wake on the sea.

Antonio Machado, 1912[1]

In the development of a human brain, in the metamorphosis of a caterpillar into a butterfly, and in the coordination of elements in a habitat, who or what is in control? How can the complex interactions of numerous independent agents and processes *all by themselves* and with great reliability produce a functioning brain, butterfly and habitat?

Biologists in recent years have claimed that such developments are not exceptional in nature, but are actually indispensable qualities of all living systems. Their claims are argued using a short list of terms of which self-organization is the most common. A longer list would include words like *self-assembly, emergence, inherency, stigmergy* and *autopoiesis,* this last term a combination of words of Greek origin meaning "self-making." Unity of purpose is what Cecil Sharp first thought he saw in starling swarming, as if someone or something was directing the whole. And yet, he saw no

[1] Machado's poem was referenced in the biologist Francisco Varela's essay on self-organization entitled "Laying Down a Path in Walking."

sign of such a central command, nor can we, only webs of interactivity and materials which change according to their own physical laws.

In biology, conversations about self-organization move quickly into the thickets of biochemistry, genetics and systems theory, the natural habitats where self-organization as a scientific concept was first applied. This conversation will happen here in only as much detail as necessary (and in as much detail as I am capable of) in the chapters in Part 3. But as a beginning, the experience of artists can carry us a surprising distance into the mystery of how, as the poet says, life is lived with no road. The only road is the one made in walking.

12

Ways of the hand

The title of this chapter is borrowed from David Sudnow's story of his ten year introduction, as an adult pianist, to the world of jazz improvisation.[1] For our purposes here, his book of the same name has implications which extend well beyond the specific habitat of mid-twentieth-century jazz practice known as bebop which he is describing and, more importantly, beyond the limits of the I-word itself. Because of Sudnow's careful avoidance of conventional musical terminology and personalities, and his rigorously experiential perspective, I am offering his memoir here as a first pass at connecting the music world with *self-organization*, the concept so central to current scientific conversations about the spontaneous creation of order which distinguishes the biosphere from the non-living regions of the planet.

From Sudnow's description of the working of his hands in the fashioning of music in jazz, it is just a short step further to consider the hands of musicians making other kinds of music, and from there to the hands of other craftsman fashioning solider materials – not sound, but textiles, metal, pigment, wood or clay. Ways of the hand can then be translated to other human organs – ways of the ear and ways of the voice, for instance – and to the extensions of the human body, to the tools for making things, from hammers to musical instruments and computers. Language too, has its self-making ways, as does thought. Sudnow's story is a genuinely musical step into the realm where, perhaps surprisingly, we can see artistic creation happening according to what Darwin saw as laws. At the same time, Sudnow provides an account which, like Sharp's account of murmuration, is still firmly rooted in behavior, in the actions of individual actors. Sudnow calls those acts "handways."

1 Sudnow, D. (1972).

Handways, melodying and foraging

As a memoir of one man's musical learning curve, *Ways of the Hand* is striking for its disruptive, even tortured, language. The author consistently avoids standard music-talk of pitches, harmonies and rhythms. Even the conventional jazz lexicon of "swing," "blues," "comp," chord names and even tune names only make an appearance in one chapter. Instead, he introduces a number of photographs of his own hands at the keyboard, and his narrative is crammed with uncomfortable made-up verb forms and one-of-a-kind hyphenated nouns. All of these expressions evoke ways of moving, ways which are only partly governed by the player as the conscious and solitary creative decision-maker. "Scaling ways, up-a-little-down-a-little ways, rocking ways and every-other-finger ways, and skipping ways, hopping ways, rippling ways, ways to go a long way with"[2] all refer to what the hand does, rather than to the musician's aesthetic goals.

There is little mention of melodies, only of "melodying,"[3] and there is no analysis of "places" on the keyboard (keys, chords, motives, tunes), only "pathways." It is the finding and digesting of these pathways which he says occupied his years of practice. "I set out on a course of notes,"[4] he says, as if on a journey with an indeterminate destination. He tosses out paradoxes to circumvent any claim he might be tempted to make that he authored the products of his setting out: "I am not using pathways to make melodies," he says, "I find places to go in the course of going to them,"[5] a kind of Zen koan prioritizing what emerges along the way rather than claiming that *he* did it. Sudnow foregrounds what his hands do the way the Shona mbira master foregrounded the work of his ancestors. Both musicians direct attention away from themselves, not out of false modesty, but out of a sincere effort to describe what they know to be true.

Each gesture, he says, "provides a platform for the next gesture,"[6] and his hands are always "hovering over the keyboard, tasting possibilities,"[7] as if the possibilities were there in the keyboard waiting to be discovered. He describes this as "foraging,"[8] a word evoking the browsing of an animal for food on a forest floor. His talk of hand ways rarely yields to the temptations of essentialized relationships such as chords or intervals which he

2 Ibid., 73.
3 Ibid., 179.
4 Ibid., 55.
5 Ibid., xiii.
6 Ibid., 56.
7 Ibid., 144.
8 Ibid., 266

as an individual artist chooses, preferring to think of them as systemic phenomena, as "classes of action."[9] It is apparent in his narrative that standard jazz terms were in regular use in the many conversations with his principle teachers and with ensemble mates over the years. It's just that here, on the page, he is choosing to rely on extraordinary language in his descriptions to keep himself focused on what he calls his "phenomenological inquiry into the structure of handwork," in accordance with the book's only theoretical reference points, the philosophies of Edmund Husserl (1859–1938) and Maurice Merleau-Ponty (1908–1961).[10] Once he has mentioned them in his preface, Sudnow takes these philosophies no further and launches into his catalogue of "ways."

In folk music, the unique actor or agent responsible for creating the song has largely been ignored in favor of the performer of the moment. It was the hiddenness of any individual creative agency which Sharp saw in murmuration that suggested to him the practice of making folk songs. Sharp's folk singers, the starling swarm, and Sudnow's hands, rather than creating particular products, are each engaged in setting out on paths, as Sudnow put it. Their ways are classes of action and more like foraging than constructing a lasting visual or musical moment. In each of these cases, there are indisputable agents of variation, but in each case, the individual agent is subsumed, even dwarfed, by a systemic interaction of other agents.

Agency from within and from without

Ways of the hand have also attracted the attention of twenty-first-century cognitive science. A direct link to Sudnow's handways at the piano can be found in a thought experiment created by the neuropsychologist Lambros Malafouris featuring the potter working with clay.[11]

When the potter sits down at the wheel, she or he does so neither as a neutral agent who contributes nothing to the final product nor as a machine designed to execute exact replicas of a pre-programmed design, as in industrial production. The potter's head is not empty but crammed full of memories of vessels created by the potter's own hands in the past, as well as the instructions of teachers, the prototypes of possible vessels accumulated through observation, and a host of social expectations of the vessel's use, beauty and social status. Now add to these factors the pressure of the fingers on the lump of wet clay turning on the potter's wheel and

9 Ibid., 160.
10 Ibid., xiv.
11 Knappett, C. and Malafouris, L. (2008), 19–36.

the resistance and malleability of the material itself which feed back to the potter's mental input. Together, these complex interactions of nerve, muscle, mind and material give us a picture of human handwork in which "no *a priori* hierarchy can be argued between the potter's brain/body/wheel/clay."[12] As in the case of musical creation, the artistic agency of the potter is without question: it is the potter's choice to sit down at the wheel to begin with, it is the potter's hands in the mud, and any credit or blame for the outcome would reasonably be acknowledged as the potter's.

And yet, says Malafouris, "Potters know more than what they can tell or explain and their hands often have reasons of which their mind is not aware."[13] He backs up this claim citing neuropsychological experiments demonstrating, for example, that there is a time lag between the neural anticipation of an act and the execution of the act. The "readiness potential" in the potter's hands precedes the potter's awareness of "the conscious wish to act" by at least 350 milliseconds. But what is more, "the conscious wish to act also precedes the final motor act by about 150-200 milliseconds."[14] The short way of saying this is that the nerves and muscles know what the hand is about to do before the brain does. One third of a second is still time enough for the potter's conscious will to intercede and approve or veto the act. In other words, the potter's agency is clear, though perhaps not quite as total as we might have hoped – more collaborative and interactive, more ecological. He is an inter-actor, less like the God of Genesis and more like the master singer in a medieval monastery.

Sudnow recalls that he would ask his teacher to explain what he had just played, or even to play it again, to which he would often get the reply, "I don't know what I just played."[15] I take statements like this as an acknowledgment that self-organization is taking place. They add a layer of complication to the more commonplace explanation of artistic activity which places the creator's mind up front, as if "the potter's [or pianist's] fingers do nothing but execute the orders of the potter's brain…the central 'executive' mechanism."[16] In the view of Malafouris, pottery-making, like jazz performance, is not all in the head of a single human actor. In fact, at certain moments in any act, it is unclear how much it would be proper to claim that the head is involved at all. In baseball, for instance, a 100-mph fastball takes 375–400 milliseconds to reach the plate, roughly the blink

12 Ibid., 32.
13 Ibid., 21.
14 Ibid., 27.
15 Sudnow, D. (1972), 31.
16 Knappett, C. and Malafouris, L. (2008), 20.

of an eye,[17] justifying Yogi Berra's rhetorical question, "Who can think and hit at the same time?"[18]

Malafouris calls these examples of agency "embodied skill," a skill which is extended beyond the individual human consciousness and distributed among head, hand and clay, a distribution which the potter "discovers each time, in real time and space." Recognizing an artistic agency which is distributed in this way can be personally disorienting. As the sculptor Sarah Gilbert describes it, "However much agency may seem to come from within," says the sculptor Sarah Gilbert, "it also comes from without."[19] Following convention, the anthropocentric creative executive, which Malafouris goes so far as to call "the illusion of conscious will," more often than not will collapse all the distributed inputs into itself and claim "*I* did it."[20] In the end, says Malafouris, the human brain is not the only place to look for the origin of physical (or sonic) objects. Causal agency "is not something you discover by going deeper into the brain."[21]

The distributed mind

What a neuropsychological view of the potter's art adds to Sudnow's descriptions is not just that mud-work is comparable to keyboard-work. Both involve handwork, but both are also cases which demonstrate the complex interaction of individual brain and artefact, of imagination and sensible objects. Lambros Malafouris, an archaeologist turned neuropsychologist, even goes so far as to recommend to those of us in the art world an unfamiliar way of working: we should look "between," he says, rather than "within." This is advice consonant with what Capra and Luisi recommended earlier (Chapter 1), that we must try to understand living processes as inseparable from ecosystems. They also recommend looking "between" by relying more on "mapping" the connections among living things than on "measuring" them, an activity which fits most comfortably with stationary objects whose attributes can be quantified [22]

In music, the particular resources and feedback loops are different, of course, consisting of sound rather than mud, the feel of the instrument or voice, the eye and ear and hand attentive to them, the interplay of sound

17 Quinton, S. (2017).
18 For this reference, I am grateful to Lewis Lockwood.
19 Gilbert, S[arah] (2018), 7.
20 Knappett, C. and Malafouris, L. (2008), 79.
21 Ibid., 78.
22 Capra, F. and Luisi, P.L (2014), 6.

with notated instructions or even with intellectual constructs based on notation, the deep memory of years of pedagogy and practice, and the moment-by-moment awareness of the musical ensemble. "There are no clean hands in pottery making," Malafouris observes – there can be no claim of "I did it" which excludes the hands' interaction with the clay. Likewise, in music there are no clean hands. There can be no musical artefact which can be defined simply in terms of its sound (recorded or live) or its abstract representation in writing. Nor can one untangle the artefact from the web of sound waves, neural pathways, memories, muscles, social inheritances and human idiosyncrasies from which the artefact emerged. Mind itself is embodied, extended and distributed, says Malafouris, making the human being an "intelligent problem-solving ensemble of brain, body and the external materials we work with."

When the anthropologist Gregory Bateson spoke in the 1970s of the claim "I did it," he refused to identify it as universally human. Instead, he spoke of it as a particular folkway, a creature of culture, what he calls "the delimited agent" characteristic of "the average Occidental."[23] My Zimbabwean mbira master's confusion at my question, "have you ever composed a piece of your own?" perhaps stems from the sense of a distributed agency which was his cultural inheritance, a conviction which left him disinclined to reach for the phrase "I did it" to explain his life-long relationship to the music he played.

Composers at the keyboard

Igor Stravinsky (1882–1971) often referred to the fact that a lot of his compositional work was done at the piano. He also acknowledged the importance of improvisation in his creative process throughout his lifetime. In his autobiography his insistence that he did not regret his creative roots in the keyboard is a bit defensive, since it is also known that in his early years he felt uneasy with this way of working and sought the reassurance of his teacher Nikolai Rimsky-Korsakov (1844–1908). "Some compose at the piano and some without a piano," the teacher told him. "As for you, you will compose at the piano." The pupil, writing years later as a celebrated composer, went a step further, insisting that he preferred "to compose in direct contact with the physical."[24] "Fingers are not to be despised," he declared. "They are the great inspirers, and, in contact with a musical

23 Bateson, G. (1973), quoted in Knappett, C. and Malafouris, L. (2008), 318.
24 Stravinsky, I. (1936), 5.

instrument, often give birth to subconscious ideas which might otherwise never come to life."[25] In his Charles Eliot Norton lectures at Harvard in 1939 he declared "a composer improvises aimlessly the way an animal grubs about...in expectation of our pleasure, guided by our scent."[26] This is a statement which sounds a lot like Sudnow's "foraging" at the keyboard and Malafouris's claim that the potter's hands "often have reasons of which their mind is not aware." The testimony of others suggests that some of the uncertainty Stravinsky felt about his foraging may have stayed with the mature composer, leaving him uncomfortable in situations where the foraging was heard by others. Stravinsky's long-time musical collaborator Robert Craft reported in 1949 that "when Stravinsky composes he must be alone, even closing the windows of his studio as if to seal himself in." It seems that the composer's need to browse at the keyboard was linked to the need to hide the act of browsing from the ears of others, the audience for his finished works. Orchestration was different, says Craft. "When orchestrating he likes companionship, often asking me to read to him."[27]

This keyboard-centeredness also seems to have been linked to the I-word in Stravinsky's thinking, beginning with his own accounts of his first years of music training as a child, when he recalls how he quickly gravitated toward "improvising," despite the reproaches of his musician parents for not practicing properly. Stravinsky's use of the I-word also recalls a somewhat different, but related example: the painter Delacroix's claim that Chopin's improvisations were like a painter's preliminary sketches. This is a view which is consistent with how such musicking in real time has been understood in art music discourse as a rough draft of the museum-grade musical work.

According to a widely-circulated legend, another famous pianist-composer, Béla Bartók, was once asked by a visitor in his studio, "Mr. Bartók, where is it that you do your composing work – over here (indicating the desk) or over there (indicating the piano)?" Bartók's answer was to point to the space in between the two: "approximately here," an answer which seems to allow room for the more distributed creative agency that Malafouris described. The story testifies to how the silent musical work in the mind, with or without the act of notating, was inseparable from the noisy foraging at the keyboard. We should not forget that the medieval and

25 Ibid., 82. For a recent examination of the importance of the piano in Stravinsky's work, see Brown, K.G. (2022).

26 Stravinsky, I. (1947/1970), 55.

27 Stravinsky, V. and Craft, R. (1978), 384. My thanks to Sean Gallagher for this reference.

renaissance term for the fashioning of music in the head, *alla mente* ("in the mind"), the work of the composer of polyphony was presumed to be rooted not in the hand but in the voice. To Johannes Tinctoris in the late fifteenth century, Bartók's "approximately *here*" would have been located somewhere between the singer-composer's writing desk and his own voice.

In a more specific case of handwork at the keyboard, Bartók's graded collections of student pieces, the six books of his *Mikrokosmos* (1926–1939), included many pieces which harness a limited array of pitches, often playable without moving the hand – in other words, no thumb under, no scales, no gregarious movement across the keyboard, just five notes for five fingers. His use of narrow-range folk tunes, many of them pentatonic, matched well this acceptance of the most severe limits of the hand. Stravinsky did the same with his *Pour les cinq doigts* (1921), which also do not require movements around the keyboard but explore the many sequences and rhythms of the notes that lie right under the fingers. Many of Bartók's 44 Violin Duets (1939), like *Mikrokosmos*, a graded series for students, extend the limits of the hand to the bow. In the less advanced duets, the left hand remains in one position, limiting movement up the violin's neck in short phrases composed of adjacent notes on the fingerboard or rocking movements of the bow to adjacent strings. In all these works designed for students, the limited range and limited note choices lend themselves to repetitive phrases with many small, surprising and continuous variations, full of both symmetries and asymmetries. The resulting texture is often surprisingly complex.

The ecology of handwork

The particular array of pitches and the particular movements of the hand which are available to the composer and performer are always limited, but that limited field provides the most elemental constraint for creative work. In the same way, the stickleback is physically shaped not by all possible interactions of its body chemistry, body armor and feeding structures with its habitat but with a limited set of interactions presented by lake bottom or running stream or ocean. These are all examples of the limits provided by ecology, by which musical or natural creativity is shaped. In both cases, limiting what is permissible in the habitat focuses available energy and resources and eliminates options that do not work. In the case of Stravinsky and Bartók, the extreme limits of the hand and a limited palette of tone colors (they each were confined to either piano or violin) proved sufficient for successful music-making, just as immobility and dependence on symbionts and the nutrients provided by a particular tree trunk or stone wall is sufficient for successful lichen. At every moment, all of the frenzied flapping movements of a given starling are limited severely by the movements of other birds on all sides, by gravity and by wind. Not a single bird in the flock is free of these systemic effects and feedbacks. Like any of the composer's five finger creations, the constraints on the bird's choice – the very fact that all but a few choices of movement and direction are ruled out at any given moment – is essential for the creation of the form of the murmuration. In its gorgeous shape and dynamism, the flock's choreography is as much an expression of limits as its unpredictable changes in shape may strike humans as an expression of freedom.

Stravinsky is known for his at times strident assertions of the executive authority of the composer, but his lifelong habits of foraging at the keyboard also make him a bio-musical exemplar. On the one hand there was his approach to recordings and to performers of his works which seemed to stress the composer's executive control. For a few years in the late 1920s, Stravinsky collaborated with Robert Craft on a series of recordings for Columbia Records which the composer believed would "serve as guides to all executants of my music," making the recordings an extension of the composer's instructions on the page (it seems that the "executants" did not comply as he had hoped. Thirty years later he lamented that "very few conductors avail themselves of them").[28]

On the other hand, in his *Poetics of Music,* he famously embraced an ecological view of his own creativity. He could have been as easily speaking for the stickleback or the lichen when he wrote, "My freedom will be so much the greater and more meaningful, the more narrowly I limit my field of action and the more I surround myself with obstacles. Whatever diminishes constraint diminishes strength. The more constraints one imposes, the more one frees one's self of the chains that shackle the spirit."[29] Constraint, as he is using it here, is an ecological concept, an acknowledgment of what would be regarded in biology as an epigenetic truism, that a unique musical artefact arises from interaction with its environment. The constraints of natural and musical habitats enable the variations that emerge within their confines, even as they limit which variations can thrive in those same conditions.

I recall one summer standing in the August heat at the edge of an open meadow in the hills overlooking Berkeley, California. It was more than 30

[28] Stravinsky, I. (1936), 150.
[29] Ibid., 65.

degrees centigrade in the meadow, which was dry and brittle, and almost white in the sun. The field was brimming with flying and crawling insects and there were signs of burrowing rodents and ground-nesting birds which fed on the insects. But I was looking around for shade. A few feet away, the edge of the meadow was marked sharply by a stand of evergreens – a few redwoods surrounded by a dense mix of spruces and pines – whose lowest branches with densely packed fronds of fine green needles were high enough to stand under. From the cool twilight I stared out at the parched brilliance of the meadow a few paces away and shivered. My skin was moist and there were droplets on my glasses. It was raining. The ground was soggy, covered with mosses and other bryophytes and fungi adapted to cool, moist and fertile environments, with no sign of the tall, thick grasses and flowering vetches a few feet away. The boundary between the two ecologies, complete with their unique climatic conditions, was stark. I was looking at an enactment of the compossibility which Scott Gilbert, citing Leibnitz, claimed was behind these contrasts: Forests full of squirrels can thrive at the edge of oceans full of whales but the boundary separating the two habitats is severe, as severe as I experienced on that summer day in the California sun. In nature and in music, the web of interconnected life forms which this limitation makes possible and even *inevitable*, is also prohibitive. More options of variation are excluded by a habitat than can be sustained by it. Only certain organisms fit for the cool shade and moisture under the pines could long survive there.

This is my own bio-musical understanding of Stravinsky's embrace of constraints, though I cannot claim that the composer would have agreed to this interpretation of them. Nor is it certain that he would have recognized the handwork of the jazz pianist or of the potter as akin to his own handwork. In the end, it is not for us to presume what Stravinsky, or any other composer, might never have had to consider – how his creative process fitted into, not just the web of musical traditions, but the web of life forms on the planet. The perspective which contemporary biology has made possible permits us now, as never before, to look beyond what Stravinsky, like Darwin, could not see.

13

The path is made in walking

Henrich von Kleist, writing in 1805 about the mundane interactions and chance happenings which contribute to the development of ideas, was also considering, well ahead of his peers, the mysteries of self-organization, though he did not use those words. His subject was not the handwork of pianist and potter, but the speech-work and word-work of his own tribe of orators, novelists and thinkers. He also had no thought of the biological self-organization which has become a staple of scientific debate about living processes in recent years. But his short essay "On the gradual construction of thoughts during speech"[1] is full of signs of struggle with conventional thinking about thinking itself, about the agency of ideas, and about how all manner of everyday conversation emerges in time from the catalytic impacts of accident and trivia.

Rational thought, he was arguing to his contemporaries, moves less in straight lines than they were inclined to assume, an idea which first led me to caption what I saw in biology in Chapter 1 as there are no straight lines in nature. He even suggests that knowledge itself is a looping process which entangles the spontaneous and messy act of speaking with the notion of a pristine idea. "For it is not we who know," he says, "but at first it is only a certain state of mind of ours that knows."[2] His insight sounds like a premonition of Sudnow's approach to jazz improvisation, of Malafouris's arguments for a distributed agency and the physicist Eric Smith's claim that "ecosystems are more fundamental than their organisms." In the science

[1] "Uber die allmahliche Verfertigung der Gedanken beim Reden." 1805–1806. For more on von Kleist's essay, see the treatment of it with reference to the concept of improvisation in light of eighteenth- and nineteenth-century German philosophy in Chapter Four of Landgraf, E. (2011).

[2] Von Kleist, J. (1805–1806), 45.

unknown to the German writer, it is entire ecosystems, not particular organisms, which are the bridge between the primordial geochemistry of the barren planet and the abundant life so familiar to us. I have already given this formulation my own biblical twist: without Eden there would have been no Adam.

The idea comes in the speaking

The hero of von Kleist's story is the revolutionary orator Mirabeau (Honoré-Gabriel Riqueti, comte de Mirabeau, 1749–1791), whose spontaneous speech in the French assembly in 1789 helped transform a submissive response to the demands of the king's representative into a defiant and unprecedented declaration of the assembly's sovereignty. As von Kleist tells it, the politician rose to speak not knowing fully what he would say and discovered his own thinking as he spoke – "the idea comes in the speaking" ("L'idée vient en parlant"), an adaptation of the bon mot's famous monologue about wine-drinking and gluttony in *Gargantua,* "the appetite comes in the eating" ("L'appétit vient en mangeant").[3] The speech proceeded, not according to a pre-conceived outline, but rather by way of the successive platforms of thought that emerged as he spoke. In von Kleist's words, "speech is not an impediment on intellect, a brake on the wheel, but a second wheel on the same axle."[4]

In von Kleist's version, the King's Master of Ceremonies insisted that the assembly respond to His Majesty's demands, followed by what might be described as Mirabeau's temporizing: a series of repeated phrases of no importance – "yes, we have heard it" – separated by silence and perhaps a bit of throat clearing, showing that, according to von Kleist, "he still did not really know what he wants." And then, abruptly Mirabeau asks the king's emissary, "but what entitles you to draw our attention to commands in this place?" launching himself into what might have shocked him to hear a moment before: "we are the representatives of the Nation. The Nation gives orders and does not take them." The passage of time, a very short time during which the speech continues, is critical in Mirabeau's transformation. In that space, under the pressure of the moment, the ideas come, *en parlant.* "The moment when he opened his mouth, many a great orator did not know what he was going to say," says von Kleist. "But the conviction that the necessary wealth of ideas would be provided

3 Rabelais, F. (1552).
4 Von Kleist, J. (1805–1806), 43.

by the circumstances and by the resulting excitement of his mind, made him bold enough to pick opening words at random."[5] In other words, just begin and your words will lead you to a meaning you are not yet aware of. *The thoughts will self-organize.*

Perhaps most provocatively of all, von Kleist argues that thought emerging *en parlant* is not limited to lofty oratory or literature, but to everyday interactions as well. He describes an interchange with his sister, sitting in the same room as he puzzles a problem, as she periodically interrupts him with casual comments and questions about unrelated topics. As much in response to her presence as to her specific questions, in the manner of Mirabeau's temporizing he says "I interpose inarticulate sounds, draw out the connecting words, possibly even use an apposition when required and employ other tricks which will prolong my speech in order to gain sufficient time for the fabrication of my idea in the workshop of reason." In these moments, he even welcomes his sister's interruptions or sudden movements, which press him to more intense reflection, just as "a great general in an awkward position, reaches an even higher tension and increases in capacity," or the constraints which impose themselves on Stravinsky's "field of action" and thereby increase his creative freedom. No less than the lofty deliberations of the general, von Kleist says (and here I would add the composer), the naïve interruptions of a servant girl in a play of Molière simply provides a human face which can affect a speaker as "an extraordinary source of inspiration."[6] He applies these same principles to a person holding forth at a party or an oral examination at a university.

Nature, tradition and the individual talent

The topic of self-organization, once it is admitted to fields beyond the biochemical and cybernetic specialties where it emerged, offers many surprises. Consider how one of the most celebrated poets of the English language in the twentieth century could defend the European literary tradition using the language of chemistry. Not long after the publication of *The Lovesong of J. Alfred Prufrock*, the poem which established his reputation in the literary world, and decades before the publication of the *Four Quartets* which earned him a Nobel prize in 1948, T.S. Eliot offered the short essay "Tradition and the Individual Talent." Eliot famously spoke for the greater-than-individual and systemic processes in which he, the

5 Ibid.
6 Ibid.

poet, is embedded. He was attempting to describe a realistic place for poetry-making in a cultural setting where the word "traditional" has a negative connotation, as if it could only stand for a remnant of a superseded civilization of the past. "We praise a poet upon those aspects of his work in which he least resembles anyone else. In these aspects or parts of his work we pretend to find what is individual, what is the peculiar essence of the man." Ultimately, Eliot says, "you can hardly make the word ["traditional"] agreeable to English ears without this comfortable reference to the reassuring science of archæology." This is the basis on which, Eliot says, much European literary criticism tends to promote the view that "novelty is better than repetition."[7]

Eliot's "tradition" resembles von Kleist's more elusive "it is not we who know, but a certain state of mind of ours that knows."[8] But Eliot has declared that the subject of his essay is "individual talent," so it should be no surprise that he emphasizes the special sensitivities which are required of a poet who wishes to access what is held in common. It is not "personality" which differentiates the mature poet from the immature, nor is it that the mature poet has "more to say." Rather, he says, the difference is in "the mind of the mature poet, which is a more finely perfected medium, in which special, or very varied, feelings are at liberty to enter into new combinations." In Sudnow's terms, tradition takes the form of the "possibilities" or the "classes of action" which are still current, waiting in the piano keyboard (or is it in the player's hands or the player's memory?) to be discovered through a process that he describes as "foraging." Sudnow's phenomenological inquiry into the structure of handwork roughly translates across differences of media to the word-work of von Kleist and Eliot. All three workers see themselves – perhaps as the Shona mbira master did – as attached through a lifeline to the workings of others. "The existing order is complete before the new work arrives," Eliot declares. It "has a simultaneous existence," a "simultaneous order" which dwarfs the novel contributions of individuals. This is a perception, he says, "not only of the pastness of the past, but of its presence."

Here is where the poet, in what he calls a "suggestive analogy," turns to chemistry. Consider, he says, "the action which takes place when a bit of finely filiated platinum is introduced into a chamber containing oxygen and sulfur dioxide."[9] When the two gases mix in the presence of the sliver of platinum, they form sulfurous acid, an unstable and gaseous

7 Eliot, T.S. (1919/1975), 37.
8 Von Kleist, J. (1805–1806), 45.
9 Eliot, T.S. (1919/1975), 39.

substance. The oxygen and sulfur dioxide combine only if the platinum is present, but the resulting acid shows no trace of platinum. Likewise, the platinum sliver shows no sign of being affected by the chemical reaction it has participated in – that is, the molecules of sulfurous acid contain none of the atoms of the element platinum. The platinum remains "inert, neutral, and unchanged."[10] This is a role well-known in chemistry, that of a *catalyst.* For Eliot, the mind of the poet is the catalyst, the shred of platinum – "the necessary ingredient which makes the chemical reaction possible, without which the reaction would not have happened, or more likely, would have created a different reaction."[11]

If we accepted Eliot's term "tradition" as the caption for all of the experiences of self-organization in playing, making, speaking and writing, we would have to be willing to accept the diverse ways that each of these observers locate tradition in the body. Together, they present a paradox: the experiences of the pianist, the potter, the speaker and the poet are commonly located in different places– in the hands, the brain, the mind and the memory. But they each describe these experiences as happening a step or two removed from the individual human executive function as we conventionally think of it.

For Eliot, the poetic process is an act of continual self-sacrifice and even extinction, but this is not how Sudnow, von Kleist, Malafouris and Stravinsky express it. Nothing in what they say suggests that, for them, the pianist, the potter, the composer or the orator would disappear in their creative acts. In each of these cases, the creative agent remains fully present and un-extinguished. They presume that the idiosyncratic ways of each agent, and the impact of accidents on them, continues intact. It's also true that in Eliot's view, the catalytic agent (the platinum) never actually self-extinguishes, but keeps a respectful distance, as it were, from the outcome of its interactions with the other substances. The shift of emphasis from poet to poetry, from singer to song, from works to processes remains in all these testimonies. We are left to wonder at the way that outcomes arise from sources other than what we normally associate with our individual personalities. Eliot's description of the mind of the poet would also fit the descriptions of von Kleist's different speakers as "a receptacle for seizing and storing up numberless feelings, phrases, images, which remain there until all the particles which can unite to form a new

[10] The element Platinum (Pt, atomic weight 78), has been referred to as a *noble metal* for its special inert or non-reactive qualities, which has made it a frequent choice as a catalyst in industrial processes. See Royal Society of Chemistry (2026).

[11] Eliot, T.S. (1919/1975), 39.

compound are present together." This is a view consistent with what the science historian Philip Kitcher described as interactionist.

Like a rabbit from a hat: emergence and the systemic view

In the general (but not universal) consensus of contemporary biology, self-replicating life on earth could only have arisen from the interactions of inorganic materials present in the atmosphere, oceans and geology of the early earth. This is another statement of the paradox that has come up in different ways in previous chapters – that the permanent presence of life on this planet was made possible by a transformation (somehow) of geochemistry into biochemistry. This so-called *abiogenesis* theory – the theory that life originated from non-living matter – is not new, only newly demonstratable. In 1924, the biochemist Alexander Oparin (1894–1980) proposed that, more than four billion years ago, an accumulated primordial stew of inorganic chemicals in the atmosphere, oceans and geology of the earth – in the presence of the energetic inputs of sun, volcanic heat and lightning – resulted in compounds found in living organisms.[12] Numerous attempts to test Oparin's idea under laboratory conditions were unsuccessful until 1953 in the now-famous Miller-Urey experiment. Water and a mixture of hydrogen, methane and ammonia gases consistent with the primitive earth's atmosphere were placed in a closed apparatus and then heated and subjected to high-voltage electrical shocks.[13] In only a week's time (on an evolutionary scale, a blink of an eye), accumulations of compounds in the apparatus contained traceable amounts of certain amino acids, the building blocks of the proteins found in all life forms.

Like Darwin's adaptive radiation (so-named by later researchers) and Margulis's theory of endosymbiosis, the Miller-Urey experiment continues to be questioned, tested and duplicated, giving the nineteenth-century idea of abiogenesis an established place in the mainstream of controversies in biology.[14] However this particular controversy eventually resolves, the general principle of *emergence* that the experiments demonstrate is here to stay. Terrence W. Deacon, whose roots are in anthropology

12 Oparin, A.I. (1938).

13 Xie, X. et al. (2015).

14 An example of a recent critique of the original Miller-Urey experiment does not dispute the concept of abiogenesis which the experiment demonstrated, but it does point out that the non-living material of the glass apparatus in the experiment actually plays a role in the experiment's outcome. See Criado-Reyes, J. et al. (2021).

and neuroscience, describes emergence from the point of view of the life sciences in a way which I believe will be recognizable to musicians:

> …major transitions in the organization of things are often described as *emergent,* because they have the appearance of spontaneous novelty, as though they are poking their noses into our world from a cave of non-existence. And while they are not exactly something coming from nothing, they have a quality of unprecedented discontinuity about them – an almost magical aspect, like a rabbit pulled from an apparently empty hat."[15]

Once again, as with epigenesis and symbiosis, emergence is an old concept more resonant with our time than its own. The English philosopher George Henry Lewes noted in the 1870s in his *Problems of Life and Mind* that "the emergent is unlike its components insofar as these are incommensurable, and it cannot be reduced to their sum or difference."[16] A generation earlier, John Stuart Mill's puzzlement over the boundaries of thought, biology and chemistry lead him to wonder how chlorine (a gas) and sodium (a metal), both of them toxic to humans and many other organisms, could combine to create sodium chloride (common salt), which is essential to life and ubiquitous on the planet. "No mere summing up of the separate actions of these elements will ever amount to the action of the living body itself."[17] Mill's observations are an early version, like Lewes's, of *the whole is more than the sum of its parts.* Both of them anticipated Oparin's abiogenesis. For Deacon, the interest in these same properties of life which he shares with Mill and Lewes is focused on the ultimately puzzling fact that "mind emerged from matter."

What follows from the problems posed by Lewes and Mills, says Deacon, is the "Humpty-Dumpty problem," a problem which might strike musically minded observers as familiar. Deacon compares a mechanical device, the wheel, with an organism: "[the wheel] can be dissected into its parts and reconstructed without loss, but a living organism taken apart suffers from the Humpty-Dumpty problem. Most of its parts are themselves unstable entities outside of the context of a living organism."[18] Molecules are not like marbles in a jar, he says, but "reciprocally synthesize each other, none are parts. *Being alive does not merely consist in being composed in a particular way. It consists in changing in a particular way*" (emphasis

15 Deacon, T.W. (2012), 144.
16 Lewes, G.H. (1874), 412.
17 Mill, J.S. (1843), Bk III, ch, 6, sect.1. Quoted in Deacon, T.W. (2012), 148.
18 Deacon, T.W. (2012), 164.

added). If this process of change stops, Deacon says, "life stops."[19] This is to say that emergence, like the variability which is a recurring theme in this book, takes place in time. It is a process which can be frozen in a snapshot, in a PET scan[20] or in an audio recording, but in the end, its origin, persistence and demise are diachronic and historical. What happens now arises from what was happening a second earlier, and there is no turning back. There is no reverse-engineering of a musical event any more than there is a reverse-engineering of burnt wood. Ashes cannot be made to reconstitute into oak, or into its original fibers of cellulose and lignin. Ashes are the remainder of an event, the after-product of spent energy. The flame which consumed the wood, Mark Bickhard has reminded us, was a process, just as human mentality is a process. Music is also a mentality, an acoustic phenomenon which also registers in human consciousness and memory by first passing through neurochemical processes in the body and brain.

Choice, with and without scare quotes

For our purposes, one of the take-homes of Deacon's point is that emergence, like the variability of music, takes place in time. Any process which can be frozen in a snapshot, in an ultrasound image or in an audio recording in the end is historical and develops diachronically. The living processes of nature and music are equally irreversible.

When patches of light gray-green lichens form on the limb of an oak tree, each patch is as unique in its details of shape, color and texture as the lines of iambic pentameter of an Elizabethan sonnet. But lichen and sonnet are recognizably what they are because they are faithful replications of the structures of poems and lichens of the past. We say that the poem is the result of the poets' choices, but what do we say about the lichen? If we were to suggest that there is a creative continuity between lichen on a branch and poem on a page, at one end of this continuity we tend to use scare quotes but at the other end, we do not. We say that the lichen "chooses" where and how to settle on the branch, but when making artistic choices, we uncritically allow the artist to simply *choose.* The pattern of lichens on the branch is dazzlingly varied, so is it proper to say that a particular alliance of cyanobacteria and fungi *chooses* this spot rather than that one, to expand in this direction on the branch and not in that direction? No two square centimeters of lichen growth are

19 Ibid., 175–176.

20 Positron emission tomography (PET) used for cancer detection, creates a picture of the inside the body in real time.

identical, which is one sign of the variability that tells us that there is life going on in the lichen. Each lichen body develops uniquely through a back-and-forth exchange with the atmosphere, the solar energy and the chemical properties of that spot on the tree limb. The tree itself – its unique assemblage of nutritive properties – participates in a conversation which shapes both lichen and tree.

The subtle variability of the Cortot-Rubenstein-Hoffman versions of a Chopin Nocturne – apparent to any attentive listener – are also signs of a living, developing and evolving practice. My eye is drawn to the infinite variety of the lichen as its separate bodies advance, shrink and bloom on the tree trunk, and I am no less drawn to the infinite variety of treatments of a particular musical phrase, with the variety of movements in the starling murmurations in the evening sky, and with the diverse adaptations of sticklebacks in the Baltic and in Canada. If I were able to gather an assemblage of finches from the Galapagos and Cincos, an assemblage many times the size of Darwin's collection, I would be dazzled by the immense variety of beak adaptations, of which the summary pictures of just four beak types in Figure 1 of Chapter 3 are merely a convenient simplification. We might as easily say that this finch here "chose" a nest-building beak which is different from the "choice" of a seed-cracking beak made by that finch over there. Certainly, both finches were engaged in responding individually and uniquely to their environment in the same way that the lichen "chose" to settle on this square millimeter of the branch over another, and a freshwater Canadian stickleback "chose" an increased belly armor over its cousins in Lake Constance. However, we say that Stravinsky and Sudnow chose their morsels at the keyboard free of quotes, presumably because they, like the rest of us reading these words, are humans who we prefer to think are neither feral creatures nor symbiotic mashups. We are sentient and sapient. We are free individuals, made in the image of God. We, unlike anything else in nature, choose.

The self-organization of everyday life

So much of life is self-organized. Beginning with our bodily functions, the established processes of breathing, digestion, waking and sleeping, sensing the world, circulating blood, staying erect while walking – the self-regulating feedback systems of the human anatomy which in biology are referred to as *autonomic*. Even further, there are those countless examples of self-organization in natural systems – cells dividing, embryos developing and ecosystems evolving – and the processes which are inscribed in the concentric swirls on the shells of snails and in the growth rings of trees described by Paul Weiss. From the perspective of individual organisms, these autonomic systems and self-organizing processes are built into the ecologies in which organisms move. They are behaviors which are beyond the control of individual organisms, including humans, whether or not we take note of them.

Moving beyond the body proper, consider our thinking and speaking, the words we choose (with or without scare quotes), the sentences that fall from our mouths hundreds of milliseconds before our conscious minds have a chance to censor. Our cart is perpetually before our horse. At times, we acknowledge that we sometimes say things "thoughtlessly," usually because what results can be an embarrassment. And yet, as we sit at the computer keyboard composing our sentences, the words sometimes tumble out faster than our fingers can execute. Each sentence, each phrase, creates a platform for launching the next. On each occasion, when we go back and correct, the same thing happens. We are now writing on top of writing, correcting corrections, rethinking thoughts, each layer delivering a fresh zone from which a new word or image can emerge, a word or image which, a moment ago, we would not have thought to use. Looking back over the page full of type we have created, complete with layer upon layer of revisions and edits, we can find ourselves wondering where it all came from. So much of what we read on the page we were unaware of when we set out. Words, and therefore thoughts, emerge. *L'idée vien en parlant.*

And yet, looking at what we have written, we still feel confident that "this is mine, this is me." Of course, this makes sense. If these words are not mine, then whose are they? But in retrospect, we can see that what slipped out of us unbidden is adopted as ours, though we had only a few milliseconds before discovered it. In this sense, it was given to us, or perhaps we might call it a lucky mistake. We may have chosen to walk on this beach, but the shiny object we stumbled upon in the sand is ours in the finding. Others had passed it by and missed it. The stochastic, the chance events, which we may experience as external to us, as found objects or as intrusions and nothing to do with us, are also personal. It is *my* accident, the accident which happened to *me*, like a banana peel which I, and no one else, happened to slip on, making me the subject of a funny story afterwards. It is not that the banana peel is mine, but the slipping – the act, the behavior – is all mine, though it would not have become mine had it not been for a chance intersection of my wayward stroll and a discarded fruit skin.

Self-organization – in the larger, umbrella sense of the word as I am using it here – comprises all of these moments. The shiny object, the poem, the banana peel, the slip and fall, the musical variation – all converge on me, which leads me to think that I am in some way their author. Perhaps a better way to say this is that *I am the locus of their occurrence*, or perhaps, more simply, *they are gifts I found on my doorstep*. From a musical perspective I might even go so far as to say that, through my own native sensitivity and discipline, developed over many years through criticism and practice, this word or act or musical gesture, the discovery of this shiny object, was *enabled*. I am indispensable to its creation, a willing catalyst.

14

Autopoiesis is never optional

"Autopoiesis [self-making], the chemical basis for the impatience of human beings, is never optional. Absolutely required at all times for any life form in a watery milieu, once autopoiesis appeared in the tiniest bacterial ancestor, it was never completely lost."[1] When biologists like Lynn Margulis use the term autopoiesis rather than self-organization, they tend to move quickly beyond the kind of everyday language quoted here to take up the language of chemistry.[2] The two Chilean scientists Humberto Maturana (1928–2021) and Francisco Varela (1945–2001), who are said to have invented the autopoiesis concept, were both biologists and medical doctors whose research drew attention to the way that all organisms sustain themselves by producing their own components.[3] On a planetary scale, the principle of autopoiesis was an active ingredient in the Lovelock-Margulis theory of Gaia around the same time, which described the essential feedback loops between the earth's multitude of life forms and the ecology which sustained them. This is a circumstance where at times autopoiesis is used interchangeably with self-organization, but at other times, self-organization – the term I have been using more generically in a bio-musical context – becomes a big tent which encloses autopoiesis as a special case. Together, these concepts make a special contribution to the tendency of current biologists to rely on "co-" concepts like co-creation, co-evolution and co-development.

The prime example of such self-making is the *metabolism* of living cells which at every moment repairs cell damage, drives cell division,

1 Margulis, L. and Sagan, D. (1995), 62.

2 For an accessible account of the interdisciplinary evolution of "self-organization" and "autopoiesis" between the 1940s and the present, see Clarke, B. (2020).

3 Maturana, H. and Varela, F. (1972).

invents modifications of organismal bodies to fit new circumstances, and processes the components of energy which power the whole. *Self-assembly* and *inherency* (about which more will be said) are additional expressions with somewhat more specialized meaning which also fit under the big tent definition of self-organization. *Stigmergy*, another terms which fits under the same tent, is assigned to patterned behavior in nature like the elaborate nest-building of swallows, the courting dances of sticklebacks, and to beehive building, firefly swarming, fish schooling and starling murmuration. Such complex and artful ordering behaviors cannot be satisfactorily explained simply by genetic programming or by the kind of planning we are familiar with in human work.

Margulis's description of how the packets of chemical energy (ATP, Adenosine triphosphate) are produced in a cell's mitochondria and Weiss's description of the growth patterns on the shell of a snail can both be talked about as self-organizing processes. They are emergent behaviors which arise from interactions among multiple independent agents and events in the ecology, rather than from a pre-designed program. Nor are they the products of an executive intelligence. At the same time, the consistent behavior registered in Margulis's mitochondria and of Weiss's snail shell are the result of distinct chemical and physical processes which no scientist would claim to be the same either in materials or in method. Neither would any scientist claim a broad consensus in their fields about how those distinct and predictable processes of self-organization originated on this planet – how biogenesis could have erupted on a rocky ball without life several billion years ago. This is, once again, the question unanswered by natural selection which kept Darwin awake at night.

The page is never blank

If we first run across the general concept of self-organization in the chemistry of habitats, living cells and bacterial colonies, it can seem like overreach to consider applying the concept to the arts, to language and to human thought. But when we approach self-organization by way of the more familiar everyday acts of handwork, voice-work and word-work, self-organization becomes more recognizable, a concept which is native not just to forests and coral reefs but to the commons of everyday human life. Coming from neither biology nor music, the anthropologist Timothy Ingold has weighed in on self-organization, though he doesn't use the term. His home turf is art and architecture, and he pays special attention to archaeological study of what he calls "makers" in human societies of the past, and this experience has led him to question the conventional image of the artist/craftsman. When he looks closely at the role of the maker, he finds that, "far from standing aloof, imposing his designs on a world that is ready and waiting to receive them, the most he can do is intervene in worldly processes that are already going on...adding his own impetus to the forces and the energies at play."[4] Ingold's words describe a world where the maker is not a demi-god who, solely through his own unique abilities, brings life to a lifeless world, but is instead a participant (Eliot's word is catalyst) in existing processes and energies.

The world as described by Ingold is a world compatible with that of music-makers, pot-makers and word-makers who browse among the tasty morsels scattered by mostly anonymous contemporaries and predecessors on the forest floor. This is a commons like that from which string quartets draw for their respectful interpretations of hallowed classics and which jazz musicians absorb through conscious imitation and osmosis. For the Rolling Stones, Robert Johnson was just such a tasty morsel – a shiny object discovered on a beach – which, in their effort to take him into their own practice, they ended up transforming. What began as imitation and replication became variation and adaptation. Circles became spirals. From there – from Ingold's "intervention in worldly processes that are already going on" – it is only a short step to the biologists' claim that the inherent properties of matter, the self-assembly of molecules, and the feedbacking loops within a cell are all components of the metabolic processes characteristic of living systems. From the point of view of a human observer, all of them seem to emerge, *on their own,* as it were, even *inevitably.*

To the naked eye, a glass of spring water can look clear, telling us that there's nothing there. When drinking is our goal, we will probably find this reassuring, since when we're thirsty, nothing-but-water is what we want. If the glass of clear fluid were an artist's easel, we would say the canvas is empty, meaning that it offers us nothing but the neutral background of water itself, a canvas to be painted on. To the naked eye, the glass of water would only become more than nothing if we added something to it – a drop of ink, perhaps, or some grains of sand, or a goldfish – *now* there is something to look at and respond to. But the swarm of living material populating the glass – the tiny "animalcules" first viewed witha microscope in the seventeenth century – now in the twenty-first century can no longer be marginalized, even though they are not seeable unaided. The human awareness which the microscope symbolizes – an extra sensitivity to even the smallest discriminations of difference – is as much a foundational

4 Ingold, T. (2013).

quality of scientific work as it is of work in music. A listener's preference for Richter's version of a Chopin Prelude over Rubenstein's, or the fierce arguments of the Juilliard Quartet about a dynamic in a Beethoven string quartet are instances of the microscopic scrutiny of variations which is native to all serious music-making.

All this is to suggest that what may appear to be an empty medium – the artist's blank canvas, the writer's empty page, or the silence in a composer's inner ear is in reality a tangled bank. The morsels which find their way into a performance of a classic piece or into a composer's head and onto the empty staff lines consist of specific gestures, fragments of tune or texture, lines of text, muscle memories of music executed in the past, snatches of rhythmic patterns, unidentifiable auditory echoes in the ear, as well as stored-up templates, models, techniques, teachings and theoretical ideas. The canvas, the page, is never more blank than the musician's inner ear and the writer's inner voice are ever really silent.

Louis Pasteur's experiments in the time of Darwin bring home this commonplace observation. The first self-sustaining life forms on this planet did not, as the Greeks famously proposed, appear "spontaneously" – that is, out of nothing. The microscopic spores of harmful molds, the fungi, bacteria and other invaders of human bodies and food which earlier generations had thought arose from nothing were, in fact, already there in the sealed chamber of Louis Pasteur's demonstration. Humans simply couldn't see them. But in Pasteur's experiment, once the invisible microscopic life was killed with heat and the chamber sealed, no new life appeared. When the seal was broken, the signs of life reappeared. Pasteur is usually credited with the memorable caption for his experiments, "Omne vivum ex vivo" – *all life comes from life*. Looking into biology from the arts, that sounds about right. In writing and speaking, it is no stretch to say *all words come from words,* nor in music to say *all music comes from music.*

The idea that primitive life might have appeared on this planet as an inevitable outcome of the physical conditions of the planet itself is still controversial in biology, but it is fair to say that the controversy has now moved out of the margins. However the controversy is resolved in the years to come, the ideas being argued suggest the outlines of an unfamiliar creation story, a story of mythic proportions on a par with the Genesis myth. The Lovelock-Margulis story of a planet which is a self-making and self-sustaining Gaia is one version of the myth. This new story of the planet earth also suggests a new context for musical creativity, one which is potentially as challenging for musicians as it is for biologists. Lynn

Margulis declared unequivocally that our planet cannot function without self-making – that is, without the self-generating and self-perpetuating processes of metabolism which exist in every living cell, in all domains of Linnaeus's phylogenetic system. Throughout the multiple extinction events that have assaulted the biome since the first appearances of replicating life forms between three and four billion years ago, the ability of life forms to carry on has never been extinguished. "Carry on" in this case means that the loops of chemical transformation in essential processes like the famous Krebs Cycle continue on their own, as long as there are steady infusions of energy from multiple potential sources – from the sun, from volcanic heat, from lightning and from the energy-generating chemical reactions catalyzed by other chemical reactions. Self-making is a foundational creative principle that proposes a continuity of everything living on the planet – from a garden of heirloom tomatoes to a pod of whales, from a human embryo to an utterance by a human orator or musician.

What the Cambrian explosion tells us

The planetary event which took place between 600 and 500 million years ago known as the Cambrian explosion,[5] famously recorded in the fossils of the Burgess Shale deposits in the Canadian Rockies, offers another glimpse into the idea of self-organization as it emerged in late twentieth-century biology. The Cambrian explosion refers to the dramatic appearance in the fossil record of an enormous variety of novel animal body types not found in the fossils of preceding periods, all within just a few million years early in the Cambrian period.[6] This is what a recent study described as the "very short burst of accelerated evolution at the start" which then "flat-lined for the rest of the Cambrian."[7] Though most of these newly-invented Cambrian body types were unprecedented in the fossil record, the majority of them have nevertheless persisted over millions of years, even into our own time. They are a strong presence in Linnaeus's phylogenetic classification system which is still the basic framework of all work in biology. These basic body types include the exoskeletons of the insect and crustacean groups *(phylum Arthropoda)*, the segmented bodies and

5 Stephen Jay Gould's *Wonderful Life: The Burgess Shale and the Nature of History* (1989) is a readable account of the Cambrian explosion viewed through a close examination of the large number of fossil forms found in one location in the Canadian Rockies since the early twentieth century.

6 Morris, S.C. (2003), 27. In Müller, G.B. and Newman, S.A. (2003).

7 Paterson, J.R. et al. (2019), 4394–4399.

jointed limbs of the *sub-phylum Vertebrata* including mammals, and the hollow cylinder-like, soft-bodied groups *(phylum Ceolenterata)* including corals, sea anemones and jellyfish.

In these Cambrian examples we confront head-on the porousness of the conventional boundary between adaptation of what is already there – the common meaning of variation – and invention *de novo.* Darwin, we should remember, embraced the general model of evolution as a series of incremental and uniform changes over very long periods, Stephen J. Gould's "stately and orderly process, imperceptible within a human lifetime." Because Darwin was also aware that, as Müller and Newman put it, "selection can only work on what already exists,"[8] natural selection is unable to explain the instances of organic invention like the Cambrian explosion with no clear precedent, which seem to arise relatively rapidly, responding to local conditions and, in Lamarckian fashion, passing on the new traits to future generations.

As early as his first edition of *The Origin of Species,* Darwin made the first of his many disclaimers of the universality of natural selection, acknowledging the possibility that certain traits "originated from quite secondary causes, independently of natural selection."[9] He was mindful of the gaps in his own theories, gaps marked by those moments in the evolutionary record when organismal forms without precedent seemed to appear suddenly, as if out of nowhere, and yet also managed to establish themselves in future generations.[10] These gaps were often referred to by nineteenth-century biologists, thinking in terms of millions of years, as "saltations" (L. *saltus,* leap).[11] During the same period, farmers and plant and animal breeders, thinking in the short term about anomalies of variation, referred to unprecedented flower colors, leaf shapes or animal coat patterns as "sports." Such acknowledgments of exceptions to the glacial pace of Darwinian evolution have kept alive attempts to recast the prevailing understanding of biological variability into something more nimble and interactive. The gradual acceptance of epigenesis and

8 Müller, G. and Newman, S.A. (2003), 3.

9 Darwin, C. (1859/1993), 196.

10 Darwin was not the only writer in his day to worry this problem. See Thomas Huxley's famous review of *Origin of Species* the year after it was published: "it is not absolutely proven that a group of animals, having all the characters exhibited by species in Nature, has ever been originated by selection, whether artificial or natural." Huxley, T.H. (1860).

11 The phrase "Natura no facit saltum" (nature does not make leaps) has been argued since the eighteenth century. For example, see Mayr, E. (1991), 46–7.

symbiosis over the past century accelerated the acceptance of relatively rapid change in the biome.

Speaking in the 1990s, one of the leading neo-Darwinists, the venerable Ernst Mayr, could easily dismiss the concept of self-organization as a form of Lamarckism which had been completely refuted by the modern marriage of Darwinian evolution and genetics.[12] And yet here was the evidence of the Cambrian explosion. From Mayr's point of view, how could the creative surge in unprecedented body forms in the Cambrian period have taken place – like a rabbit from an apparently empty hat – and yet with such lasting effects? As the nimbleness and interactivity of nature has been increasingly observed and accepted in science, it has become more reasonable to consider the possibility that nature and human sound art – music as made and heard in real time – exist on an unbroken continuum.

The inherent properties of oil, water, snow and sand

Beginning in the 1980s, the biologists Gerd Mueller and Stuart Newman, among others, proposed that the formerly unknown body types such as jointed limbs, hollow bodies and hard exoskeletons did not arise simply because of recipes transmitted in the genes. Instead, they argued that the early eras of evolution were characterized by "an archaic soft stew" of protein molecules joined by relatively loose bonds and susceptible to the influences of surrounding conditions. "Tissue forms emerged early and abruptly because they were physically inevitable," says Newman. "They were not acquired incrementally through cycles of random genetic change followed by selection."[13] This is the *inherency* defined by Newman and Bhat in 2009 as "the idea that biological forms...are expressions of inherent dispositions of living materials" which need no direction from a command center such as a genome or a brain to take shape.[14] To bridge the theoretical gaps offered by the Cambrian explosion, Newman says, "there are now credible alternatives to scenarios of adaptation by increments."[15] The alternatives now offered by many other cell biologists and geneticists push past gradualism to question long-dominant assumptions about the behavior of matter itself. They argue that, at this early stage of evolution more than 500 million years ago, some three billion years after the appearance of the first primitive forms of microscopic

12 Mayr, E. (1991), 179.
13 Newman, S.A. (2003b), 221.
14 Newman, S.A. and Bhat, R. (2009).
15 Newman, S.A. (2011), 3.

life, *metazoan* (multi-cellular) body forms were malleable and susceptible to morphological creativity. They were more malleable than those more highly evolved forms we now know which have been hardened through still more millions of years of natural selection.

In other words, the rapid changes in living organisms that were possible earlier in evolution are less possible later on. This is an interpretation of evolution itself which has relevance more generally, outside of biology. It raises the possibility that novel structures, novel relationships and novel energy infusions will continue to emerge in later epochs, but those novelties will increasingly have to find a place for themselves among all the structures which have already proven their durability over time. They are like newcomers moving into an established neighborhood who find the streets are already constructed, the power, water and sewer lines laid, houses built, and the web of real estate investment and taxation in the neighborhood deeply linked to the surrounding webs of city, region and nation. As the history of any neighborhood or city demonstrates, some new structures, new populations, and new interdependencies can still establish themselves, but the available options for novelty at any given moment will be more limited and the alterations – whether incremental or radical – can be expected to take longer to accomplish.

Newman's example of what happens when oil and water are mixed illustrates how the differing capacities for intermolecular *adhesion*[16] in two substances can produce characteristic and predictable structures. At first, the small, round, island-like clusters of oil molecules on the water's surface take shape, but these islands are ultimately dynamic and even mobile in their behavior. Shortly after the oil and water are stirred together in a petri dish, small circular islands of oil form; after a longer period, these small islands come together to make fewer, larger islands. After a still longer period, they are likely to transform into a single unbroken continent. The consolidation of oil molecules predictably continues until an equilibrium is achieved between the forces on both sides of the oily membrane separating the two substances. The oil molecules amassed into a continent-like formation are capable of holding together as long as the equilibrium state remains undisturbed – that is, until the container is stirred again, heat is applied, or another substance with different adhesive properties such as alcohol is added.

16 *Adhesion* is the term given to a situation where substances with dissimilar adhesion properties interface – for example, water droplets on a spiderweb. The interface of substances with similar adhesion properties is usually referred to as *cohesion*.

In a similar way, when taking a shower, water vapor condenses in a cloudy coating on the cool surface of a bathroom window. When the windowpane is at a freezing temperature, the vapor just as dependably crystalizes into a frosty coating. If we were to look closely at the frost, we would see the gorgeous fractal forms of snowflakes. These shapes, these movements, are said to self-organize due to the inherent properties of oil and water according to the physical laws which govern the behavior of all gases and fluids in the presence of temperature changes. These generic, not programmatic, processes are what researchers like Margulis over the past 40 years have seen in the naturally occurring segregation of lipids (oils and fats) and water to form the cell wall, the structure common to all complex life forms. Cells themselves can also behave like the molecules of oil and water, says Newman, "when differential adhesion exists between two cell populations. Not only will each type of cell keep to its own side of the interface, but, when dissociated and randomly mixed, the two populations will 'sort out,' much like the shaken mixture of oil and water, and for the same thermodynamic reasons."[17] Complex multi-cellular organisms emerged from early aggregates of cells which had the properties of "excitable, viscoelastic soft matter," properties "which made a profusion of multilayered, hollow, segmented forms virtually inevitable early in the history of metazoan [multi-cellular] life," of which the Cambrian explosion is an example.[18]

Surprisingly, the idea that organic matter can assemble into functioning structures without the directions of an all-controlling genetic code, is not a new idea but one of several strains of evolutionary thought which were taken seriously through much of the nineteenth century alongside the slow adaptationism which eventually came to dominate biology. We have already seen how the studies of Johannes Reinke and Albert Frank in the 1870s describing the mutualism of algae and fungi in the lichen would eventually be mainstreamed as symbiosis by Adolf Meyer-Abich in the 1940s and Lynn Margulis in the 1990s. The studies by Abraham Trembley in the mid-eighteenth century, and of H.V. Wilson in the early twentieth had called attention to the ability of the multi-armed photosynthetic *hydra* and the marine sponges to reassemble themselves from dissociated cells.[19]

17 Newman, S.A. (2003b), 225.
18 Ibid., 233–234.
19 Steinberg, M., in Müller and Newman (2003), 138.

Newton's matter and matter which is protean and self-organizing

What is now known about organic matter itself provides a perspective on novelty and adaptation in nature which is quite different from the view provided by the nineteenth-century evolutionary consensus. Stuart Newman's review of the range of available physical theories which Darwin could have chosen to serve as the basis of evolution is revealing. Three of the four principle theories known to eighteenth- and nineteenth-century physical science – chemistry, thermodynamics, and electricity/magnetism – were understood to be capable of, in Newman's words, "qualitative transformations of matter: metals and gases into salts; solids into fluids and back again; moving charges into magnets." However, although this "protean side of the natural world" was the starting point for the biological theories of several other scientists and natural philosophers of that period,[20] this was not the theoretical model adopted by Darwin, and the "Darwinist" school of thought forming around him. They chose instead the fourth of the physical theories then available, the science of mechanics, in the philosophical lineage of Aristotle, Galileo and Newton and "the oldest and most established of the four physical theories." But mechanics was also, according to Newman, the one "least concerned with qualitative change."[21] For all of its shortcomings in explaining the behavior of organic matter, says Newman, mechanics seemed to offer Darwin the most authoritative materialist basis for arguments against the general cultural resistance to evolution, especially as it was embodied in biblical interpretations of creation.

In the end, Darwin's choice was fateful for the development of biology, since it hitched evolution's wagon to the tried-and-true methods of classical physics. "We now know a great deal about the actual chemistry of cellular life and the physics of embryonic development," Newman points out, "and they have turned out to be nothing at all like classical physics."[22] In the physics of Newton, "matter was inert, altering its trajectory only when acted on by external forces…Things moved when pushed and then stayed put unless pushed again." Writing elsewhere in 2011, Newman said that the incompatibility of these assumptions with the conditions of living

[20] Newman, S.A. (2011), 6. Newman cites Johann Wolfgang von Goethe (1749–1832), Etienne Geoffroy Saint-Hilaire (1772–1844) and Lorenz Oken (1779–1851), among others.

[21] Ibid., 6.

[22] Ibid., 8.

molecules and cells had become a hindrance to biological understanding. "Aristotelian and Newtonian mechanics...successful as they were in their respective realms [for example, in the building of steam engines], were incapable of capturing the dynamic, self-organizing and self-realizing qualities of living systems, or their capability of undergoing abrupt transformations in form and behavior, particularly during embryonic development."[23] Burkhardt, looking back at a century of thought in biology, had observed in 1977 that Lamarck's more plastic view of evolution had implied that behavior itself could be "an agent of organic change" (see Chapter 7). But for over a century, the more protean, Lamarckian views of matter remained in the realm of belief rather than fact, sidelined by the excitement over the tidal wave of data coming in from genetics research.

Those of us in the music world are left to wonder what the effect on the young science of musicology in the nineteenth century would have been if the early ideas of self-organization, Lamarckian evolution, epigenesis and symbiosis had been able to compete with Darwinian gradualism. We will return to this question in Part Four.

The stable, the unstable and the metastable

Matter which is capable of self-organization can be expected to exhibit the qualities scientists would associate with *metastability,* another concept used in biology which, I believe, suggests musical thought. Most simply defined, metastability is a state which is *stable,* provided it is subjected to no more than small disturbances. Part of the charm of this concept is that it has an opposite definition which is equally true: metastability is a state which is *unstable,* but which is so persistent over time as to be considered stable.[24] It is a liminal and dynamic state, in other words, neither stable nor unstable in the everyday sense, but perhaps both, or perhaps something altogether different. (A musical concept if ever I heard one.)

As a general concept, metastability originated in such remote fields as metallurgy and particle physics. The term describes the dynamism of Newman's oil and water mixtures, but its classic examples are the sand pile and the avalanche. An immense pile of sand such as a sand dune may achieve a stable form which can hold for an indeterminate duration, secured by the friction acting on trillions of separate grains. As individual grains are added or shifted – for example, by wind – the whole pile is

23 Ibid., 6.
24 Lamm, E. (2011) in Gissis, S.B. and Jablonka, E. (2011), 349ff.

gradually reshaped. A sand dune is therefore inherently dynamic in form. Each gradual disturbance contributes to a new equilibrium, until the next gust of wind. In a much stronger wind or in a heavy rain, the dune might abruptly alter its shape, or even disappear altogether.

With the avalanche, the disturbance is likely to be a change in temperature – the colder the snow, the less the individual flakes tend to bind with each other and the greater the likelihood of an abrupt and drastic shift. The equilibrium of forces keeping the grains of sand and flakes of snow stable is inherently temporary, though it could, under the right conditions, remain undisturbed indefinitely. What is also important is that each of these forms of matter – sand and snow – behaves in accordance with its own inherent properties: the binding qualities of sand are the result of friction and gravity, while with snow what holds back the avalanche are the chemical bonds created or loosened when water changes its state from fluid to crystal.

If we apply the concept of metastability to the DNA codes of living systems, we get a *metastable genome*. It is an idea which a generation ago would have appealed to geneticists like Lynn Helena Caporale (see Chapter 4) who were feeling the inadequacy of the mid-century neo-Darwinian view of DNA as the code of codes. Metastable genome is a term which the science historian Ehud Lamm calls "a Lamarckian concept in a Darwinian world,"[25] and a crucial piece of the argument for rethinking the accepted view of the genome "as a static container of genetic information."[26] "Various environmental stresses, such as temperature shocks and pathogen attacks...can cause anything from small-scale genomic changes to wide-ranging and repeatable genomic patterning," says Lamm. "The genome is constantly dynamic and in constant danger of unintentional switching from one metastable state to another." It would not be going too far to say that all living matter – cells, tissues, organs, genes – may be viewed as *essentially* metastable, kept in a provisionally unchanging state by a complex matrix of interactions among highly evolved structures in the cell, in the genome itself, and in the environment.

Artistic sovereignty and the stable thing-ness of music

Natural systems – cells, organisms and whole ecosystems – are variable, even though the world we live in and the bodies we inhabit are as a rule

[25] Ibid., 345.
[26] Ibid., 347.

invariant *enough* to allow us to act as if they are invariant. Likewise, in music, individual pieces and events are dynamic, continually inclining toward variations small and large. But we can treat whatever we are experiencing now in music as stable enough that we can act as though it were stable, at least for the amount of time it takes to make an impression on us. Metastability is, in this way, a useful model for music, but I don't believe that this concept resolves the perennial tension between our experience of music as both dynamic and fixed. The *thing-ness* of music – the perception, usually transient, that music exhibits the qualities of an object – remains ambiguous and changeable with context.

The Irish singer Joe Heaney would always tell a bit of a story about the song he was going to sing – an anecdote about a famous singer from whom he learned it, or something about the words. Then he would pause a moment and say, "well now, here it is, I guess." The preamble and the reference to "it" communicated to his listeners that this was a *thing* that he was passing on. He was stating his own conviction that what he was about to perform had substance and durability. It was not just the singer's song, but rather a piece of shared cultural property, like a recipe, a historic building, a poem, or a gravestone. On the other hand, we should also bear in mind that, among traditional singers like Heaney, references to musical objects were always mingled with expressions of admiration for a particular rendering of a song by someone else. Such singers often seemed unconcerned or even unaware of the variants they were bringing to their performances. The eight different versions of just one line from a song which Heaney provided at the request of an ethnomusicologist, mentioned in Chapter 1, were unusual in folk song research. The versions acknowledged in folksong what the hypothetical performer of the Bach Chaconne acknowledged in art music when she took offense at the suggestion that her rendition was exactly like the score. The concept of metastability could reasonably apply to both cases, but in naming this dynamic condition, the dynamism is not made less dynamic. It is just that, by calling the dynamism of music by the same name as the dynamism of nature, we are drawn another step closer to acknowledging the unbroken continuity of the natural and musical ways of working.

Looking back at the medieval psalm-antiphon practice, we can see that the objectness of the antiphon and the variability of the psalm recitation merge into a single fluid practice. There is a clear advantage gained for musical thinking when we accept the stability of a musical object as a kind of metastability. By seeing an example of medieval song practice as a metastable phenomenon, we can leave room in our thinking for self-organization. The shiftiness of the materials in the sandpile and snowbank are the result of systemic forces which "all by themselves" contribute to the behavior and the shape of the sandy or snowy aggregate. Assembling, connecting, adapting and re-inventing the many components of the psalm-antiphon is a compositional task requiring skill and musical intuition. Some humans can be expected to accomplish this task in distinctive and personal ways which call special attention to themselves. The master singers in the monastery were those who had demonstrated competence and a quality of performance which inspired emulation, a combination which we can easily call artistry. This is the same artistry which was at work in the versions of Frere's Salisbury cathedral chants where patterns of melodic behavior show tendencies toward, on the one hand, conformity to family characteristics and, on the other, continual variation of those shared features.

Once again, this is what I would consider the most important take-home from the comparison of a particular melodic practice and the behavior of living matter: the fact that the comparison calls into question the role of the creative executive – what might be called *the artistic sovereignty* of the singer-composer. His or her skill and experience are clearly indispensable to the outcome, but the singer-composer's role is also reactive and interactive as much as it is authorial – he or she personally *composes* the melody (Lat.: *con-* [with] + *ponere* [to place] = "place-together"), but their melodic choices are drawn from a vocabulary and a grammar which are communal and systemic. There is no top-down programmatic direction for what happens. The received musical materials themselves – like the inherent qualities of sand, snow, oil or water – are as determining of the musical outcome as the singer's individual choices. When the singer-composers of Frere's Salisbury Cathedral community created melodies for their texts, they didn't assemble new melodies note-by-note. Instead, they created as nature creates, by recombining small conventional structures, clusters of a few notes or of many notes which were already proven over time to be memorable and graceful enough to warrant reuse.

Frere's melodic families recombine and modify their conventional elements to match circumstances, but they leave behind in their notations patterns of similarity that reflect successfully repeated behavior. The lobed leaves of the oak tree are examples of deeply established patterns in the tree's metabolism. Paul Weiss called these patterns precipitates and transcriptions – fixed structures in cellulose which trace the preferred bio-chemical pathways of the tree. The serrated leaves of the chestnut are precipitates of a particular expression of the same metabolic process, resulting in a different physical appearance, honed into its conventional shape by repetition over time in a particular habitat. In the non-living circumstances of the sand dune, the rhythmic patterns of ripples on the surface of the dune, as well as the overall shape and metastability of the dune, are also transcriptions of wind patterns. But these inherent behaviors of sand grains are the result of gravity and wind which operate by different rules than the biochemical churning of metabolism. The interactions of living materials and the interactions of non-living materials, though they are very different, are each agents of self-organization. Metabolism and inherency have this in common: neither of them happens by authorial choice, but by the consistent behavior of the living and nonliving materials provided by the planet. Translating this directly to musical terms, the singer-composer is, once again, a musical actor who is a mediator, an inter-actor in an epigenetic exchange with the ecology in which he or she is embedded. For T.S. Eliot, the poet was also an inter-actor, a catalyst, an enabler.

The notion that the entire biosphere is perched on a wall of snow waiting for the cold snap which could convert that wall to an avalanche is perhaps unsettling. Such precarity feels like a mismatch with the confidence humans have placed in the solid works of civilization – the edifices and institutions and art works which are the product of countless lives over centuries and which assure us that some human products endure. Perhaps even more unsettling is the fact that metastability also seems much less certain than we have believed is appropriate to the material world and to real science – that quantifiable concreteness which has been science's defining feature, supposedly distinguishing it from the mentalness of the humanities, what Mark Bickhard described as a flame state. The predictable comfort of a Newtonian view of matter – matter which is stable, neutral and inert, which will not move unless acted on externally[27] – seems like a lot to give up for a metastable system. But recent biology assures us that there can be no life at all without the indeterminacy which, so dependably for the past 3.7 billion years, has arisen from "the dynamic, self-organizing and self-renewing qualities of living systems," as Stuart Newman put it. They are protean and full of surprises. To the extent that human sound art is metastable, self-organizing, self-renewing and full of surprises, I am inclined to give it the benefit of the doubt and place it in an unbroken continuity with a forest, bird and microbe.

27 Newton's first law: An object at rest remains at rest, and an object in motion remains in motion at constant speed and in a straight line unless acted on by an unbalanced force.

Part Four
The macroevolution of species and musical practices

At every step we are reminded that we by no means rule over nature like a conqueror over a foreign people, like someone standing outside nature – but that we, with flesh, blood and brain, belong to nature, and exist in its midst, and that all our mastery of it consists in the fact that we have the advantage over all other creatures of being able to learn its laws and apply them correctly.

Friedrich Engels, 1873–1886[1]

Evolution has long been another way of talking about history. On reflection, the implication of this usually casual way of speaking is surprising since it presumes that there is an equivalence between human actions and natural events. In everyday conversation in both public media and academia we refer to the evolution of personal relationships, the evolution of hip hop, and the evolution of democratic institutions. This metaphorical habit betrays our instinctive acceptance of the continuity of biology and human history. It seems to cause little disruption in our thinking about evolution to allow the concept to stand for the stories of dinosaurs and the stories of human civilization, as if the changes in a fish's feeding structures over thousands of years and the changes in the form of a sonata over decades were the same kind of story.

[1] Engels, F. (1873–1886).

15

The single road of history

For Darwin, music remained a relatively minor theme which he attempted to flesh out only in the latter part of his career as a factor in what he called sexual selection, a subset of his central theory and dear child, natural selection. In *The Descent of Man*, published a dozen years after *The Origin of Species*, he argued that music evolved as part of the vast repertoire of plumes, horns, ritual behaviors, cries and birdsongs common to the mating practices of many species.[1] Darwin's contemporary and self-appointed public spokesman, the British social scientist Herbert Spencer, had his own view of what the concept of evolution meant and, more importantly, how it could be harnessed to the writing of history, including music history. It was Spencer who coined the infamous term "survival of the fittest" which became the popular caption for Darwinism in general as well as for its derivatives in social science gathered under the term "social Darwinism." In 1857, two years before the publication of *The Origin of Species*, Spencer articulated a universal "law of all progress" in a highly influential essay entitled "Progress: its law and cause."[2]

Looking at Spencer from a musical point of view, Ruth Solie describes Spencer's essay as an argument for an inexorable evolution in society as in biology from the simple (homogeneous, primitive, instinctual, aesthetic) to the complex (heterogeneous, sophisticated, intellectual, architechtonic).[3] This is a view which had important implications for history writing in

1 Darwin, C. (1871).

2 Spenser, H. (1857). A later version published in 1881 in *Popular Science Literature* included a section on "The Origin and Function of Music."

3 Solie, R.A. (1982), 297. For more on various aspects of evolutionary thinking in music and the critique of evolutionary thinking, see also: Burnham, S. (1992); Graziano, A.B. and Johnson, J.K. (2006).

Figure 15.1 "The March of Progress," an illustration by Rudolph Zallinger (1919–1995), first published in Howell, F.C. (1968). This famous illustration by American artist Rudolph F. Zallinger is a mid-twentieth-century version of a venerable lineage of nineteenth-century march-of-progress illustrations. The author of the volume F. Clark Howell later commented on its oversized impact and the misleading impression it gave: "The artist didn't intend to reduce the evolution of man to a linear sequence, but it was read that way by viewers....The graphic overwhelmed the text. It was so powerful and emotional." Barringer, D. (2006).

general and for the writing of music history. Although Spencer's polemics on Darwin's behalf and their absorption into popular discourse may certainly be considered an example of the steady misrepresentation which Darwin regretted, no one could say that, as a twit-nerd coalition, the polemics were ineffective. Spencer, like Darwin, was in no sense a music specialist, but he supported the idea that music had a rightful place in an account of biological evolution, but offered an alternative theory – that music arose within the evolution of human speech.[4] A later version of Spencer's 1857 essay on progress included a section on "The Origin and Function of Music."[5] Both Darwin's and Spencer's approaches to music in biology continue to be debated and explored today using experimental protocols, often focused on the links between genetics, brain development and the evolution of music and speech.[6]

Progress backwards?

The role of music in Darwinian thinking and the occasional public tension between Darwin and Spencer on the subject of music remained a sideshow next to much larger controversies. Questions surrounding the place of

4 Solie, R.A. (1982), 298.

5 The later version was published in 1881 in *Popular Science Literature*.

6 The continuation of Darwin's sexual selection thesis in recent studies is surveyed by Wilson, D.S. (2007), 182–193. See also: Perry, J. (2000).

human art in evolution were no competition for the widespread religious resistance in Europe and the United States which viewed evolution as a heresy seeking to invalidate the biblical Genesis.[7] More widespread at the time, and also more lasting in effect, was the prevailing view of history which was embedded in the evolution debate. This was a story of inevitable advancement which for many, including Spencer, fit comfortably with the European colonial outlook of the time. Over the course of his career, Darwin's role in the development of this view is mixed. Beginning with the publication of his *Voyage of the Beagle* in 1839, his initial publication as a naturalist, he at times seemed to support the idea of a single macroevolutionary road to a superior stratum of organisms which was the scientific consensus of his time. But by 1859, in *The Origin of Species*, his preference for the tangled bank as an image of evolution bore little resemblance to either a straight and stately tree or a single road, and privately he expressed abhorrence at these ideas. In the margin of a later edition of *The Origin*, Darwin scribbled a note to himself – "Never use the words 'higher' and 'lower,'"[8] as if he were painfully aware of the need to separate himself from the linear view of evolution which, in the eyes of the public, his own work had at least appeared to support.[9]

Nevertheless, despite Darwin's disclaimers and increasingly robust evidence against the single road doctrine, the idea persisted in public discourse, and ways were found to adapt Spencer's universal law of progress to music. Most commonly, this was accomplished by equating "simple" with monophony and "complex" with polyphony. The generations of musicians and music historians who grew up in the shadow of Spencer's version of Darwin fleshed out Spencer's principles with references to specific repertoires and practices which seemed to illustrate the progress from simple to complex. As W.J. Henderson put it in his *How Music Developed* in 1898, "in the beginning there was only melody: and that was like the earth before the creation, without form and void."[10] When he reached beyond the monophony-polyphony issue to describe the development of music between Scarlatti and Beethoven, there is no mistaking the ethnocentrism of Henderson's language. It is worth quoting at length:

7 Solie, R.A. (1982), 298.

8 Quoted in Mayr, E. (1988), 62.

9 The pros and cons of biological thinking in music reaching back to the Enlightenment were efficiently outlined by Warren Dwight Allen in 1962. Since then, the critique of Spencer's law of all progress has been continually amplified and sharpened by others, including Kivy, P. (1959); Solie, R.A. (1982); and Tomlinson, G. (2015).

10 Quoted in Allen, W.D. (1962), 291.

> Like other evolutions, [musical evolution] was marked by a constantly increasing *differentiation*, so that elements that in its early stages were simple, as animals low in the zoological scale – for instance, mollusks – are simple, became in the end highly complex...The first movement of Beethoven's *Hammerklavier* Sonata, op.106, is to Scarlatti's Pastorale as a dog or a horse is to an oyster.[11]

Ruth Solie makes the case that many nineteenth- and twentieth-century music histories shared the same spirit as Spencer. J.M. Fischer's *Die Grundbegriffe der Tonkunst*, published more than 20 years before Darwin's *Origin*, divided western music history into an early "melodic" period through the eleventh century, followed by a "polyphonic" period to around 1750, followed by a "harmonic" period through the nineteenth century. As the musicologist Karl Stumpf expressed the Spencerian evolutionary view in 1911, any other movement in history would be an oxymoron, an example of "progress backwards."[12] Traces of the presumptive "Darwinian" melodic-polyphonic-harmonic scheme can be found in the introduction to volume I of the *Oxford History of Music*, second edition, in 1932.[13] As late as the 1980s, Solie considered views like those of Anderson, Stumpf and Fischer to have a continuing "subliminal influence on musical thought."[14] In 1992, Scott Burnam felt the need to critique a similar lingering influence in music theory in his description of Hugo Reimann's influential harmonic theory from the 1890s in which "earlier stages of development were seen as the primitive yet fundamental seeds from which later stages could grow and develop."[15]

No matter how we read these views, they profess a world historical perspective in which the inevitable way forward for humanity – inevitable because it is part of the natural order embodied in European history itself – is toward the polyphonic textures exemplified so prominently in western art music. Clearly, much of the rest of world music culture, including the modal systems of the Middle East, North Africa and the Asian subcontinent, has remained vibrantly monophonic. This fact, when read against the background of Spencer's law of universal progress, would inevitably be interpreted as a symptom of a culture's

11 Ibid., 291.
12 Solie, R.A. (1982), 302.
13 Allen, W.D. (1962), 294.
14 Solie, R.A. (1982), 305.
15 Burnham, S. (1992), 3. For a twenty-first-century review of the persistence of the single road view of history, see Perry, J. (2000) on the continuing use of musical genealogies which uncritically accept the ladder of progress.

backwardness. As a teacher of music history and ethnomusicology in an American music conservatory for more than three decades, I can recall numerous occasions when universal progress surfaced in a conversation. In a particular classroom exchange some twenty years ago, a freshman music student, already an accomplished classical pianist, confidently questioned the importance of studying Gregorian chant because, after all, "monophony is something we've outgrown," betraying an unambiguous but naive affirmation of a subliminal Darwinian (Spencerian) single road.

Families of melodies and the single road

Much has already been said in earlier chapters about what the study of oral tradition melodies in English peasant villages and the study of song-making in Medieval Europe reveals about musical variability. But against the background of Spencer's single road, for all the enthusiastic collectors, editors and publishers of melodies in Darwin's century it was natural to assign folk songs and songs of earlier periods a place closer than art music to music's origins. For European observers beginning in the early nineteenth century, the idea that English folk monophony was underdeveloped seemed to match what they saw in the material underdevelopment of a rural hamlet. Polyphonic development instinctively correlated with the level of built environment in a city like London. It should therefore be no surprise that discussion of folk tunes by scholars and collectors was also saturated with evolutionary language. Tropes of inheritance and family traits were part of the evolutionary air Cecil Sharp breathed. Notions of kinship commonly evoked to explain similarities and differences among humans and animals were unhesitatingly applied to tunes with similar characteristics. In 1855 in his *Ancient Music of Ireland,* Sharp's predecessor, the Irish folk song collector George Petrie (1789–1866) remarked that certain groups of tunes like the three in Figure 2 related to one another as "of the one family."[16]

In even the most casual hearing or reading of these three melodies, the ear and eye are drawn less to the diversity of their many details of pitch and rhythm than to their generally similar contours. Writing five years before the publication of Darwin's *Origins,* Petrie declared that the resemblance of folk melodies such as these "could not have been wholly accidental," he was voicing evolutionary assumptions of his time, the belief

[16] Petrie founded the Society for the Preservation and Publication of the Melodies of Ireland, and spent a lifetime assembling a collection of 1,582 traditional melodies which was finally published in 1905, 39 years after Petrie's death.

Music Example 15.1 Three tunes which relate to one another "as of the one family" according to George Petrie (1855). Based on Petrie, G. (1855).

that an organism's gross external features provided a sufficient basis for any comparison, and more importantly, that common external features betrayed common origins. For Petrie, similarities of contour among a song's variants were sufficient proof that the songs were members of one family. He was not alone in this view. Shared traits, to another collector, Lucy Broadwood in 1893, were proof that the tunes were "versions of the same original,"[17] and for another song collector, Father Richard Henebry in 1927, that they were "descended from a common prototype."[18] Perhaps it was because of the shared understanding of origins embedded in the Eden creation story that helped persuade folk song scholars, music historians and Spencerian Darwinists that, behind the familial commonality of features in a set of organisms or melodies lay a common single ancestor.

Talk of family and inheritance among students of folk music began as an informal concept in the time of Petrie. It was another sign of the presence of evolutionary thinking that was in the air in Europe even before Darwin pushed it closer to the mainstream. In folk song scholarship, the idea of *tune family* formalized in the middle of the twentieth century in the work of the influential scholars Samuel Bayard and Bertrand Bronson. It was Bayard in 1950 who summarized what had been up to that point the concept's default definition:

> A tune family is a group of melodies showing basic interrelation by means of constant melodic correspondences and presumably owing their mutual likeness to descent from a single air that has assumed multiple forms through processes of variation, imitation and assimilation.[19]

17 Broadwood, L. and Maitland, J.A.F. (1893), cited in Shapiro, A.D. (1975), 31.
18 Henebry, R. (1928), cited in Shapiro, A.D. (1975), 23.
19 Bayard, S. (1950), 33.

Bronson took this definition a step further into both abstraction and statistics by first reducing tunes to what he regarded as their essential pitch structure – usually a sequence of pitches on the first, fourth, fifth and eighth scale degrees – and then feeding this streamlined data via punch cards into the first Fortan computers which were at that time finding their way into academic research. What emerged, in Bronson's opinion, was "the favorite path" of a given tune based on statistical evidence which he felt supported Bayard's tune family definition. There was, in Bronson's view, a fixed number of "originally distinct" tunes constituting "a canon of our melodic tradition."[20] Bronson was providing one explanation for the endless variants which Gavin Greg had considered the despair of the conscientious collector.

The three tunes in Figure 2 which Petrie had said related to one another as a family all exhibit Bayard's key feature, *constant melodic correspondences*. These are the general features of the tune which are most likely to evoke in any listener the sense of identity with another tune having similar features. In the same way, in everyday speech we may for a moment choose to look past the distinctive qualities of individuals and comment on the wavy hair, willowy stature and aquiline nose possessed by most of the members of the Jones family. In the eighteenth century, Carl Linnaeus, using only the naked eye, had easily grouped warm-blooded vertebrates with furry bodies which nurse their offspring as one family, *Mammalia*. Even though taxonomic emphasis has shifted since the mid-twentieth century to the molecular level of similarity, many of the eighteenth-century groupings based on gross features are still considered useful in contemporary science.

The story of starling murmuration in Chapter 4 served as an introduction to the practice among folk song collectors of grouping folk tunes in families according to overall contour, what James Cowdery was to call in 1985 *outlining tune families*. Two other families which Cowdery called *conjoining* and *recombining* add two additional types to Petrie's family concept, illustrating how musical thought has continued to take steps toward refinements that correspond to refinements in biological thought. These are refinements in musical thinking which occurred independently from biology in response to internal debates within the music profession. For Cowdery, outlining relationships conform to Bayard's classic definition – constant correspondence of melodic contour in which beginnings correspond to beginnings, middles to middles and ends to ends (Figure 2).

20 Bronson, B. (1950), 1079.

In conjoining relationships, a single segment of one tune will resemble a single segment of another tune.

Recombining relationships, on the other hand, according to Cowdery, draw attention to a level of identity which biologists would call more systemic. In the musical case, this is an identity which is less readily apparent and only perceptible with special scrutiny of smaller melodic bits. Before Cowdery, earlier students of folk song had already noted the possibility that relationships might exist among tunes which were less obvious than constant melodic correspondence. Petrie, for example, while foregrounding the outlining model, had also noted that other tunes displayed convincing, though less obvious levels of family identity. "It was only through a careful analysis of their structure, aided, perhaps by a knowledge of their history and the progress of their mutations," he said, "that they could be recognized as being essentially one air."[21]

Cowdery's method of achieving a more careful analysis than the outlining model was to compare tunes in a matrix of smaller melodic gestures which recur in each melody, though not necessarily in the same order. In Music Example 15.2 two songs which Cowdery transcribed from performers singing in the Irish *sean-nos* or old style are shown cycling through a repertoire of six short melodic gestures.

Music Example 15.2 Two examples of a recombining tune family according to James Cowdery. Cowdery, J.R. (1985).

Both tunes employ variants of all or most of this shared repertoire of melodic moves, but the sequence of the gestures and their organization into phrases of text are strikingly different, creating strikingly different outlines. It takes analysis under a higher magnification for this kind of detailed resemblance to emerge.[22]

21 Petrie, G. (1855), xv, quoted in Shapiro, A.D. (1975), 21.

22 Cowdery, in personal communication, attributes the first use of this analytical matrix or "barber pole" notation of melodic recombining to an unpublished transcription of an *alapana* in *Ritagaula Raga* by Harold S. Powers dating from the late 1970s. Both

Lord Gregory 156 123 345 56
Sail Og Rua 345 561 1235 56

Cowdery's recombining tune families are another way of describing the modular construction which scholars of Gregorian chant have identified over the centuries. Frere's comparison of melodic modules in "Nos alium" and "Domine Rex" in Chapter 10 parallels Cowdery's comparison of "Lord Gregory" and "Sail Og Rua":

Nos alium: O F1 D2 d1 C D3
Domine Rex: O D1 d1 C F D3

Bronson and Cowdery inherited and embraced the bio-musical term tune family, and yet they still diverged in their use of that term. Cowdery's matrix of recombining gestures reflects the method of "mapping" the behavior of living systems in their habitats which the biologist Fritjof Capra favored over "measuring." Bronson's emphasis, by contrast, is on large surveys of melodic pathways, on the reduction of performed melodies to pitch skeletons, and on statistical data, all of which are closer to Capra's paradigm of "measurement."

Following the single road beyond Europe

Looking outside of European art music and folk music, the degree to which the single road paradigm also spread beyond its original Anglo-American habitat should not be underestimated. The appearance of those ideas in the regions of the former Ottoman empire in the Middle East is a case in point. In the Turkish language liner notes to his 1987 recording of *Yunus Emre Oratoryosu*[23] for symphony orchestra and mixed chorus by the Turkish composer Adnan Saygun (1907–1991), composed in 1942, the Turkish conductor Hikmet Şimşek (1924–2001) provided an example of the global adaptive radiation of Spencer's universal law. In his telling, this grand adaptive process is given a deterministic spin:

> Turkish traditional music occupies an important place in the world's musical heritage because of its subtleties of rhythm and melody.

Powers and Cowdery offer examples of much larger arrays of tunes which, in the interest of legibility and space, are impractical to print here. See Cowdery, J.R. (1985) for a longer and more detailed example of recombining in his "Blackbird family".

23 Saygun, A. (1987).

> However, because Turkish music did not keep pace with the innovations of the Renaissance it lacked polyphony until the reforms of Atatürk.[24]

In other words, according to Şimşek, the Renaissance was not a local European occurrence, but a mythic event for the whole planet, like the Flood or the birth of the Prophet Mohammed. Polyphony, in this view, is a normal attribute the lack of which would be judged a deformity, as in the case of a human born without one eye. Most important of all, to Şimşek and many of his Turkish colleagues in the period of western-leaning reform initiated after 1923 by Mustafa Kemal Atatürk, polyphony *replaces* monophony as inevitably as adulthood replaces childhood. This is an idea which remains a staple item in Turkish thought today as the so-called "polyphonic-monophonic debate" *(çoksesli-teksesli tartışması),* though the debate has in recent years slipped to the margins as a subcurrent in the perennial public arguments over the pros and cons of westernization *(batılılaşma).*[25] As a conversation about westernization, the replacement of monophony with polyphony has for more than a century functioned as just one among other inevitable replacements – for example, the replacement of religion with secularism as the foundation of society, and the replacement of Ottoman law based on the Koran with a European-style legal system based on private property.

The German composer Paul Hindemith (1895–1963), commissioned by Atatürk to help design a modern system of music education for the new Turkish Republic, had voiced confidence in the single road from monophony to polyphony during the years when Şimşek would have been a student. In his 1937 report to the Turkish government, Hindemith wrote that the Ottoman urban tradition exemplified by the modal system of *makam* had "reached the apex of its development, perhaps even passed it. Through its concentration in a single voice its technical potentialities have...been exhausted. All that can be done now is repeat past discoveries." The fatal flaw of Turkish makam, according to Hindemith, is that, because of its microtonal scale system with intervals smaller than those on the piano, it is "unsuited to polyphonic treatment – and in all attempts to develop music, an ordered combination of various voices is essential."[26] As my gifted freshman conservatory student put it three generations after

24 Şimşek, H. (1987). Recording liner notes.
25 For example, see: Çetinkaya, Y. (2019).
26 Skelton, G. (1975), 137.

Hindemith, melody-making is for civilizations in their infancy. Grown-up civilizations make harmonies.

Both Hindemith and Şimşek were passing on the wisdom of their own teachers who were either European or who had European educations and who were born in the period of Darwinism's ascendance.[27] But we have already seen that neither the tree of life nor the single road of history sprang full-blown from the head of either Darwin or Spencer, but were ideas already percolating below the surface of European thought. Long before Atatürk, before the teachers of Hindemith and Şimşek, and even before their teachers' teachers, Europeans were responding in writing to the music they heard on their travels to the east of Vienna by comparing them to what they knew.[28] While the subject of so many of the travelers' anecdotes was the music of the Turks (the word commonly applied at the time to any of the inhabitants of the vast Ottoman domains), the substance of their narratives was as much the travelers' own musical beliefs and assumptions. The historian D.R. Howard concisely captured the reflexive quality of early modern European travel accounts in this way: "things are other than we expect at home, and the contrast turns us back upon ourselves."[29]

Most of these commentators between the fifteenth and twentieth centuries were not musical specialists, so they relied, largely unconsciously, on aural and visual analogies to what they knew at home as well as on the contemporary conventions of European musical thought they had picked up by osmosis. Travel put them in a comparative frame of mind, provoking the travelers' ongoing question, "how is the music (or food or dress or religion) of the Turks different from my own?" None of these travel writings covering some five centuries make any mention of biology. However, taken together, their observations trace a familiar macro-evolutionary pattern, a narrative of a single road from simple to complex leading to the familiar European present.

In their struggle to understand what was unfamiliar, these travelers turned back on themselves, as Howard put it, bringing to bear their own

27 Hindemith was a student of Arnold Mendelssohn (1855–1933) and Bernhard Sekles (1872–1934) in Frankfort. Şimşek studied in Ankara with the German theorist Eduard Zuckmeyer (1890–1972) who was educated in Köln, and also with Ferit Alnar (1906–78) who was Turkish, trained in Vienna, and with the composer Adnan Saygun himself, who was educated in Paris.

28 For a fuller treatment of travel writings on music in the east see Labaree, R. (2019).

29 Howard, D.R. (1971), quoted in Campbell, M.B. (1988), 149–150.

orientation to the stories of classical antiquity and the Bible which was their inheritance. The Italian Pietro della Valle, traveling in Turkey, Iran and India in the seventeenth century, saw the pan flute *(miskal)* being played in Constantinople and thought of lines from Vergil.[30] Later on in that century, the English physician and clergyman John Covel observed, "I'm inclined to think this musick old, and mention'd in scripture," citing chapter and verse in the book of Chronicles.[31] Lady Mary Wortley Montagu, the wife of the English ambassador to Turkey noted "there being not one instrument among Greek and Roman statues which is not found in the hands of people of this country?"[32] In 1767, Charles Henri de Blainville, an aristocratic French intellectual who was also a musician, added a touch of scholarly perspective, declaring that "the music of the Turks is none other than the remnant of that of the ancient Greeks."[33] This was the pastness of Ottoman music, a theme more often voiced by travelers than any other. The idea matched the travelers' perception of the contrast of the Ottomans' industrial capacity with that of Europe was comparable to the English urbanite view of rural English folk music.

European travelers in the east shared with folk song collectors in Europe a cluster of linked assumptions about music which they instinctively applied to what they were witnessing: orality and monophony were relics of the past, whereas music writing and polyphony represented the reality of the present. In the experience of educated travelers from cosmopolitan centers in the west, monophonic forms and oral transmission had been superseded by centuries of European polyphony, musical notation and music publishing. Orality and monophony were on a par with nakedness and belief in magic, signaling to their readers back home that these eastern practices were relics of the past. Blainville, one of the small handful of music professionals and scholars who have left accounts of their travels in the Ottoman lands,[34] cites a conversation with a Turkish informant (an "intelligent man of this Nation") which sounds like what is probably Blainville's

30 Della Valle, P. (1614). Della Valle (1586–1652) was a classically trained humanist from an aristocratic Roman family who wrote many letters to a correspondent in Italy during the 12 years of his travels.

31 Covel, Dr. J. (1670–1679), 19. Covel (1638–1722) was chaplain for the Levant Company in Constantinople, 1670–1679.

32 Montague, Lady M.W. (1717), 25.

33 Blainville, C.H. de. (1776), 4. Blainville (1711–1771) was a published composer of instrumental music, and writer on music theory and history, who spent the years 1737–1738 in Constantinople.

34 Other musically educated writers about music in the Ottoman domains besides Blainville include Giovanni Battista Donado (1627–1699), Franz-Josef Sulzer (1727–1791),

own macroevolutionary credo. This anecdote resembles a similar tale in the writings of an earlier traveler in the previous century, M. de Thévenot in 1686, and so it may be taken as a trope all on its own.[35] According to Blainville, "the Turks" harbor a little-known respect for "the true beauties that the fullness of our harmonies gives." At first, the informant told him, "the mélange of sounds which you find marvelous, only made a confused noise in my ears; but at length, I sensed that you sang with much grace and dignity, and this assembly of sounds of the voices and instruments seemed to me to form an admirable harmony."[36] In this tale Blainville finds support for his faith in polyphonic music as a manifest destiny – that is, what the evolution of chant to symphony in European history had clearly demonstrated. Throughout these fragments of testimony, mostly casual and informal, it is assumed that history leads inevitably to where the writer now stands. Cosmopolitan, urban Europe represents the destination of history into which rural people and non-European people inevitably will fit. The history of Europe's music (the only music history most of these travelers knew) seemed to lead from simple to complex. The logic was obvious even before intellectuals like Spencer had articulated it. It was not just a historical observation; it was an article of faith.

Compared to their own music, in the ears of early European travelers, the music of Ottoman Muslims was old, resembling the music of the ancients, and its monophonic character made it analogous to European folk music, which is the product of peasant culture, the most under-developed segment of culture in the travelers' habitat of origin. Few Europeans appear to have been aware of the more than 300 years of music theory writings by Ottoman musicians, including extensive notation collections using the Arabic alphabet and Greek and Armenian notation systems.[37] The mistaken belief that Ottoman music was lacking in notation

Charles Fonton (1725–1793), Giambattista Toderini (1728–1799) and Guillaume André Villoteau (1759–1839).

35 Thévenot, M. de. (1686), 22. Thévenot (1633–1667) was a French linguist and traveller who spent the years 1655–1667 in the Levant, Persia and India.

36 Blainville, C.H. de. (1776), 6.

37 For a discussion of the notated tradition of Ottoman music, see: Wright, O. (2000). The introduction of western staff notation is generally associated with the commissioning of Giuseppe Donizetti ("Donizetti Paşa"), the brother of Gaetano Donizetti, to lead the palace orchestra of Sultan Mahmud II in 1828. But music publishing was slow to follow. With the exception of several publications of Ottoman music by the Orthodox Patriarchate in Istanbul using Byzantine neumes (see Sanlıkol, M.A. (2023)), it was not until the last decade of the nineteenth century that the dissemination of Ottoman music through printed notations became more widespread, and it was not

further suggested to them that it lacked a rational basis, a view which was reinforced by what seemed to be a lack of stable compositions which are the creations of individual composers.

The role of Christianity in this belief system cannot be ignored. For most of the travelers, whether in the fifteenth or the nineteenth centuries, their shared conviction of the universal truth of Christianity and the Bible lent an additional measure of certainty to the belief during those centuries that music writing and polyphony were on a single road from simple to complex. The repertoires which mapped the law of progress were all products of Christian Europe. Both the folkishness and backwardness of Ottoman music made sense to European travelers during the six centuries of Ottoman rule because of their unfounded belief that music was prohibited in Islam. The English doctor James Dallaway in 1779 expressed a widely held opinion in the travelers' accounts: "When music was expelled from public worship, and held to be profane in society, it soon became silent and consequently forgotten. The dronish chant, unassisted by instrumental music has high antiquity only to recommend it; for the dullest ears would be disgusted by such grating monotony."[38]

We should not be surprised that for European travelers, religion had offered a ready explanation for the backwardness they heard in Ottoman music. Europeans were, after all, heirs to the many constraints on music over centuries exercised by Roman Catholic councils, by the Inquisition and by Protestant puritan rule. In fact, looking to the east and to the west of Vienna, both Islam and Christianity had acted to shape music-making through both approval and disapproval. In both cultural regions, the shaping process waxed and waned in intensity, and the two faith communities were frequently torn by internal debates and conflict between competing interpretations of biblical and koranic authority. In the accounts of European travelers, the very sound of Ottoman music even at times confirmed the linking of faith and evolutionary inevitability. Among European travelers, the word "Christian" seems to have stood for an aesthetic sensibility as much as a confessional category. At the end of the sixteenth century, the Prussian Protestant Reinhold Lubenau summarized what he heard in his travels in the east in this way: "strange music when we Christians hear such music."[39] This was a sentiment echoed by Salamon Schweigger, another German, in 1608, who described

until after the establishment of the Turkish Republic in 1923 that it found its way into popular education.

38 Dallaway, J. (1797), 14.

39 Lubenau, R. (1573–1589), 7.

Turkish music as having "no charm or grace in it; on the contrary, it is quite violent and hostile and furthermore it is entirely contrary to the art of true music which is practiced in Christendom, and everything is dissonant."[40] Religion is what Darwin was running up against in the nineteenth century, and it is also what was a controlling factor in the travelers' descriptions of non-European civilization as they found it in the Ottoman region extending from Vienna around the Mediterranean littoral as far as Morocco. At many points in European history, religion had been an impediment to the inevitable progress from simple to complex – in music, slowing and constraining the acceptance of polyphonic innovations over the centuries by church leaders and theologians is a case in point.

The macroevolutionary bio-historical view that was developing incrementally in the eighteenth and nineteenth centuries finally crystallized in what we now know as Spencerian Darwinism. According to this view, the monophonic forms of the Ottomans, when compared to the compositions of Handel and Beethoven in their own day, would have been easily interpreted as superseded forms of the past – at best, mollusks, or at worst, worms or insects. While Darwin was regretting his own use of "higher" and "lower" in his writing about evolution, the observations of these travelers, music professionals and music amateurs alike, provide a rich archive of the hierarchical discourse which Darwin wanted to avoid.

Travelers were instinctively measuring what they witnessed against what to them was an unquestioned truth, an evolutionary certainty that the simplicity of melody-making inevitably lead to the admirable harmony of European music of their own day. These ideas were expressing themselves in bits and pieces long before Darwin and the Spencerian doctrine of evolution.

It is worth wondering whether this assumption – that the civilization of the European travelers is the pinnacle of evolution – might be best understood as *preceding* biological evolution as an idea. Perhaps Spencerian Darwinism could be viewed as an heir to these Eurocentric convictions which found support in the enduring certainty of the biblical story of Eden, the first chapter of European civilization's founding document. On the sixth day of creation, before He rested, God created humans and made them the custodians of nature. So, they began to cut down forests for fuel and to replace those trees with grain and cattle. So began the story of human civilization and – if we choose to see it this way – the straight road of history.

40 Schweiger, S. (1608), 8.

16

Biology and music listening to each other

So much of the back and forth between biology and music over the centuries has taken the form of metaphors of one sort or another. Even Herbert Spencer's "law of all progress," demands scare quotes because it presumes to thrust his conjectural doctrine of universal development from simple to complex into the arena of hard science alongside Isaac Newton's Laws of Motion. It cannot be said too often: anyone attempting to move between the materiality of cells and habitats and the immateriality of sounds and ideas will need to spend a lot of time stretching the meanings of words.

The term *macroevolution* in the lexicon of natural science has become a comfortable enough metaphor for the transformations in nature and in civilization taking place on a grand time scale. This being the case, then the equivalence which Eliot saw between the work of an individual poet and the catalytic interaction of platinum and nitrous oxide might reasonably fit into *microevolution,* another of biology's metaphors. In the *Encyclopedia of Evolution,* microevolution is defined narrowly as "... genetic change within the lifetimes of populations and species,"[1] that is, changes occurring on a time scale shorter than the glacial pace of macroevolution. Applying this concept to the human arts, any variations that occur in a piece of literature or in a piece of music within the lifetime of those works – for example, in different versions of a poem in the poet's notebooks or different performances of a Schubert song – could also qualify as microevolution.

1 Pagel, M. (2002).

The fit of Eliot's chemical metaphor into microevolution becomes a bit less comfortable when we recognize that his focus was on the relationship between the individual talent and tradition, which inclined him to insist that his chemical analogy also had a historical dimension. For Eliot, the chemical reactions of the moment – the memory of the catalytic agent (the poet) interacting with the accumulated poetic capital of the poet's time – merge their time scales. The individual artist's choices made at a particular moment in time, and the artist's inheritance of words, models and ideas from artists of the past are on a continuum which is unbroken. In Eliot's view, these micro- and macro-time scales were never separate to begin with. They are certainly not separate in science, at least not in principle. When biologists speak of the endemic variability of individual organisms on the one hand, and of species on the other, the two scales are never considered absolute or unconnected. Adaptations in a symbiotic microbe in the stickleback's gut or in a single cell in the stickleback's body can occur in fractions of a second – a level of microevolution which makes the fish's three-to-eight-year lifespan look macro. At the same time, neither of these adaptations can be understood apart from the evolution of the genomes of the fish and the microbe over millions of years.

Finding music in planets and human bodies

It was not until Darwin's own lifetime (1809–1882) that what we now understand to be modern biology emerged from its niche within *natural history*, where it had existed alongside *natural philosophy* for centuries. While music historians were struggling to create authoritative editions of music of the past, natural historians were struggling to assemble an evolutionary timeline for all species. The conversation between the two groups of historians was less direct than osmotic, the result of cultural seepage through the soil of shared assumptions. This is the soil from which the evolution metaphor had been sporadically springing as an invasive species since the eighteenth century, a weed among the dominant monocrop of Christian belief in a natural world which was essentially unchanged since Eden. If there was conversation, direct or indirect, between biology and music in this period, it was decidedly one-way, since it was science that was doing most of the talking and music, the listening.

There had been a time in the past, we should remember, when these roles were reversed. The "harmony of the spheres" *(harmonia universalis)*, the medieval European metaphor for the solar system based on a mashup of astronomical observation and the mathematical proportions of the musical scale, was just such a case of science listening, with at least one ear, to music.[2] The relationship of the movements of the planets since creation were understood as a perfect harmony, and the most persuasive model for harmony was music itself, expressed in the mathematical proportions of the common musical intervals. It was a harmony which had been observed in nature and immortalized in the theory associated with Pythagoras in the sixth century BCE: a sounding body of any kind (a taut string or a slab of wood or of iron) divided in half (2:1) gave a pitch an octave higher; if the division was at a proportion of 3:2, it produced an interval of a fifth; at a proportion of 4:3, a fourth; and if at a proportion of 5:4, a third; and so on, mathematically locating the seven pitches of the diatonic scale using the simple integers 1–9. The scheme in Robert Fludd's 1617 treatise *De Musica Mundana* shows a monochord whose intervals are marked by the orbits of the planets, with the hand of God tuning the tuning peg of the string. It is a late representation of that conceptual synthesis which uses a musical yardstick for understanding the constancy of the solar system. The order which kept all the planets in their orbits was natural and ultimately divine in origin.

But in Fludd's day, these roles were already beginning to shift. By the seventeenth and eighteenth centuries, writers on music like the composer-theorist Johann Mattheson (1681–1764) were systematically linking the emotional impact of various intervals, melodic types and the major and minor modes to the physiological theory of the emotions based on the flow of fluids in the human body. This was the so-called theory of the humors put forward in the generation before him by the mathematician, naturalist and philosopher, René Descartes (1596–1650).[3] In his famous *Passions of the Soul* in 1649, Descartes' new psychology was in fact a refashioning of a much older theory of the humors, the bodily fluids of blood, yellow bile, black bile, and phlegm which had been a feature of western thought since Aristotle. Like the classical theory of epigenesis, the theory of the humors was recast and given new life by recent science. The increased knowledge of those fluids, as for example, in the studies of the circulation of blood by Descartes' contemporary William Harvey (1578–1657),[4] gave new credence to the old idea that a variety of natural fluids were essential both to bodily functions and to emotional states.

2 For a summary of the tradition of linking science and music from Pythagoras to the present, see James, J. (1995). For the enduring role of musical imagery in English nature poetry 1500–1700, see Hollander, J. (1961).

3 Mattheson, J. (1739/1958).

4 Ribatti, D. (2009), 3.

As we now know, Descartes' new psychology did not survive later scientific discoveries and the findings of modern psychology. But Descartes' humors theory shared with Harvey's work a growing tendency among European thinkers to look beyond the material determinism rooted in biblical beliefs about creation and human behavior, and to search for explanations in natural processes. For Mattheson, natural systems provided the benchmark for understanding how melodies and harmonies were created and how it was that they so reliably triggered human feelings of sadness, love and joy. Mattheson's language is often as vague as Descartes's, but he believed that specific emotions or *affects* of music were not representations of human emotions, mere symbols of joy and sadness created through tradition which were subject to changing social ideas of taste and beauty. In Descartes' theory, the emotions aroused by music were causally connected to the natural movements of the body's fluids or "spirits": "Since, for example, joy is an expansion of our vital spirits, it follows sensibly and naturally that this affect is best expressed by large and expanded intervals. Sadness, on the other hand, is a contraction of the those same subtle parts of our bodies. It is, therefore, easy to see that [to express sadness] the narrowest intervals are the most suitable."[5]

Viewed from the twenty-first century, we can understand that Mattheson was not so much stating laws of nature based on empirical evidence as he was observing the human sound art he knew and noting that joyful melodies tended to make use of wider intervals and sad melodies tended to use smaller *(chromatic)* intervals. The association of "feminine" qualities of sadness and sensitivity with the interval of the minor third and "masculine" qualities of strength and forthrightness with the major third was a further link to the natural state of human bodies in Mattheson's theory. His observations were not original with him or with his generation, but echoed similar patterns of musical affects codified since the Renaissance by composer-theorists like Zarlino in Italy and then re-stated by Thomas Morley in England. By Mattheson's day, these natural states of music were considered to be in tune with the laws of nature which were acted out at every moment in the necessary functions of the human body. Like the eternal mathematical proportions of the musical scale, the human responses to music, and therefore the craft of musical composition, were extensions of natural systems.

All these claims were consolidated by Mattheson and others with another reference to human bodies, the common social dance forms of the time,

5 Mattheson, J. (1739/1958), par. 56 and 57, quoted in Lenneberg (1958), 51–52.

into a comprehensive science of musical emotions or Affektungslehre. Mattheson's characterizations of each of the dance forms so central to European social life between the sixteenth and the nineteenth centuries are surprisingly specific and were given detailed choreographic forms in the voluminous dance publications of the period, which featured detailed diagrams of the dance movements, often based on the choreographies of popular operas of the time.[6]

> Now it may be difficult to believe that even in simple little dance melodies the passions must be distinguishable as light and shade can always be. To give just a small example, in a *chaconne* the affect is more noble and proud than a passacaglia; in *courante,* the feeling is directed toward a tender hope (but I do not intend the Italian *corrente*); in a *sarabande,* to a public, formal seriousness; in an *entré,* to pomp and vanity; in a *rigaudon,* to pleasant joking; in a *bourée,* to contentment and a pleasing air; in a *rondeau,* to a cheerfulness; in a *passepied,* to fickleness and inconstancy; in a *gigue,* to passion and fervor; in a *minuet,* to moderate gaiety; and so on.[7]

As dance, the rhythmic component of Mattheson's science of musical emotions is kinesthetic, embedded in the muscle memories of human beings, and it is social as well, embedded in European dance culture rooted in everyday human rituals of courtship and class. In this system, rhythm is not abstract. This science of affect was not limited to the simple contrasts between metrical divisions of two and three, for example, but incorporated the emotion and social meaning embodied in the elaborate choreography of the dance movements. Bearing names like those which Mattheson mentions, the dance forms migrated globally, following the paths of European empire in a curlicued path from village habitats in Europe, North Africa and the New World to courtly and aristocratic habitats throughout what would have been unblushingly referred to from the Renaissance to the twentieth century as "Christendom."[8] The dance forms, in particular, inhabited and adapted to the ecologies of secular and

6 *Recüeil de dances contenant,* a treatise on dancing practice, featuring the choreographies of Louis-Guillaume Pécour (Feuillet, R.-A. and Pécourt, G.L. (1704)). Many dances in Pécour's treatise first attracted popular attention on stage in the operas of Jean-Baptiste Lully.

7 Mattheson, J. (1737). Cited in Allanbrook, W.J. (1983), 334.

8 Wye Jamison Allanbrook's descriptions of how this science was embedded in the arias of Mozart's operas reads like another example of adaptive radiation in music. See: Allanbrook, W.J. (1983).

sacred vocal music, symphony and opera and continued their spiraling path back into villages, public houses and bourgeois salons, and then out again to still wider habitats of church and, ultimately, of twentieth-century cinema.[9]

A bio-musical metaphor is a two-way street

Ultimately, any conversation between biology and music depends on how humans process their common sense observations of the mutual otherness of nature and music. Leaves, water, rocks, hurricanes, bacteria and dinosaurs are all physical phenomena which tell our senses that they are different from both the invisible sound waves of music and music's internal registrations of meaning for humans. Metaphor-making comes in handy in just such boundary-crossing situations where we suspect our sensory information and our empirical understanding is limited, so the very word "life" or "lively," even the musical (Italian) terms "vivo" and "vivace," fill a conceptual zone lying between the domains. Hanslick's description of Beethoven's music as developing like a blossom from a bud has nothing to say about either chains of proteins or relationships between pitches, and so it may still be considered of more interest to philosophers and poets or the general public, with no practical interest to either the study of botany or the composition of musical works.

Nevertheless, the adaptations of biological concepts to music history and analysis over the past two centuries are many, from *organicism*, the enduring philosophical concept originating in the late eighteenth century, to *meme*, the more ephemeral concept at the center of a late twentieth-century movement which applied the idea of genetic inheritance to all things cultural.[10] Through this whole span, there has been the deep and continuing reliance on the concept of evolution itself in musical discourse. In conversations about the western classical canon, and no less in jazz and in a range of other global urban and rural practices, music is routinely considered to "germinate," "put down roots," "blossom," "grow," "develop," "evolve" and "die out." We commonly speak of this or that musical "hybrid," of musical "genealogies," and the concept of the "transmission" of musical traits flourishes alongside the transmission of viruses and genetic codes. I

9 For a concise account of the transmission of the affective codes of European music to early Hollywood cinema in the work of Max Steiner (1888–1971), see: Gorbman, C. (1989), 70–98.

10 For "organicism," see: Hubbs, N. (1991) on Heinrich Schenker and Solie, R.A. (1980); for "meme," see Blackmore, S. (2000) and Aunger, R. (2000).

cannot fail to mention as well the ubiquitous and conversation-stopping undergraduate explanation for any similarities between the work of any two artists in any medium – "influence" – a term with a more than coincidental etymological link to *influenza*.[11]

However, with the image of the blossom, Hanslick's image also confirms that the metaphor can be read in the opposite direction, as well. If anything in the natural world makes sense to humans, it is because it presents itself in, so to speak, human form, or at least in forms resonant with human aesthetics.[12] This is, in part, what Paul Weiss was saying to his biology colleagues in the 1950s when he insisted that the order and the beauty we see in tree rings and snail shell patterns is the result of a developmental process that humans consider orderly. It is also worth wondering if metaphors of nature remain irresistible in the arts because human artifice itself pales in the mind of artist and non-artist alike against the orderliness of nature. Compared to natural beauty, human artifice can look like an add-on, a decoration, icing on the cake, and therefore inherently unnecessary – all expressions which may come uncomfortably close to describing the common view of the non-essential nature of the arts in modern civilization. The medieval vision of the orderliness of the musical scale in the movement of planets and the Enlightenment idea of musical passions as an outward sign of systems in the human body can also be understood as ideas which confirm that music participates in systems which are necessary.

For musicologist Holly Watkins, the problem in our own day is not with bio-musical metaphor itself but with the outdated biology which was the point of departure for Romantic metaphors. Here, I will be siding with Watkins' call on scholars of western art music in our own day to rely on current life science to update nineteenth-century thought, making room for "a species of organicism that dispenses with humanism's anthropocentric conceits."[13] Her goal of creating "a renewed organicism" has much in common with music in vivo.

After two centuries of work by biologists, an increasingly confident consensus has emerged that the conceptual boundaries and the material

11 In mid-eighteenth-century Italian, 'influence,' from medieval Latin *influentia*. (*influentia* 'inflow,' from *influere*, *in-* 'into' + *fluere* 'to flow.') The Italian form of the word *(influenza)* was also applied specifically to an eighteenth-century epidemic in Italy, and the name of the disease was later adopted into English.

12 Matthew Rahaim identifies two broad categories of evolution metaphors in music: the "progressive" and the "situated." See: Rahaim, M. (2006).

13 Watkins, H. (2018), 18–19.

boundaries separating humans and nature are more porous than has been thought, therefore altering the actual function and content of at least some bio-musical metaphors. In important ways, the reimagining of these boundaries has resulted from technological innovations which help us see what had never before been seen, but only imagined. Since the introduction of the electron microscope and the computer to biological research in the second half of the twentieth century, the ability to witness and model microbial life and the movements of molecules and even to scrutinize them in vivo has increased exponentially, and the cloak of invisibility covering this swarming global stew has been removed. An important border wall between human activity and the activity of nature has been transformed by the growing recognition of the central role played in the planetary life system by the formerly unseen horde of microorganisms a horde which includes both bacteria and the largely underground networks of fungi.

In a recent study supported by the United Nations, the percentage of living mass on the planet represented by plants is 83% and by animals 0.4%, while human beings occupy just 0.01% (see Figure 16.1). Even if we were to leave aside for a moment the contributions of microorganisms to functions crucial to human life – functions such as the digestion of nutrients, the production of atmospheric oxygen and the fixing of life-supporting elements in the soil – bacteria and fungi together represent 15% of all living mass, dwarfing the actual physical presence of human bodies on the planet. What had been for all of human existence unknown and unknowable was becoming as knowable as a volcano or a dinosaur bone.

We can now confirm that the sound waves battering our ear drums, and the feverish activity of our gut bacteria impinge on us similarly: they are both invisible under normal conditions and at the same time undeniably physical. Moreover, while music's impact on human affect is undisputed, recent research has even begun to record a similar impact on the emotions by gut microbes and their role in creating cognitive conditions such as autism as well as their effect on mood.[14] The boundaries between the seen and the unseen, between the real and imaginal, and between the material and the mental – for centuries the exclusive domain of metaphor – have turned out to be further examples of a porous membrane, as much conduit as barrier.

In 2016, the developmental biologist Scott Gilbert and the immunologist Alfred Tauber noted that biology was undergoing its own shift of emphasis "from a molecular-centered science towards ecology,"[15] that is,

14 Schuller, K. (2018).
15 Gilbert, S.F. and Tauber, A.I. (2016), 849.

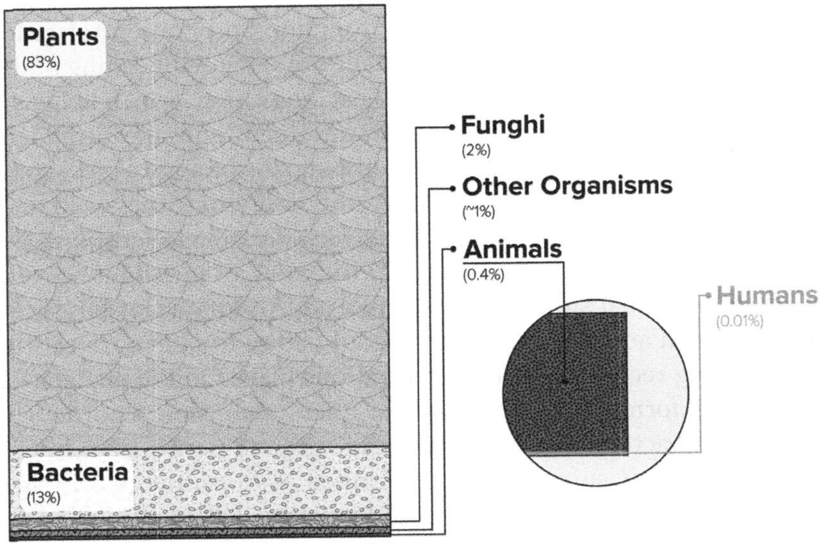

Figure 16.1 Humans make up 0.01 percent of all living things by mass (Bar-On, Y.M. et al. (2018)). According to Rosenberg, Y. et al. (2023), "...the total biomass, or collective weight, of terrestrial arthropods is about 1 billion tons, which is roughly the same as that of all people (about 400 million tons) and all farm animals (about 600 million tons) combined." This study suggests that soil is likely the most biodiverse single habitat, home to 59 ± 15% of the species on Earth.

away from the emphasis in the twentieth century on chemical reactions and genetic recipes toward an emphasis on how those processes and structures originated and evolved in relationships with the swarm of life around them.

The centuries-old boundary between humans and the natural world has been gradually melting in the presence of all these findings based in experience outside of music. Coming out of the philosophy of science, Mark Bickhard's depiction of mental states as flame states is one such example. Flames can only be said to exist so long as they continue to consume energy and to change. Music is no less a flame state. The mouvance within musical transmission is plain to see, a systemic norm – like the Cortot-Rubenstein-Horowitz versions of Chopin and the Juilliard and Guarneri versions of Beethoven and Schubert – rather than an exception which must be explained away simply as error or as unique authorial choice. In

1991, Nils Wallin, the musicologist and neurophysiologist credited with coining the term biomusicology (no hyphen), saw a direct link between biology and music in human perception, calling music "a real homologue of brain processes."[16] In other words, music in important ways enacts in sound, and thereby embodies, structural relationships of the brain. The brain, we now know, has also been shaped over the millennia by music (and language and toolmaking) as much as music has been imprinted by the brain's own structural demands. These different brain-music interactions are an example of the so-called "Baldwin effect" described by the psychologist J.M. Baldwin at the beginning of the twentieth century, an early version of the view that all development is co-development.[17]

If we accept that brain and music have co-evolved side-by-side, the relationship between brain biology and music opens another door to a re-positioning of music – for all of its mentalness – closer to the world of matter. "Ways of the hand" conveys an orientation to the physical contact between hand and medium in pottery and piano playing and the idea of distributed agency as recognized by neuropsychology. Forest ecologist Suzanne Simard has recently moved the boundary of metaphor even further by ushering plant life into the exclusive inner sanctum of nature, *intelligence*, for centuries a domain reserved for humans alone. The hypersensitive electro-chemical communications among forest vegetation through the dense underground web of roots and hair-like *hyphae* connecting trees, shrubs, fungi and the vast bacterial colonies associated with them (*mycelium*) occupies about two per cent of the mass of the planet's living matter. This is, according to Simard, approximately the percentage of the human brain's mass to the rest of the human body.[18] The mycologist Merlin Sheldrake describes the fungal mycelium in the soil as a series of "fantastically complex networks of electrically excitable cells." Brains too, he says, are "fantastically complex networks of electrically excitable cells."[19]

It is worth mentioning an example of bio-musical metaphor that may be at least one answer to Holly Watkins' call for an updated organicism. It reminds us that, for all that metaphors are not the same as hard science, they are important, and not just for those who don't know enough about science to do more than poach its lexicons for figures of speech. In his immensely readable little book *The Music of Life: Biology Beyond Genes*

16 Wallin, N.L. (1991).
17 Quoted in Deacon, T.W. (2012), 156.
18 Simard, S. (2022).
19 Sheldrake, M. (2020).

(2006), the distinguished English physiologist, cardiologist (and guitarist) Denis Noble lays out a post-Neo-Darwinian view of transmission and change comparable to that of the biologists in the interdisciplinary writings examined here. He does this by methodically exploring as metaphors certain standard topics drawn from western classical music: *pipe organ, score, mode, key, conductor, composition,* and *orchestra.* Coming from decades of immersion in the complexities of systems biology (a field for which he wrote the first textbook) and the electro-chemical systems which regulate the human heart, Noble's robustly detailed use of bio-musical language embodies ideas which are as challenging for biology as they are for music. "The score: is it written down?"[20] and "the orchestra of life works without a conductor,"[21] for instance, may be metaphors, but they are metaphors which require both twits and nerds to stop and think. In both disciplines, the many individuals who are more comfortable with the clear lines of causality conventionally implied by "score" and "conductor" would find Noble's language controversial. Much of Noble's book is devoted to applying each level of metaphor to specific examples drawn from his own research in cardiorhythms. He then links them to other bio-musical metaphors on his list, always reminding the reader of the limits of each one, both as music and as science. In the end, through a matrix of interwoven bio-musical terminology, music gains by its association with the knowable and at times quantifiable material complexity of biological feedback systems. Though Noble doesn't say so, I would suggest that biology, for its part, gains by its association with music's intuitive wholeness (harmony in the largest sense) and its emotional impact, what the liberal arts tribe refers to as *meaning.*

Darwin's sensible pluralism

Perhaps it is Darwin's awareness of how little of his theory he was able to prove that leaves his writing such a complex mix of modest conjecture, detailed observation and far-reaching assertions spanning geological time. This is a quality which links his work with that of our own day, but also with others in the natural sciences within Darwin's own orbit, including predecessors like Darwin's famous grandfather Erasmus Darwin (1731–1802) and the celebrated poet and naturalist Johann Wolfgang von Goethe. These figures mark the beginning of the end of natural history,

20 Noble, D. (2006), 33.
21 Ibid., 113.

the term inherited from Aristotle's *Historia Animalium*, when the word *historia* stood for an explanation of phenomena. Like his predecessors, Darwin's science was based on observation – both of his own collection of specimens and in vivo – but less on experiment and mathematical calculation, with his own modest experiments in dove-breeding and botany as an important exception. In this approach to scientific study, he was not alone among his contemporaries, both mentors and peers. Charles Lyell (1797–1875), the man considered to be the father of modern geology who was an important supporter of Darwin's thinking about evolution, and Darwin's colleague/competitor in biology, Alfred Russell Wallace (1823–1913), shared his emphasis on observation. All three contrast sharply with their contemporaries in physics like Joseph Fourier (1768–1830) and Michael Faraday (1791–1867), and in biology, Gregor Mendel (1822–1884) and Louis Pasteur (1822–1895). Compared to Mendel's tracing of plant inheritance through statistical data sets and Pasteur's study of bacteria through chemistry and the rapidly advancing techniques of microscopy, Darwin the biologist was as much a member of the liberal arts as of the sciences, and he shared the historian's and novelist's instinct for filling in missing details with a plausible story.

Stephen Jay Gould recalls how Darwin at times referred to his famous theory of natural selection, a staggeringly ambitious account of how all living forms on earth evolved over millions of years, as his 'dear child." At the same time, Gould says, Darwin also frequently expressed his discomfort with the way that his beloved theory had been granted, by supporters and detractors alike, exclusive explanatory status for the evolution of all life forms. In the preface to the sixth edition of *The Origin of Species* in 1872 he wrote:

> I may be permitted to remark that in the first edition of this work and subsequently, I place in a most conspicuous position...the following words: 'I am convinced that natural selection has been the main but not exclusive means of modification.' This has been of no avail. Great is the power of steady misrepresentation.[22]

It is significant that, no matter how much Newtonian assumptions about the neutral and passive nature of matter set the limits to Darwin's explorations of heredity and variation, in the end, he still declined to press

[22] Quoted in Gould, S.J. (2007), 444. In the first edition of *The Origin of Species*, Darwin acknowledged the possibility that certain traits "originated from quite secondary causes, independently of natural selection." Darwin, C. (1859/1993), 196.

for the establishment of natural selection as a kind of biological law of gravity which pulled, à la Newton, all life uniformly toward a predictable end. Darwin's willingness to accept an explanatory universe which is open, and his willingness to allow for the possibility that matter might be less determined in its behavior than he was able to describe with the tools at hand, is what Gould referred to as his "eminently sensible pluralism."

Cecil Sharp's metaphorical leap from a flock of birds into music claimed something disruptive – that biological activity and human activity are not just similar, but somehow the same. Looking at the mass movement of birds in the evening sky, Sharp saw humanity in nature because the murmuration behaved like the music that humans make. The insistence that music exists in a biological continuum may not be a solution to our predicament or a rallying cry for calling musicians to the barricades of the environmental movement. But it is an indispensable perspective, like the Apollo earthrise photo, another step closer to positioning ourselves in the universe.

17

Root, stem and branch

Darwin's sensible pluralism, his combination of openness and empirical exactness, is appealing, in part, because of the possibility it holds out of a natural alliance between Alfred W. Crosby's mutually suspicious tribes of the liberal arts and the sciences. We now know that macro-evolutionary ideas about music were not unique to Herbert Spencer, for all that he was the loudest voice in the room articulating them at the time, and they cannot all be traced to Darwin's thinking. It is more accurate to say that Darwin and Spencer and the writers of music histories in the same century were all drinking from a common stream of ideas, some of which predate the nineteenth century.[1] In his five volume *De la Nature,* French naturalist Jean Baptiste Robinet (1735–1820) first declared that all the myriad forms of organisms in the world "constitute natural variations of a prototype that may be regarded as the generating element of all beings." At the center of his description was not an organism, but a prototypical cell which differentiated infinitely, a concept which suggests something like the stem cell of twenty-first-century biology.[2] A few years later, Goethe's term *Urpflantz* ("plant archetype") stood for a hypothetical primordial ur-plant whose structures served as the starting point of the complex and diversified forms of the present. For Goethe this was the paradigm for other ur-concepts – *Urtier* (animal archetype) and even *Urtype* (archtypal forms of all biological phenomena).[3] In the *Urpflantz* all the basic elements of a modern plant are there, only in elemental form: root, stem, leaves,

1 In his 1879 book *Evolution, Old and New,* Samuel Butler, in his persistent claim that Darwin had stolen the idea of evolution, credited Jean-Baptists Lamarck, Erasmus Darwin, George Louis Leclerc Buffon and others. Margulis, L. and Sagan, D. (1995), 184.

2 Morrison, S.J. et al. (1997), 287–298.

3 Hubbs, N. (1991), 153.

branches and reproductive organs. In Goethe's later thinking, the simple-to-complex paradigm of the Urtype evolved to become an archetype more like a Noah's ark which contained representative structures of all the different forms of all future plants.

The philosopher R.G. Collingwood once suggested that evolutionary ideas became attractive to historians at the turn of the twentieth century in part because a link between biology and history seemed to offer the possibility of historical writing which was authentically "scientific."[4] The attempt to classify all life forms by Carl Linnaeus (1707–1778) established itself as a constant during the next two centuries of thinking about nature. Among professionals and the general public alike, his scheme became the starting point of most natural history discussions in Darwin's day. In Linneaus's *Systema Naturae* of 1735, which went through twelve increasingly enlarged editions in his lifetime, the system was represented in a carefully categorized chart which featured two main branches of life based on their gross observable characteristics: those which have breathe (*anima:* L. breathe) and can move, and those which do not have breathe and are stationary (*plantae:* L. *plantare,* to fix in one place). In 1866, the polymathic naturalist Ernst Haeckel's "Stammbaum der Organismen," (family tree of organisms) in his famous Generelle Morphologie der Organismen was based on Linneaus, but added a third domain consisting of microscopic creatures, the *protista* which, because of continued advances in microscope technology, could no longer be ignored or marginalized[5] (Figure 17.1). His Stammbaum helped to popularize the arboreal root, stem and branch as the central visual metaphor of inheritance and development in biology.

The tree evolves

The nineteenth-century tree of life consisting of three main branches was incrementally revised by researchers throughout the twentieth century. By the 1990s, the most widely-used scheme, largely the work of the legendary geneticist Carl Woese, consisted of five *kingdoms* which are grouped, not according to their gross structures – the presence of segmented bodies or hard exoskeletons, for example – but according to the RNA genes of their chromosomes. *Monera, Protists, Plants, Fungi and Animals* – these five kingdoms are distributed by Woese among three overarching *domains* – *Bacteria, Archaea* and *Eukarya.* The first

4 Solie, R.A. (1982), 297.
5 Pearce, J.M.S. (2022).

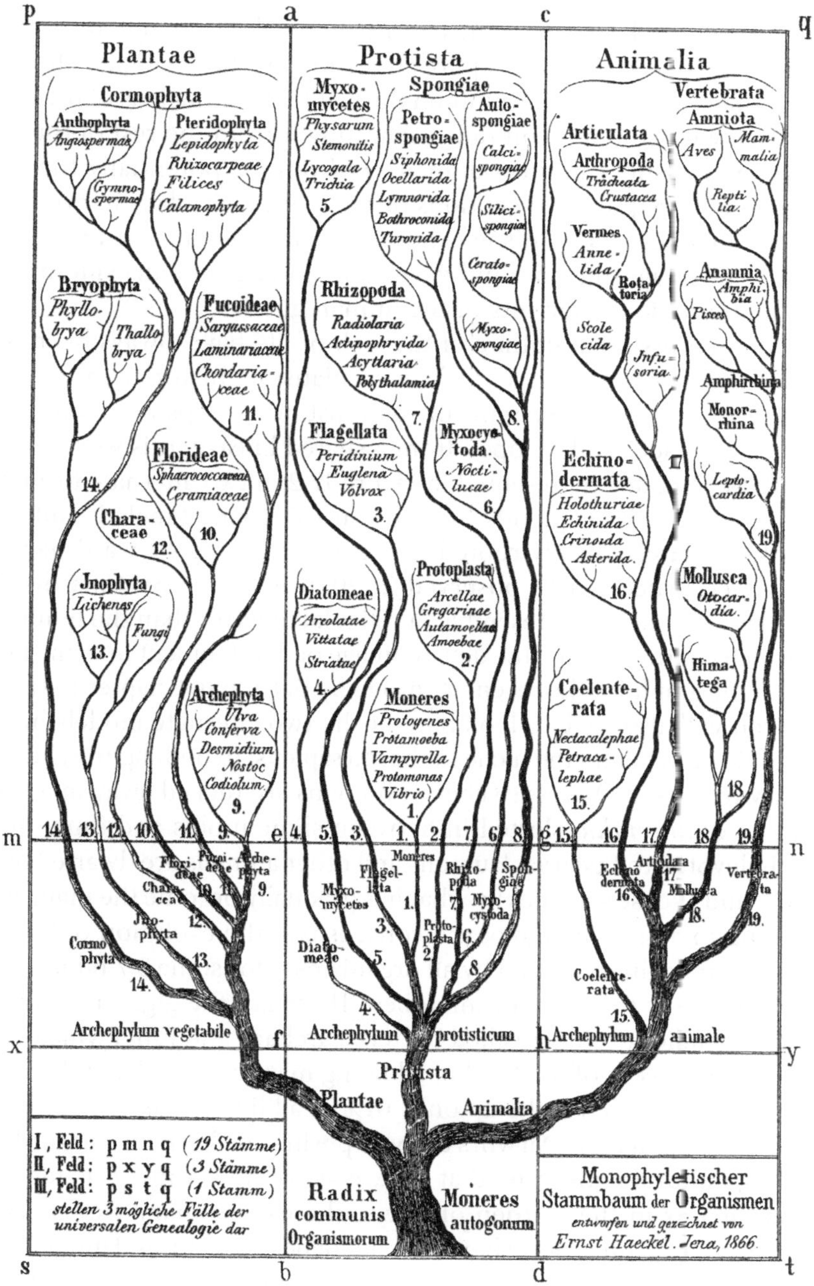

Figure 17.1 The "Stammbaum der Organismen" in Haeckel's *Generelle Morphologie der Organismen* (1866). This is Haeckel's original conception of the three kingdoms of life, including the new kingdom Protista. Public domain, Wikimedia Commons.

two, bacteria and archaea, consist of single-celled organisms with no enclosed nucleus – that is, they are all organisms (both single-celled and multi-celled) which have in common the fact that the nucleus of their cells floats freely without a membrane separating it from the rest of the cell's watery cytoplasm. The third domain, *Eukarya,* is potentially the most challenging to popular conceptions of homo sapiens' place in evolution. Eukarya includes all organisms whose cells have an enclosed nucleus, a category of staggering size and variety, including humans side-by-side with microscopic inhabitants of the human gut.[6]

Such attempts at an outline of the web of life have always been both a taxonomic scheme and a *cladistic,* a plausible historical account of how the life-sustaining systems of such distinct groups of organisms as humans and bacteria developed from a shared lineage. It is worth noting that Woese's model, represented in Figure 17.2 in two alternate formats – rooted and unrooted – tends more toward Darwin's tangled bank than toward the stately, deeply rooted tree trunk reaching skyward pictured by Haeckel. Bruce Clarke observed in 2020 in *Gaian Systems* that the latter-day schemes more resemble "a reticulated web"[7] than the branching tree which continues to be the default image of development in many fields.

As in the graph for the recent study led by Bar-On in Chapter 16 (Figure 16.1), you will also have to look hard at the two alternate models here in Figures 17.2a–b to find the niche for homo sapiens. In both figures, humans are easy to miss. In Woese's scheme, humans are tucked away in a single off shoot of *animalia,* slotted in between slime molds and fungi in the branch devoted to Eucarya. Humans and other animals are dwarfed in the Bar-On and Woese schemes by what has gradually become the main event according to a growing biological consensus – the evolution of the life-sustaining machinery shared by all organisms – for some 3.7 billion years has been largely the work of microbes. These are the gigantic, formerly invisible families of single-celled bacteria, archaea and fungi which make up some 15 percent of the total of all living mass on the planet. Humans, we should remember, make up only .01% of all living mass.

Scientific thought in Darwin's day was posing an alternative to the long-standing cultural consensus that nature as we now know it is identical to what it was in the Garden of Eden. The evolutionary view injected dynamism into Linnaeus's exhaustive catalogue of creatures. Human civilization would have to adapt to the idea that the biome was in a state of

6 Woese, C.R. and Fox, G.F. (1977). 5088–5090.
7 Clarke, B. (2020).

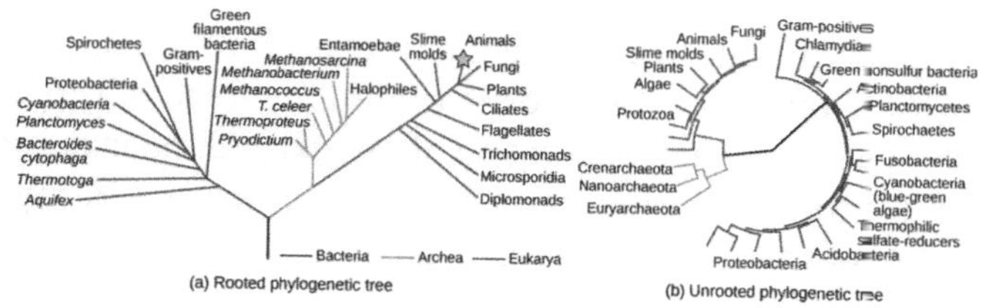

Figures 17.2a–b Two current representations of the universal phylogenetic tree in three domains (*Bacteria, Archaea* and *Eucarya*) as rooted and unrooted. The rooted tree (Figure 17.2a) attempts to identify approximately at what point in evolution various species diverged from a single ancestor. The unrooted tree (Figure 17.2b) does not. Graphic: modification of work by Eric Gaba, NASA. https://courses.lumenlearning.com/wm-biology1/chapter/reading-phylogenetic-trees-2/ (accessed 2/22/26).

perpetual modification. No plant and no animal today exactly resembles its forebears of a million years or even a hundred years ago. This raises a question for music: if biological consensus has been gradually shifting attention from the cataloguing of gross morphologies to bio-chemical processes, and from there to ecologies, as Gilbert and Tauber claim, how would a shift of attention from gross musical morphology to musical processes and to ecologies effect our understanding of how music is made and heard? The familiar catalogue of musical forms comparable to the adaptive radiation of Darwin's finches discussed in Chapter 3 – structures like aria, mass, symphony and popular song – are plain to see as both taxonomic categories and as cladistics. They are items in a catalogue and they are also nodes in a web of historical relationships. Each of these musical genres can be understood as –to take one familiar example – the creative variants of basso continuo, a harmony-regulating process shaped by the distinct demands of the various habitats of theater, church, concert hall and public house within different cultural settings across Europe.[8]

Both biologists and musicians think in terms of families of behavior and both make use of arboreal thinking – family trees which link descriptive features and inheritance. In 1927, in *The Etude,* a popular magazine whose readership was primarily piano teachers, Carl Czerny is positioned at the root of a tree on whose branches are the names and faces of more than a century of famous pianists. The evolving practice of piano is

8 The continuity of the practice of basso continuo between Baroque era Europe and twentieth-century jazz was perhaps first articulated by European observers. See: Schmitz, H.-P. (1979), 75–78. See also: Gooley, D.A. (2018).

represented with the bio-musicological terms "forefather" and "family tree." The passing on of genetic information in the form of protein chains is implicitly equated with the passing on of mental and muscular technique from teacher to student. In this reductive popular image, Czerny becomes the great progenitor, the pianistic equivalent of the biblical Adam.[9]

Root, stem and branch reconsidered

Consider the arboreal metaphors of root, stem and branch. How would these concepts shaped by biology advanced during the lifetimes of Darwin, Spencer and Stumpf shift in meaning if they were anchored in twenty-first-century science? What changes would we have to read into these commonplace borrowings, so stubbornly convincing and useful in everyday speech, despite their basis in outmoded science? Perhaps attempting to make over the bio-musical assumptions buried in these terms makes little sense now, no more than would poking at the weaknesses in Descartes' theory of the humors and Matthesson's reliance on them as musical pedagogy, as if recent physiology and psychology had not intervened. Perhaps it would be better if we simply transferred the discussion to a new venue altogether, furnished with contemporary terms, assumptions and views of life forms.

But it would also be senseless to advocate a purge of all musical descriptions employing, for example, life-based words with origins in nineteenth-century science like "root," "grow," "inherit," "evolve" and "influence," if for no other reason, because of the trans-generational continuity of meaning which might be lost. I believe that if we allowed that continuity to absorb some of the expanded meanings of these words developed over the past century, we could approach a bio-musical perspective matched to our time. These expanded meanings may build on the biologist's greatly increased ability to analyze the microscopic processes by which we now understand that plants actually do establish roots, by which plants and animals grow, inherit, evolve and influence one another. From there, we could begin to experiment with how these fresh meanings might continue to illuminate human activities like music.

"Root" is a useful place to begin thinking about the fresh meanings that could be applied to music. It has been a term commonly harnessed to the study of language and culture as well as to the plant kingdom, but in current science, the word acts as no more than a generic placeholder for

9 At the top of the page, connected neither to root, stem or branch, hovers the picture of Beethoven, the teacher of Carl Czerny, suggesting that in this Edenic scheme, if Czerny is Adam, then Beethoven is God.

a host of more recent terms. Words like *mycelium* and *rhizome* now stand for where the real action in the root system is considered to take place, in soil structures far more delicate and more interconnected than what the naked eye or the microscopes of either Darwin or Pasteur could see.

With scientific discussion now favoring the ecological communications among organisms – what Philip Kitcher and others saw as examples of an interdisciplinary trend toward interactionism – focus is shifting to an array of root functions that could not or did not rise to the same level of attention in earlier generations. It is a focus which emphasizes systemic relationships over individual organisms and structures, the same thrust of recent biology which Gilbert and Tauber called ecological. The reassuring solidity and singularity of a root no longer adequately encompasses the complex exchanges between plant chemistry and soil chemistry through the *mycelium,* the vast underground networks of microscopically fine hair-like *hyphae* that extend far beyond the individual plant to create a chemistry laboratory which also functions as an ecological communication system.[10]

The web of mycelium underground is a far less fixed, less unitary and more dynamic image of a root. The root is still there in the new understanding, but root is now a region of exchange, less a thing than a locus of activity. The physicist Eric Smith, for pushing the concept of biology – and with it, our bio-musicological efforts – into the realm of behavior in real time: "the biosphere is only understandable as a state of activity."[11] It is a claim which resonates with Marc Bickhard's flame metaphor. Like mental states such as music, flames exist only as states of activity. When combustion ceases, any semblance of existence the flame might have ceased as well. The cloud of root hairs in the mycelial web are physical structures which are shaped by the chemical exchanges they transmit. Like flames, they too flicker and flare in response to the ebb and flow of energies and nutrients to and from the soil and the soil's bacteria, with whom the plants are symbiotically partnered. When those exchanges increase, diminish or cease, the web tendrils grow, wither or die.

Rhizome rising

It is worth mentioning in passing that even beyond the diverse meanings of root, these interlocking functions of soil, plants and microbes are activities which we now read against a still larger ecological backdrop, the distress of

10 Perlin, J. (1989/2023), 492–505.
11 Smith, E. (2015).

our planetary habitat. This is a point of reference which had hardly begun to appear on Darwin's conceptual horizon. The cycling of water through soil fed by symbiotic relations of fungal mycorrhizae and soil bacteria, through leaves and through atmosphere, is the much studied transpiration process which we now know creates the rain that life depends on, and which cools our overheating planet.[12] Compared to the current level of knowledge about photosynthetic life, continued use of the term root by anthropologists, musicologists and literary scholars which did not adapt to the current understanding can be expected to sound less and less relevant. This is one more way that the current derangement of the biome enters our bio-musical discussion uninvited, this time giving a seemingly academic concern for terminology a fresh urgency.

Even without taking into consideration latter day planetary anxieties, the term *rhizome* is an example of a biological concept which offers an alternative, or at least a supplement to the traditional image of root, stem and branch. Rhizome describes a particular relationship of plant structures above and below ground which is found in a diverse array of plant species, including potatoes, gingers, bamboos, poplars and certain aquatic varieties, in which stems grow horizontally underground, sending off multiple roots and surface structures. The underground structures can also serve as nutrient storage, as in the case of potatoes and ginger. The rhizome has no central root, but spreads in multiple directions. In this century, the densely tangled, uncentralized growth characteristic of the rhizome has made it an appealing alternative image to the more familiar metaphor of the rooted plant. Rhizome also offers an alternative to the hierarchical distinction usually read into the above and below ground image of the tree in which the root plays a subordinate, supporting role to what is the main event, the visible vegetation favored by humans.

In the rhizome plant, each growing structure above ground appears to be unique and separate, but below ground they are a continuity of structure and energy transfer. In the image of a rhizome, where is the center of the tangle of connectivity? Where does the tangle begin? Where does it end? Instead of clear answers to these questions, we see only a maze of branches with many nodes of intersection.

Since the 1980s, the rhizome has attracted considerable attention among thinkers outside of biology as a metaphor of systemic complexity and it has even entered popular conversation as a kind of meme. As a contemporary version of Haeckel's tree metaphor, *rhizome* and *rhizomatic* are

12 Hawkins, H.J. et al. (2023).

now also applied to analyses of neural networks in the brain, to the transmission of viruses, to learning pathways, and to philosophy, psychology and electronic media.[13]

Bacteria in the tree of life

The recent acceptance of the rhizome as a popular expression is reminiscent of the widespread use of the tree metaphor since Darwin's day. Haeckel's arboreal image of root, stem and branch has been a long time in dying, though the rhizome continues to erode the tree's power to inspire questions or to explain.

As a tool for understanding how life began and then became fruitful and multiplied, the three-branched *Stammbaum* of plants, animals and protists (Figure 17.1) has been gradually replaced by subtler arboreal schemes, at least among biologists. By the mid-twentieth century, the preference for familial schemes of the eighteenth and nineteenth centuries based on the similarity of visible features was swinging toward genetic relationships, most of which were unknown and unobservable in earlier centuries. Focusing on the common features in the chemical machinery of the cell provided a different picture of the relatedness of plants and animals and of microbes and mammals than the picture based on the presence or absence of jointed limbs, reproductive organs or lungs. With this change of focus, Haeckel's three large groupings are now viewed as intersecting, recombining and diverging, pushing the tree icon to its limit. Simple organisms which had so naturally found a place earlier in the trunk of the tree are now represented as crossing and joining with later evolutionary developments, as in the biologist W. Ford Doolittle's diagram from 1999 in Figure 17.2. These organisms are both single-celled and multi-celled, but they are largely microscopic, requiring sophisticated tools to see them at work. What is more, they are no longer seen as isolated in the past, superseded by the familiar bulky, hairy and leafy structures visible in our everyday lives. The two versions of Carl Woese's shrubby

[13] The social philosopher Gilles Deleuze and the psychoanalyst Félix Guattari, who are usually credited with coining the term *rhizomatic*, have applied it to an extremely diverse array of social and cultural phenomena. See: Deleuze, G. and Guatari, F. (1987). For an example of rhizomatic applied to music, see the discussion of the transmission of technique and affect in Turkish makam practice in Gil., D. (2017), 98–100. For an example of rhizome as a popular meme in online culture see: https://www.reddit.com/r/philosophy/comments/a21qv1/the_web_as_rhizome_in_deleuze_and_guattari/ (accessed 2/24/26).

late twentieth-century diagram of three domains and five kingdoms in Figure 17.2 are, for now at least, textbook staples. But there are also other versions of evolutionary development, built on five kingdoms with three empires, or three domains with five (or six or seven) kingdoms, arrayed in cornucopic swirls and bushes or in diagrams resembling irregular crystals or jets of multi-colored paint, all of them emanating from a single center or base with Linnaeus's Latinate labels still attached. Many of these images are computer-generated from large statistical data sets.

For the specialists with the most intimate knowledge of the tree of life, there were many blows of the ax which weakened the tree's value as a heuristic tool, but the final blow may well have been delivered by the revelations of *lateral gene transfer,* a process by which an organism transfers genetic material to another organism that is not its offspring. This is an ax wielded by, once again, the humble microbe. The vast world of single-celled organisms without a nucleus, consisting of bacteria and archaea (together known as *prokaryotes*), has been in the vocabulary of scientific observers ever since the seventeenth century. That was when the first microscopists Antonie Leuwenhoek and Robert Hooke observed tiny animalcules in myriad shapes darting through pond water.[14] Haeckel, revising Linnaeus around the middle of the nineteenth century, gave microbes or *monists* a branch all their own alongside their eminently more observable cousins in the plant and animal kingdoms. Louis Pasteur's experiments in 1850 showed that humans live in an environment saturated with microbial life, a fact which to those of us living now with the new normal of global pandemics may sound like simple common sense.

Lateral (or horizontal) gene transfer (LGT or HGT), is an idea that surfaced in the 1960s and 70s, but, as was the case with Lamarckism, epigenesis and symbiosis, it was also a much older idea that enjoyed a renaissance thanks largely to new developments in twentieth-century technology. In 1928 the British bacteriologist Frederick Griffith had demonstrated experimentally that certain nonvirulent pneumococcus bacteria could become virulent simply through contact with virulent bacteria. This rapid mutation, the change from non-pathogenic to pathogenic forms of the bacteria, was accomplished by skirting *meiosis,* the much slower multi-stage process of splitting and sharing chromosomes in sexual reproduction. The simplicity and speed of lateral gene transfer partly explains how bacterial infections can mutate and spread rapidly.[15] By the turn of the new millennium, lateral gene transfer was being widely recognized as

14 Lane, N. (2015).
15 Griffith, F. (1928), 113–159.

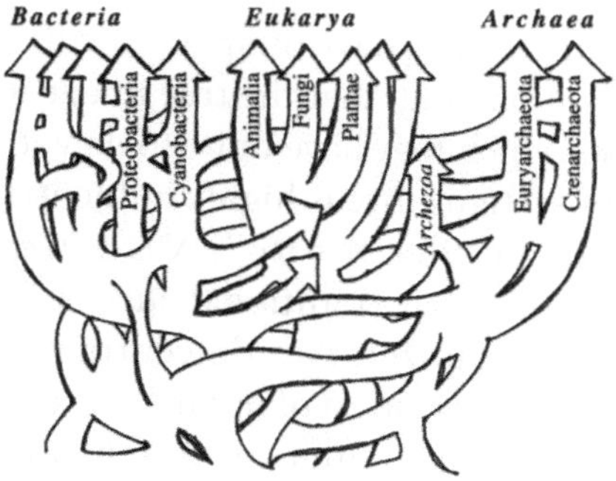

Figure 17.3 W. Ford Doolittle's scheme of evolution in 1999. Vertical arrows indicate the transmission of genes through sexual reproduction; horizontal arrows indicate the direct transmission of genes through lateral gene transfer. Doolittle, W.F. (1999). Fig. 3, 2127.

accounting for much more of inheritance than had been thought. In 1999 W. Ford Doolittle offered the example of the common Escherichia coli bacteria, twenty percent of whose genome had been acquired by lateral gene transfer.[16] In 2017 Denis Noble summarized LTG as the "highly promiscuous" behavior of a substantial number of bacteria which enabled them to perform "their own forms of multiple-partner 'sex,' and they do it all the time," Noble says, "not just when they reproduce."[17]

Lateral gene transfer adds a dimension of variation that the more linear, Neo-Darwinian view of genetic inheritance excludes: among bacteria and archaea, the most populous families of organisms on the planet, change can happen on a dime, simply by bringing two organisms near each other. The membranes that enclose the working machinery of a cell can be transiently and strategically penetrated, making possible the rapid passing of genetic material, and with it, rapid adaptation to changing conditions. This is a view of the defining boundary of the living cell (see Chapter 6) which complements the shift in the life sciences from immunity to immunitas around the same time. In this revised point of view, the cell wall is less an impenetrable militarized zone across which nothing passes, than a zone of exchange and adaptation which behaves differently depending on surrounding conditions.

Haeckel's mid-nineteenth-century Stammbaum at the time represented a substantial step closer to the current image of evolutionary complexity

16 Doolittle, W.F. (1999). Three types of lateral gene transfer are now identified by biologists, all three of which transfer traits directly between bacteria through their cell walls. Through *transformation*, bacteria absorb free-floating DNA directly from the surrounding environment; through *conjugation*, bacteria directly transfer genes to another cell; through *transduction*, bacterial viruses *(bacteriophages)* move genes from one cell to another. See: Ravenhall, M. et al. (2015).

17 Noble, D. (2017), 103ff.

than the rectilinear relationships of Linnaeus's chart of species in the previous century. But it is worth noting, in passing, that the tree image in many disciplines has persisted for reasons which may have more to do with human cognition than with the reality of systems. In his highly influential challenge to conventional urban planning in 1965, *A City is Not a Tree,* the architect Christopher Alexander argued that traditional successful cities (he calls them "natural cities") are organized as a de-centralized "semilattice," foreshadowing Bruce Clarke's description of natural systems as reticulated webs. Nevertheless, most modern attempts to design cities from the ground up ("artificial cities," according to Alexander) tend to be structured as a tree, with resources and populations "rationally" channeled through branches from a centralized trunk. "For the human mind," he says, "the tree is the easiest image for complex thoughts."[18] Whether or not this is universally true, for non-biologists the tree may nevertheless remain intuitively indispensable as an aid to thought. Still, it is striking to see how, in Figures 17.2 and 17.3, current research reinterpreting Linnaeus's table of species now resembles, not a tree, but Darwin's prescient tangled bank.

Biological thinking has changed, but the essential principles of Darwin's version of evolution are still regarded as foundational, though lacking in sufficient detail, by the majority of scientists of many persuasions. This is a view, we should remember, which Darwin arrived at through an unlikely combination of hand-written correspondence with scientists and horse and pigeon breeders (mostly in England), a five-year globe-circling field trip on the *Beagle,* close study of the plants in the hothouse of his estate in Kent, and his personal collection of barnacles, fossils and snail shells, all with equipment no more sophisticated than a small microscope.[19] In general, it is the tidal wave of details made possible by new technologies which have changed, requiring that recent generations of biologists learn to live with increasingly complex, fluid and interactive explanations for how organisms on the cellular and molecular levels manage to both stay the same and vary. It is a potential model for what is required of musicians seeking to re-connect musical thought with nature.

18 Alexander, C. (1965).

19 The microscope Darwin used for his barnacle studies, in the collection at the Whipple Museum at the University of Cambridge, had a maximum magnification of about 1100, though its optimal resolution was at a maximum of about 800. See: Jardine, B. (2016). By the late twentieth century, electron microscopy has achieved high resolution with magnification on the order of 10 million.

Part Five
Emergent order in two musical ecosystems

A living thing is any living entity, no matter what it may be. A living thing is thus not necessarily an organism. There are many examples of this. A cell that belongs to a multicellular organism is a living thing without being an organism. A species, a taxon, a population, a group, etc. can also be considered living things without, of course, being organisms. A living thing is simply an entity that lives.

Thomas Pradeu, 2012

The extended comparison of two musical examples in these final chapters will begin in the way Part 1 began – by looking closely at families of living systems. Here, those families will not be birds, fish and symbiotic micro-organisms but two song-making practices, widely separated in geography, history and habitat. The task will be to look in the fixed records of these songs for signs of change in them, both in their developmental histories and in their micro-adaptations at a given moment. Ecosystem, the scientific concept which stands for the continual replication and transformation of the constituent elements of any living thing, will be the tool for examining these two practices side-by-side.

The habits of thought which the professions of music and biology share have been recurring themes in the foregoing chapters. Before launching into the two musical events being compared, Chapter 18 returns to this theme with further examples of how biology and music both struggle with the unique complexities of their respective materials.

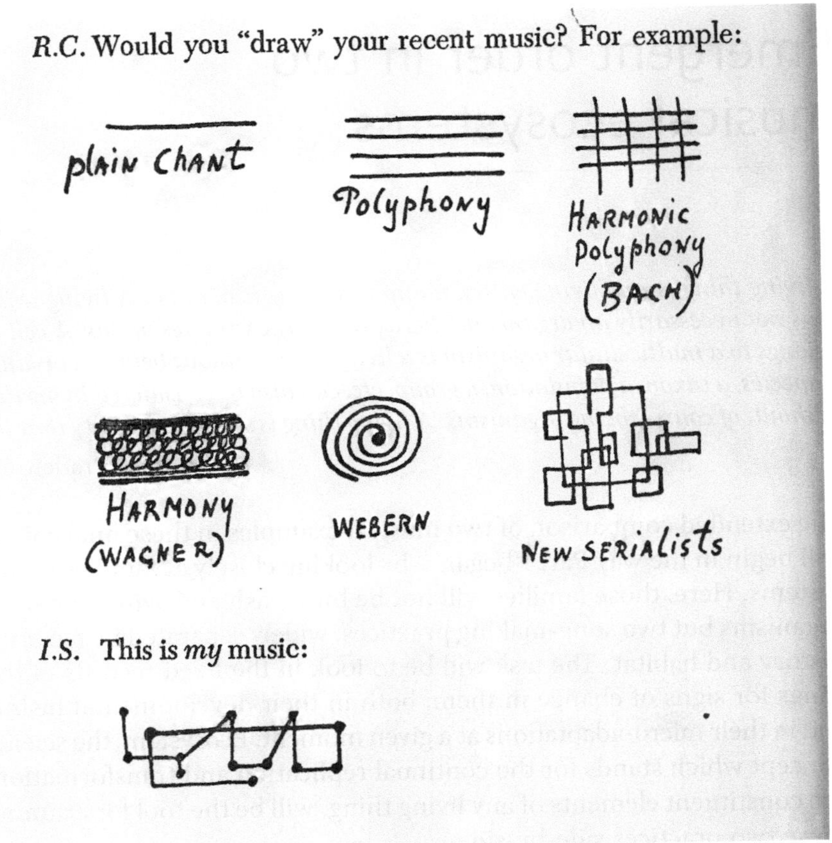

Figure 18.1 Paradigms of music-making in Europe, as recorded in Robert Craft's personal notebook. Craft drew some diagrams to illustrate the structures of various historical musical practices. Stravinsky then added a diagram of his own music. Stravinsky, I. and Craft, R. (1959).

18

A repertoire is an ecosystem

Sometime in the late 1950s, the ongoing musical conversation between Igor Stravinsky and the conductor Robert Craft, the composer's long-time collaborator, took the form of the rough sketches of musical repertoires reproduced here (Figure 18.1). In such a minimal, abstract form, these little drawings recorded in Craft's notebook cannot easily be mistaken for snapshots of complex structures which, when they are blown up, reveal all their details. They are not, for example, comparable to NASA's iconic earth-from-space satellite photos which can be enlarged many times to reveal finer and finer details of coastline and terrain. Rather, they are paradigms, models, best guesses, or even, as Capra and Luisi recommended for the study of life forms, maps of dynamic processes. Blowing up these little sketches would reveal nothing you haven't already seen in the simple design. They contain no specifics of time, form, pitch, instrumentation or function. They are not what we commonly understand to be musical scores.

The phylogenetic trees representing the relationships among species may be taken as more serious attempts at describing dynamic processes than Craft and Stravinsky's back-of-an-envelope drawings. Those trees were dramatically redesigned during the century between Haeckel and Doolittle according to fresh research (Figures 17.1, 17.2, 17.3). By the twentieth century, each new version of Haeckel's famous nineteenth-century Stammbaum was being steadily altered with new details from the latest genetics research. Increasingly, those trees have been generated with artificial intelligence and computerized data imaging (Figures 17.2, 17.3). And yet, compared to the total depth of genetic information in this pool of millions of data points, they are still sketches, like Craft and Stravinsky's – approximations, best-guesses.

In all these cases, both biological and musical, detailed topography is generalized to create typology. Craft's lines are purposely spare, so if there is any hidden detail in them it would have to be the reader who would fill them in. Renaissance polyphony is represented simply by equally spaced horizontal lines. As anyone familiar with the music of Josquin des Prez and Giovanni Pierluigi da Palestrina would immediately recognize, the lines represent the predominance of intervals of a third and a fifth (and their inversions, a sixth and a fourth) between individual parts in the multi-voice music of the Renaissance. For the polyphony of the Baroque period, captioned parenthetically by Craft with the name of J.S. Bach, those parallel lines are locked into a tight grid of horizontal and vertical lines, suggesting the blocks of built-up-from-the-bass triadic harmonies of the basso continuo model. For anyone with music school training, the mere mention of Bach's name would be enough to evoke the detailed paradigm of the Bach four-part chorale settings which in music schools are still considered the indispensable textbook of what was until only recently called the "common practice" of European music between the seventeenth and twentieth centuries.[1]

All these qualities, none of them explicit in Craft's drawings, are readable as markers of musical families, along the lines of Linnaeus and Haeckel's families of organisms. All plants, for example, fall into two massive families according to how reproduction is accomplished – by seeds (*Gymnospermae*) or by fruits and flowers *(Angiospermae)*. The groupings are large and rough, with a range of diverse forms which push the strict boundaries of the families, but the groupings remain useful today alongside the far more detailed computer-generated schemes. The Vatican-sanctioned modal groupings of melodies of Gregorian chant were also familial (also with many exceptions[2]), as were the tune families of nineteenth- and twentieth-century folk music research. (See Part 4) As maps, the rough scribblings in Craft's notebooks are also representations of families – families of musical development and families of musical

1 The widely held idea of a common practice period extending from about 1600 to 1900 testifies to the existence of a multi-generational musical ecosystem based in Europe. In recent years, objections have also arisen in music theory to the common practice idea because it imposes a Euro-centric bias. See a recent account of the controversy over the teaching of music theory in London, J. (2022).

2 The *Liber Usualis* shows how the largest of its eight modal groupings, Mode I, encloses several subgroupings with variant behaviors, including atypical openings and even finals which at moments can look more like Mode II.

behavior. Their implicit content is a finely detailed musical experience, but too much detail renders the image less useful.

Taking a crude look at the whole

In their simplicity, the lines and squiggles in Figure 18 1 might tempt us to think of them as reductionist versions of complexity – that is, as one-size-fits-all cookie cutters rather than as maps, although there is no reason to believe that cookie-cutters were at all what Craft and Stravinsky had in mind. For my own part, I am choosing to take these little sketches as moments in a much longer conversation between the two men, a conversation with many moving parts which evolved over several decades. Though the celebrated composer is clearly the center of gravity of these conversations, we can assume that both men were subtly changed in the process. Understood in this way, these musical sketches might qualify for the expression offered by economist and systems theorist Murray Gell-Mann, "a crude look at the whole."

First mentioned in an address to a conference on complex adaptive systems at Boston University in 1989, this phrase has stuck and become a kind of meme (like so many scientific propositions, controversial) in the often rarified world of systems thinking in biology, physics, economics, sociology, engineering and computer science.[3] These are conversations which often require fluency in the languages of mathematics, cybernetics and chemistry. Within music, any attempt by artists of Craft and Stravinsky's stature to make an approximate but useful description of a complex phenomenon would also not seem out of place. Such pithy non-technical summaries are the very kind of utterances based on experience which would be remembered from a lesson or master class, retold and passed on. The teacher of Arnold Steinhardt, the Guarneri Quartet violinist, had advised him decades earlier "to eat the fingers" of his quartet mates – a crude look at the inexpressibly complex problem of ensemble playing. Its graphic specificity conjures up a level of sensitivity that extends to the whole act of performance – any performance. The teacher's phrase suggests for performance what Craft's drawings suggest for composition. They are each best guesses at an overall musical approach which is too complex and changeable to break down into parts, while

3 See, for example, Miller, J.H. (2015) and the NTU Complexity conference.

they each avoid details which might obscure the principles governing the wholeness of the image.

For Gell-Mann and others after him, the idea of a crude look at the whole has, in part, served a special purpose as a tool in the ongoing argument against reductionism, determinism and specialization in the sciences. It might serve a similar purpose in music. But in any serious line of thinking in both fields, the crude look raises a common question in both fields: how can a biologist or musician stay focused on the whole while doing the necessary close work under high magnification which full understanding requires? Lynn Margulis, late in her career, castigated her colleagues for losing the thread of this line of questioning. Had they forgotten the sense of the whole – that biology is the study of life? Ultimately, she said biology is not primarily the study of the individual bits of the whole that lend themselves to specific experiments and clear outcomes.[4] For Gell-Mann, in the end, thinking about both bits and wholes is necessary. I have already spoken of the changing micro and macro perspectives on evolution over the past two centuries. My own metaphor of looking through both ends of a telescope at a distant object attempts to capture a common sense understanding of what Gell-Mann's phrase and a macro-evolutionary and micro-evolutionary approach to living systems are trying to capture. Judging from what scientists of many disciplines say when they meet to discuss any crude look at the whole,[5] this is what it takes to comprehend the wholeness of anything complex.

From a bio-musical point of view, Craft and Stravinsky's layered sketches of Bach and Wagner are typologies but not universals. The sketches are explicitly matched to particular habitats. They make no attempt, for example, to leave a place for other polyphonic repertoires in the Mediterranean, in Africa or in Asia which differ in the internal relationship of their parts from those pictured by Craft and Stravinsky. The ethnomusicologist Claudia Schmidt, for example, has created comparable schemes for *gbona*, the many-layered vocal repertoire of the Kpelle people of Liberia (Figure 18.2). Her two schemes of this repertoire are more analytical and specific than those in Figure 18.1, but they are still a crude look at the whole and illustrate how different the hierarchies of multi-voiced practices on the planet can be.

4 Margulis's criticism was at her sharpest in her article "Big Trouble in Biology" (1998) where she identified the bits of the whole as those upon which careers can be built, tenure earned and funding gained.

5 See NTU Complexity conference.

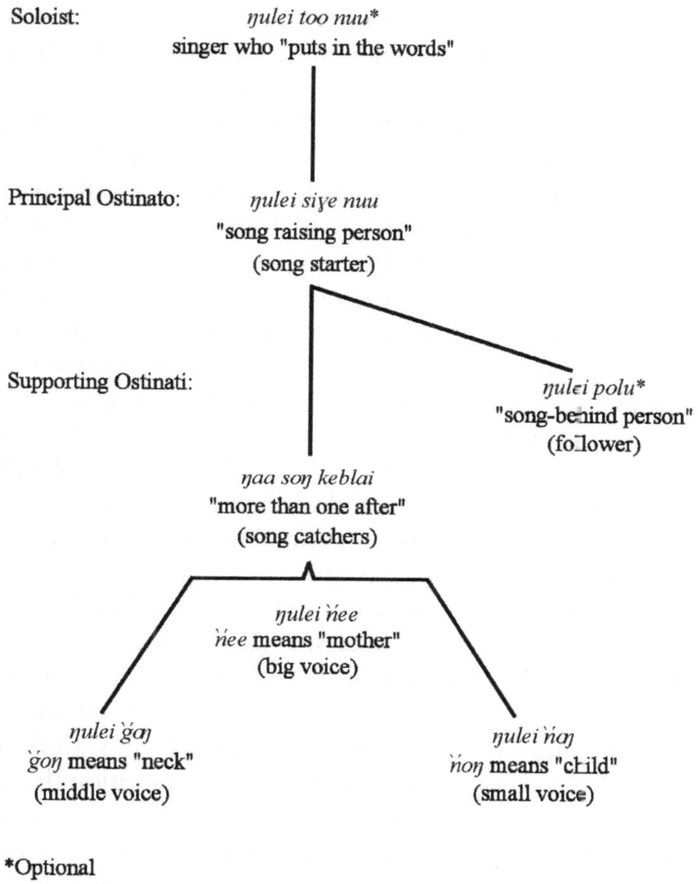

Figure 18.2 A paradigm of the *gbona* repertoire of the Kpelle people of Liberia. Based on Schmidt, C.E. (1984).

If we think of the presence of polyphonic textures in musical practices across the globe as the phylogenetic class *mammalia,* then twentieth-century Liberian gbona and seventeenth-century German basso continuo might be examples of Leibnitz's compossibility acting in music. They are both mammals, but one is a whale and the other is a squirrel – unique living systems with shared characteristics which are only possible in their distinct environments. It is worth noting that in neither Craft's nor Schmidt's model are pitch relationships between the separate parts specified.

Nevertheless, key words such as "Renaissance," "Bach," "Wagner," and "gbona" are enough to remind a person familiar with these repertoires what is explicit and what is implicit in each model. The triadic principle in Craft's horizontal lines doesn't need to be spelled out to a western musician, since in any music associated with the Renaissance, Bach or Wagner, triads are understood to be the basic building blocks, even though the use and form of the triad paradigm in each is distinctive. Craft takes the rectilinear uniformity in the Renaissance and Bach images and in the Wagner image turns it into rows of tight curlicues, suggesting the kind of dense chromatic movement through triads in Wagner's famous *Prelude to Tristan and Isolde*. In the case of Schmidt's model of Liberian polyphony, terms like "mother," "neck" and "child" used by Kapelle musicians make familial relationships and body parts paradigms of musical behavior. Schmidt's scheme is even laid out in a form suggesting a human body. Like the other examples of crude looks offered here, Schmidt's image is a sketch of a living system from which replications and variations are understood to emerge. We can expect that each musical event generated by these West African and European systems will be in some way unique, while at the same time they will bear strong resemblances to others in the family.

The internal controversies over specialization versus the big picture are important for the professionals in these two disciplines, but I believe that the crude look idea also attempts to preserve something more as an approach to complexity. I have come to value Gell-Mann's concept for the way it enables us to see beyond the facile assumption that complexity consists in nothing more than adding on *more* parts. To say, as Spencer did, that simplicity inevitably gives way historically to complexity invites a monophony-to-polyphony shortcut in music, that glib freshman view of how grown-up civilizations evolve. It is a shortcut which enables a musician or historian to leap over the paradigm of organismal complexity which for Darwin, and even perhaps for Spencer, provided the depth of detail and dynamism which the simple-to-complex idea contained. In a musical perspective one hundred fifty years ago, a notated score of a folk melody was so obviously a simpler image than a symphonic score of eighteen staves that no further thought was necessary – especially when this paradigm so closely matched the prevailing assumptions that the urban habitat of European art music was more developed than the rural habitat of folk song, the bygone habitat of a medieval monastery, and the alien habitat of the neighboring Ottoman empire.

Spencer's knowledge of biology was quite sophisticated, but in the end, his law of progress is an example of the real risks of any attempt to arrive at a crude look at the whole. In Spencer's law, the idea of an inevitable evolution from simple to complex was too tempting as a generalization for many nineteenth- and twentieth-century music historians to resist. It fit too comfortably with prevailing social views of the time to be questioned further. At the same time as it acted as a good-enough functional tool for glimpsing the whole, for several generations of scholars the simple-to-complex paradigm was also a pair of blinders. With the benefit of self-reflection, what looked one hundred years ago like a picture of the broad scope of human music-making has now turned out to have been a snapshot of just one part – Europe's view of its own role as the goal of all historical development, as in one historian's mollusk-to-mammal comparison of Scarlatti and Beethoven (Chapter 17).

The crude look comes with hazards, as do metaphors of all kinds. But keeping these perennial risks in mind, there is another potential benefit of Gell-Mann's idea to a biology-oriented view of music. In the sciences, it is widely recognized that wholes can behave in a manner which little resembles the behavior of their parts. John Stuart Mill, we now know, had puzzled over the fact that the elements sodium and chlorine act as poisons to most forms of life, but when they have interacted chemically under the right conditions, the whole which results is somehow an essential ingredient of life. The form and behavior of the whole transcend the parts which make it up and could not have been predicted simply from a study of the separate parts themselves.

Ecosystem: Humpty Dumpty in Yellowstone

The term *ecosystem* qualifies as another such crude look at the whole. Between popular speech and hard science, the term has many uses, but regardless of the specific meaning we intend when we say the word, it cannot help but evoke an image of complexity, if for no other reason than the presence of "system" in the term. With a little more familiarity with ecosystem as a biological phenomenon, the term takes on greater depth, conveying the sense of a self-sustaining living entity, like a coastal marshland or the bacterial flora in the gut of a human. An ecosystem is a whole which can surprise any observer expecting it to look like or behave like the separate organisms that make it up. With our eyes fixed on a mangrove forest in a South American estuary, most of what comprises its wholeness is invisible, submerged in the salty tidal water, soil and dense roots, in its shellfish, insect and fish habitats, and in the carbon-rich chemical exchanges in its sediments. Scrutinizing a single mangrove tree or shrub standing 4-8 feet tall at the edge of a slow-moving body of water would not give us much with which to predict how the whole Mangrove system would look or behave. The whole is not simply any one of its constituents blown up large or multiplied by a thousand.

Ecosystem suggests itself as a useful term in music because of its complexity, not its size. A coral reef such as The Great Barrier Reef off the coast of Australia, extending some 1400 miles and hosting thousands of species of organisms, is spoken of by biologists as an ecosystem of many ecosystems (more Russian dolls). This is also how James Lovelock referred to Gaia, as the planet's largest ecosystem encompassing many millions of species over billions of years, each one of them an ecosystem composed of ecosystems. The qualities which make the tiny bacteria alive and autopoietic are the self-sustaining, self-feeding and self-replicating interplay of its parts – an interactionist quality which is the earmark of any ecosystem. At the same time, while an ecosystem can strike an observer as a stand-alone entity – a particular coastal marshland unlike any other and the unique bacterial flora of a particular human – an ecosystem is lawful, in Darwin's words, governed by principles of self-organizing energy exchanges throughout the whole complex.

In an ecosystem governed by laws, no one participant is in charge. This is a fact which is a potential obstacle for anyone thinking of applying the term to music, where attention tends to be placed on actions of individual human creators and performers. For example, it may be useful to survey the life forms of a dense African habitat and say that "the lion is king of the jungle," emphasizing a key role for one actor in the system. But as a crude look as it is understood by Gell-Mann's scientific colleagues, the phrase would only make sense if we were referring to the concept of an *apex predator species*. This is a concept which incorporates a depth of knowledge about what happens to an ecosystem when the population of predators at the top of the feeding chain of a habitat diminishes or increases. The examples of apex predator species most often cited are large mammals like whales, mountain lions and humans.

The famous case of the grey wolves in Yellowstone Park in Wyoming is an example of the role played by an apex predator in an ecosystem.[6] In response to many generations of complaints by cattle and sheep farmers in the American west, in the 1890s the government-sanctioned systematic extermination of the grey wolf population began in Yellowstone. This

6 For a concise summary of the role of apex predators in Yellowstone and their role in the region's restoration, see: DeLoss, J. (2024).

initiated a cascade of destructive events in apparently unrelated parts of the park. With the rapid reduction of the region's apex predator from the habitat, over the following decades elk herds which fed on the moisture-loving willows lining the streams and rivers quickly over-populated and stripped the trees of bark and leaves. The normal transpiration of water vapor from the tree leaves and the trees' ability to hold water in the soil deteriorated, contributing to a net loss of groundwater, rising surface temperatures due to diminished evapotranspiration, and a disruption of the small water cycle between soil and rain clouds.[7] Rainfall diminished. As the streams dried, beavers (another key species, though not a predator) abandoned the region, their regular flooding of the landscape disappeared, and with it, the many species of plants and animals which depended on it. As the soil deteriorated, the landscape was transitioning to desert.

By the 1920s, scientists reported that because of this chain of consequences from the loss of its apex predator, Yellowstone was in steep decline. This ecological cycle could have been interrupted in other ways, of course. After all, it is the way of ecosystems that they evolve – they are interactional and therefore sensitive to disturbance. We may therefore speak of a king of the jungle in order to acknowledge that the impact of large mammals like the wolf and the human on the biome of a habitat can be outsized, but only because of the part which the wolf or human plays in the interplay of forces of the whole. Attempts at restoration of Yellowstone since the 1970s have demonstrated that ecosystems are not simply programmed to enact a fixed set of winning procedures indefinitely, and that the cascade of changes as I have summarized them above cannot simply be unwound or reverse engineered. Beginning in the 1970s, a program of transporting grey wolves back into the park did reduce the elk population, and by the 1990s the tree species had begun to recover, and streams were slowly coming back.

But changes imposed on Yellowstone as a system are not so easily reversed. The risk of decline is still present (as are the protests of cattle ranchers). The unwinding of the ecosystem took a generation, but the path to restoring the interactivity of the system is now projected to take many times that, and what is evolving from this process will not be a duplicate of what was there before the 1880s. If it does come back, it will be a different Yellowstone.[8] A recent study summarized this ecosystem

7 For a recent survey of the current view of small and large water cycles, see: Ellison, E. et al. (2017).

8 The same has been true for damage to mangrove forests, where attempts at restoration have proceeded slowly with only partial success. See the 70% failure rate

destruction-restoration process by cautioning against the assumption that there is a "quick fix" for the loss of predators from a habitat: "the ecosystem state that emerged following the loss of large carnivores from the food web resisted reversal after they were restored,"[9] another case of what Terrence Deacon called the Humpty Dumpty problem. You cannot unwind the ecosystem damage of Yellowstone any more than you can put a broken egg back together again.

Tonality is not a jar of marbles

In musical experience, the sound and behavior of individual tones can also interact chemically, as it were, creating a musical phenomenon that is greater than the sum of the tones in it. For example, if we were to play the pitch sequence F A G F A Bb C on an instrument, it would likely resonate in the western ear as it has for centuries, as a wholly distinct phenomenon, "F major." This is what would be called in current biology an example of *emergent order*. It is order which was not perceptible in the separate pitches F, G, A, Bb and C. The wholeness of this *tonality* of F major cannot be described simply as a collection of notes, but as an *emergent property* of the interaction among those notes in the ear of the hearer. As in everything musical, human senses, cognition and memory together form the hosting platform which allows the incoming sounds to come together meaningfully. The tonality of F major emerges in the hearer's ear from the interaction of the notes in the way that, as the systems theorist John Holland has put it, the wetness of water emerges from the interaction of hydrogen and oxygen in the water molecule.[10] There is nothing inherently wet about those elements any more than a series of tones is inherently coherent and meaningful.

To any experienced listener, the tonality or *key* of F major also demands more of the listener than the individual pitches do. In an ear sensitized through a combination of osmosis and pedagogy, the F major tonality becomes Deacon's mysterious rabbit. Its wholeness is so much more dynamic and encompassing than what we would hear if we were able to simply broadcast a random sequence of notes into the air. A random collection of organisms lacking any relationship to each other and dropped

---

of newly planted mangrove trees in the restoration of El Salvador's Jiquilisco Bay in Quarto, A. (2025).

9 "There are good reasons to restore apex predators to food webs but rapid ecosystem restoration cannot be assured among them." Hobbs, N. et al. (2024).

10 Holland, J. (2012), 45.

into a sunny meadow or dense marshland does not add up to an ecosystem. Molecules in a functioning cell, or cells in a functioning organism, or organisms in a functioning habitat are not marbles in a jar – no more than functioning poetry is a list of words or images. T.S. Eliot insisted on the emergence which occurs in a chemical reaction as the proper description of poetic process, a description which fits the chemistry of F major and its embodiment in myriad F major compositions, large and small.

Compared to the whole of global music history, the concept of tonality would stand for a small subset of the total number of ways that multipart music can be meaningful within a specific habitat. Kpelle gbona is another subset, alongside the subsets of Japanese, Korean and Indonesian orchestral ensembles and the so-called folk polyphonies of the Mediterranean basin and eastern Europe. In the whole domain of planetary polyphonies, F major is a subset of a subset, a very particular family of musical behavior which operates according to internal principles evolved over centuries which, like all ecosystems, foster certain variants and exclude others. Within the F major family, musical structures can be expected to move in ways that favor the particular gamut of seven pitches that, over many centuries of evolution, have come to constitute the tonality F major – F G A Bb C D E. To embody that tonality, the music must move through those seven pitches in a way that preserves the polarity between the so-called tonic (F) and dominant (C). Although other pitches outside of this set may play a part, in the end, they must maintain F as the gravitational center around which other notes rotate.

However, with slight changes of emphasis in the melody, these same seven notes can excite in the ear an entirely different tonality, D minor: D E F G A Bb C, with a very different gravitational polarity between a different tonic and dominant, D and A. F major and D minor employ the same pitches, but they move through those pitches in distinct ways, which is to say, they differ more in behavior than in substance. They exhibit different families of behavior in the way that a squirrel and a whale, both of them mammals, are possible only in the specific environments from which they emerged.

The relationship between a major tonality and its so-called *relative minor* – the tonality employing the same basic array of pitches but rooted a whole step and a half step below – has a long evolutionary history in European music. On the one hand, the pitches that the two tonalities share encourage the listener to hear them as if they were the same thing, only with a different emphasis, on either F or D. On the other hand, the reorientation of those pitches from F to D triggers a reorientation from the *major* to the so-called *minor*, a distinction in quality which has been identified by writers since the Renaissance as carrying with it an emotional meaning. The presence of an interval of a large (*major*) third above F (F–A) and a small (*minor*) third above D (D–F) are important identifiers of two families of music which came to be linked with conventions of affect. Even in discussions of music in popular and professional settings today, the formula "major = glad," and "minor = sad" continues to be assumed and explicitly affirmed.[11]

The widespread acceptance of the simple glad-sad/major-minor formula in no way proves it to be "natural" in the sense that it is established in nature like the inherent properties of oil and water. The major-minor musical formula is a social arrangement, a folkway. It is an adaptation established within a particular habitat, on the order of the finch's many beaks or the stickleback's variant body types, or the lichen's diverse alliances between specific algae and fungi. Over time, the links between human affect and intervals of pitch harden (as biologists would put it) through natural selection. By the end of the sixteenth century, composer-theorist Thomas Morley in England could declare (restating what was already said by Zarlino in Italy two generations earlier) that the interval of the minor third, along with the minor sixth and minor second, were most appropriate to "grief, affliction, sighs, [and] tears."[12]

The ways that this web of accumulated factors of sound and sense evolved over time is illustrated in Music Example 18.1. The excerpts from a famous aria by George Frederick Handel show how the major-minor distinction was absorbed into compositional practice and was linked with human affect. The emotional polarity of major and minor tonality commonly reinforced the very form of a piece, as in the so-called *da capo aria* which seventeenth and eighteenth-century composers frequently employed in their operas, cantatas and oratorios.

In Handel's *Giulio Cesare in Egitto* (1724), Cleopatra is thinking out loud about her love for Caesar. In the first half of the aria, which begins and ends in F major, she invites the "arrows of love" into her heart. The second half begins immediately in D minor, with no transition, and now she is begging her "woeful heart" to "take pity." The aria then returns (*da capo*, "to the head") to a repetition of the music and text of the first part in F major, creating an overall form of *aba*, and a tonality structure of major-minor-major. It is an example of how the da capo form over more than a

11 Rigg, M.G. (1964), 427–438.
12 Zarlino, G. et al. (1968), 180–181.

Music Example 18.1 The contrast between major and relative minor in a da capo aria, "V'adoro pupille," from Act I of *Giulio Cesare in Egitto* by G.F. Handel (1724): the opening and closing phrases of the A section in F major and the beginning of the B section in D minor.

century provided composers and singers an on-the-other-hand moment in which mixed emotions can be explored. Cleopatra feels both the joy and the sorrow in her love of Ceasar, and the closeness, yet distinctness, of the F major-D minor relationship is an established convention for embodying this emotional volatility. Add to this formula the fact that in Handel's day the audience would have expected the singer playing Cleopatra to create her own variations on the F major melody in the repeat. Through tonal ambiguity her unbridled love for Caesar in a major tonality is complicated by her pitiful reflections in minor. The performer's own variations showed that, in the course of the song, she had changed (as any living thing does) while the audience watched.

The close-relatedness of a major tonality and its relative minor in da capo arias like that in Music Example 18.1 is a connection so deeply established that we might even be tempted to think of it as a form of inherency, like the qualities of oil and water. It was a ready-made contrast which through many centuries of repetition in European music established webs of customary musical behavior and a deep archive of compositions and fleeting musical moments, each of them a variant of existing models. We now know that such prefabricated structures are found wherever we look in nature. All of this learned and archived behavior provided the pool for self-organization. For a musician of Handel's capacity, refined through years of teaching, imitation and trial and error, the limits of the major and minor families provided a platform where self-organizing behavior was possible, even inevitable. For the human ear saturated in western art music behavior, the major-minor contrast begets more of itself in the way that, in nature, slime mold begets more slime mold, and for the Kpelle people of Liberia, gbona begets more gbona. Da capo arias structured on the major-minor polarity beget more da capo arias like them, each one differing in ways that we recognize and find beautiful.

Ecosystem is a crude look

The term ecosystem surfaced early in this exploration of biology and music, but it is also where I have chosen to end up. In recent years, probably because of increased concern about threats to the natural environment, the word ecosystem has found its way as easily into conversation among economists and urban planners[13] as among conservationists, so its insertion into a musical discussion might not pose much of a challenge. In a review of a performance or in a biography of a musician, music-oriented readers might let a phrase like "musical ecosystem" or "musical ecology" pass without question by mentally applying scare quotes to the term, pushing these phrases into metaphor territory.[14]

Looking back at a crude look at the whole, his casual-sounding, free-hand-sketch-of-a-concept, Murray Gell-Mann applied it in 2013, not just as a descriptor of complex phenomena, but as a tool for integrating multi-levelled solutions to problems on a planetary scale, problems such as

13 For example, in economics, see: Daly, H.E. (2000).

14 For example, see two very different uses of ecosystem (without scare quotes) in public discussion of music, one applying the term to the classical western orchestra (Wolkowicz, T. (2020)) and the other applying it to global repertoires in a discussion of world music (Schippers, H. (2017)).

unequal wealth distribution, overpopulation and biodiversity loss. "For an integrative effort to succeed," Gell-Mann said, "some kind of simplification is naturally required. Certain things have to be treated in a cursory fashion, and others in more detail."[15] Engineers, environmental scientists and computer scientists around the room nodded in agreement and even those who dissented from his view stayed to argue, attesting to the role the concept was playing as a magnet for productive dialogue about complexity.

The crude look is another instance of scientific street-talk, an attempt to describe complex events on a continuum with the tendency of writers in the sciences and the arts over the centuries to rely on bio-musical metaphors. Hanslick's bud-to-blossom image of Beethoven, the Stravinsky-Craft sketches of historical European repertoires, and Lovelock and Doolittle's pithy one-liners about living systems are examples of how music and biology both grapple with complexities that overwhelm vocabulary.

15 See NTU Complexity conference.

19

Two premieres

Within the model of an in vivo study of music, the conventional concept of an artistic premiere is an easy fit. When the Juilliard or Guarneri quartets performed a piece they had played a thousand times they still looked forward to hearing it in the next performance as they had not heard it before. The next performance was, even for them, a premiere. We are told that Pablo Casals each morning would play one of the Bach unaccompanied cello suites. As the story goes, he would sometimes call out to his wife that he had just discovered something new. "Marta, come in here," he would say. "Listen. I never heard this before!" This is a story which would apply as easily to a Shona mbira composition or to a song in the repertoire of Joe Heaney or to a courtly song of medieval troubadour. In none of these cases would it be proper to say that as events they were simply texts "read in a linear fashion," as the geneticist Lynn Helena Caporale described the Neo-Darwinian view of genetic inheritance. Alfred Lord's study of Homeric epic offered an alternative to the conventional view of a poem by treating it, not only as literature, but as an event in real time. Each telling of a well-known tale before an audience was a premiere.

The Casals anecdote suggests that it was the musician's perception of a particular passage that was new, which might imply that somehow it was not "the music itself" which had changed. This is perhaps less clear in the case of a jazz performance where musicians are more likely to acknowledge the variations they participate in. But looking across the spectrum of the repertoires discussed in the previous chapters, all instances of mouvance leave us questioning the line we conventionally draw between the creation and perception of variation. The two quartets' disputes over "interpretation" of canonic works of the past resulted in subtle variations in tempo or in the pressure of bow on string. Though they were small, these were palpable alterations to a body of music in the same way that the sculptor's chisel removes another fine bit of marble from the surface of a statue. The fluttering of a single starling adds to the murmuration we see in the evening sky and the individual chemical bonds between algal and fungal micro-organisms and their habitat determine the precise shape of the lichen we see on the stone wall. Bow on string, chisel on stone, wing in the air, microbial interaction – all these variations end up registering on our senses and they all take place in time, now, at the moment we perceive them.

To keep music and nature simultaneously in view, it is useful to imagine a musical event as the split-second adaptation of a stem cell to the functional demands of its neighboring cells in an embryo. These biological events are happening in this precise way for the first time, though we also know that they have been happening for thousands, even millions of years. Casals' acute awareness of the minute details of the Sarabande from Bach's D Minor Suite was the result of lifetime of paying close attention to the many forms the piece had taken in his own hands. A bio-musical perspective urges us to pay attention to all such premieres. It takes a special degree of attention – maybe even a special kind of discipline – to resist the temptation to simplify our task by treating it as a text that is read in a linear fashion.

Premiere one

Consider the following musical event, a premiere of sorts, rooted in the culture of pre-twentieth-century Istanbul, though it takes place in Boston, Massachusetts in November of 1994. The event consists of a Turkish devotional song form known as *kaside*, executed by a 28 year-old Turkish singer, Hafız Ahmet Çalışır, while on a concert tour with a U.S.-based ensemble.[1] About the singer, the audience (most of them American) knows only his title, *hafız* (from the Turkish word for memory), which Turkish or Arabic-speaking listeners would be aware testifies to a rigorous early training in the chanting of the Holy Koran, the entire text of which he has committed to memory during his adolescent years. The event's only documentation is an audio recording on a tape cassette, from which the

1 Hafiz Ahmet's November 1994 New England tour was with the Boston-based *EurAsia Ensemble*, consisting of Beth Bahia Cohen (*yaylı tanbur* and violin), Robert Labaree (*çeng* and *bendir)*, Feridun Özgören *(tanbur)*, and Frederick Stubbs (*ney*). This was his first visit to the United States and one of his first trips out of Turkey. He was 28 years old at the time, extremely young in the world of the Koranic chanter *(hafız)*. Since then, he has travelled widely and is currently the director of the *Tasavvuf Korosu* (Sacred Choir) in Konya, Türkiye.

transcription in Music Example 19.1 was made. The text of the kaside is "Derd-i mendim ya Resûlallah," a poem, well-known in Turkish Sufi circles, by Ali Ulvi Kurucu (1920–2002), another hafiz, and the singer's mentor. It is a meditation on the Prophet Mohammed as the source of relief for human sorrow.

Derd-i mendim ya Resulallah devâ ol derdime	I have lived in sorrow, Oh Prophet, be my remedy
Dest-i gîr ol yâ Habîballah bu âsî mücrime	Take in hand, Oh Beloved of God, this disobedient offender
Sen şefâat kânı varken yalvarayım ben kime	When you, a mine of intercession, are here, to whom shall I plead?
Ben Resûl-i Kibriyânın bülbül-û nalânıyam	I am the lamenting nightingale of the great prophet
Mücrimem gerçi cemâl-i Mustafa hayranıyam	I am guilty, and yet I am a lover of the face of the Chosen One.[2]

Every line of the poem is packed with literary tropes recognizable to its Istanbul, if not its Boston, audience – the *habiballah* (beloved of God), *aşık* (lover), *pervane* (moth) and *bülbül* (nightingale) of Turkish Sufism, and the prosody and formality of its diction evokes an earlier age of Ottoman court poetry and religious discourse. An intensely emotional quality characterizes the performance, obvious to even the uninitiated, and anticipated by any in the audience familiar with the kaside form: the insistent repetition of a monotone phrase "Lâ ilâhe illallah" ("there is no god but Allah") by the ensemble, accompanied by drums, plucked and bowed strings, and end-blown flute; the high male voice, rich in overtones, moving in free rhythm in a mix of full-voice declamation and passionate bursts of rapid runs, trills and passage work. To an Istanbul audience, even the singer's posture would have been anticipated – eyes closed, body motionless, hands folded on the lap, and only an occasional movement of the head in support of a particular vocal accent or melisma.

To a Turkish audience, the similarity of the performance to two other forms – *tilavet* (the chanting of texts from the Holy Koran in Arabic) and *gazel*[3] (the musical rendering of Turkish texts about secular love) is unmistakable. The recorded document is also full of interruptions from

2 Translated by Feridun Özgören.

3 Throughout most of the Islamic world, tradition excludes the use of musical instruments of all kinds from renderings of the koranic text. It is considered improper

Music Example 19.1 A transcription of a performance of the kaside "Derdi mendim ya Resulallah," (first stanza of text only) by Hafız Ahmet Çalışır in Boston, Massachusetts, November 22, 1994. (Transcription by R. Labaree.)

coughing, the tuning of strings, and stage business muttered *sotto voce* in Turkish and English among the musicians. In this performance of "Derdi mendim ya Resulallah," notation has played no role in either creation or execution. On these pages, notation will play a role, but as an afterthought, an aid to reflection on the ephemeral event.

Premiere two

Now consider another musical event, also a premiere of sorts – that is, what could have been the first public performance of Franz Schubert's "Der Jüngling am Bache" at the home of the poet Franz Schober in Vienna in January of 1821, two years after it was written down. On this particular occasion when some 30 songs by Schubert were performed, the composer

---

to apply the term "music," a word of Greek origin, to tilavet. For a thorough discussion of the centuries-old polemic within Islam about music and Koran, see Nelson, K. (1986).

himself is at the piano and is also the main singer, his clear tenor reflecting his boyhood years as a soprano with the Vienna Hofkapelle.[4] All of this was a common enough occurrence at such informal musical evenings where Schubert's music was heard during his lifetime. The poem's four verses are set in a strophic repetition of the same tune, enhancing its effect as a kind of parlor folk-song, an effect which other songs that evening no doubt shared.[5] Just as familiar would have been the poet (Friedrich Schiller, 1759–1805) and the poem's literary tropes (a young peasant lad, a stream, an impossible love across social barriers). Even the poem itself could well have been familiar to the listeners that evening, since it was the work of a celebrated figure, considered at the time second only to Goethe, and since Schubert had most likely first encountered it in a collection published and first distributed in Vienna in 1803.[6]

An der Quelle saß der Knabe,	By the stream sat a youth
Blumen wand er sich zum Kranz,	weaving flowers into a wreath;
Und er sah sie fortgerissen,	he saw them carried off
Treiben in der Wellen Tanz.	and swept along on the dancing waves.
"Und so fliehen meine Tage	"Thus my days speed by,
Wie die Quelle rastlos hin!	relentlessly, like the stream!
Und so bleichet meine Jugend,	And my youth grows pale,
Wie die Kränze schnell verblühn!	as quickly as the wreaths wilt!
"Komm herab, du schöne Holde,	"Come down, gracious beauty,
Und verlaß dein stolzes Schloß!	and leave your proud castle!
Blumen, die der Lenz geboren,	Flowers, which the spring has borne,
Streu ich dir in deinen Schoß.	I shall strew in your lap.

4 There is no record of the actual premiere of this song. I have chosen this date, January 30, 1821 at the home of Franz Schober in Vienna, which is considered to be the first "Schubertiad." This is an occasion described by Josef Huber, who attended the event: "A lot of splendid songs of Schubert were sung and played by himself, which lasted until 10 o'clock in the evening. After that punch was drunk, offered by one of the party, and as it was very good and plentiful, the party, in a happy mood anyhow, became even merrier; so it was 3 o'clock in the morning before we parted." Deutsch, O.E. (1947), 162.

5 For the years 1826–1827 (the final productive years before his death in 1828) one study (Kuljurgis, C. (2010)) has estimated that Schubert composed some fifteen strophic songs.

6 Kramer, R. (1994), 26.

Horch, der Hain erschallt von Liedern,	Listen! The grove echoes with song
Und die Quelle rieselt klar	and the brook ripples limpidly.
Raum ist in der kleinsten Hütte	There is room in the tiniest cottage
Für ein glücklich liebend Paar."	for a happy, loving couple."[7]

The poem's familiar elements contribute a literary as well as a musical quality to the otherwise social occasion. But whether literary, musical or social, it is still as a live event – as a potentially unique performative moment rather than as a unique written artifact – that I am choosing to approach this song here. If a recording of this premiere could have been made in Schubert's day, it would have no doubt produced an impression of the event's immediacy comparable to the tape of Hafiz Ahmet's Boston appearance – with the clinking of punch glasses, the murmur of German conversation in the salon of the host, the sound of horse-drawn carriages on the street outside the window. But we cannot know with any certainty if the notation in Schubert's own hand to which the performer was very likely referring to create this event was the source for the printed version below, written in 1819 and published for the first time in 1889, more than two generations after Schubert's death and eventually listed in the Schubert works catalogue as D 638a.

Music Example 19.2 "Der Jungling am Bache" (1819), the opening line of the song with prelude and piano accompaniment.

The "Darwinian" otherness of kaside and Lied

Viewed as notations employing roughly the same writing system,[8] the differences between these Austrian and Turkish examples are what first

7 Stanzas 1 and 4 only. Translated by Richard Wigmore.

8 In the European usage of staff notation, the interval of a whole tone is understood to be divisible in two equal parts, reflected in the western system by the sharp and the flat. In the Turkish usage, a whole tone is understand to be divisible in nine equal parts *(koma).* In Turkish practice, pitch treatment is fluid and contextual, but the notation reflects only a few selected pitches of the nine: 1 koma (♭♯), 4 komas 1s (♭♯), 5 komas (♭♯).

impress us. Leaving clear language differences aside, the first song is sacred, monophonic and largely melismatic in its setting of text, while the second is secular, multipart and syllabic; the performances of the vocalists and instrumentalists in the recording of the first example quite obviously do not conform to the equal temperament of the piano, unlike the performance of the second; in the first, voice moves in free rhythm against a metrical ostinato of voices and instruments, while in the second, a piano accompanies voice in a strictly metrical setting; the first song dates from the late twentieth century, the second, from the early nineteenth century. The mere association with the cities of Istanbul and Vienna is enough to put a further gulf between our perceptions of the two: the first is a product of the Turkish-speaking Muslim Middle East, the second, of German-speaking Roman Catholic Europe. Even as musical texts, the two exemplify different types of phenomena: the first notation is a description of a particular performance, the second, a prescription for any number of performances. And finally, the impression of difference is confirmed in the fact that one song is a performer-controlled performance, an event to which the I-word would conventionally be attached. The other is an example of composer-controlled performance, a composition.

These examples seem to embody a mutual otherness which discourages any expectation that their respective identities would eventually intersect instructively. With apparently so little in common – neither history, nor culture, nor compositional structure, nor gross musical features – what is there to compare here, except perhaps the sizable accumulations of religious, social and historical baggage accompanying each one? In fact, what purpose is served by placing these two items side-by-side at all? For example, what would an intercultural and trans-historical perspective on Schubert's music add to what can be discovered through a comparison with Schubert's other songs or the songs of his contemporaries? Or, as another example, what if we chose to compare the treatment of text in each? Can we be sure that the kind of text-music analyses customary in a century's worth of Schubert scholarship and teaching would be valid in the world of Turkish *makam* where tonic-dominant harmony and a tradition of interpreting the relationship of words and music since the Renaissance cannot be presumed?[9] Where could a comparison of the lied and kaside

9 Regula Qureshi begins her study of text-music relationships in the Pakastani ghazal tradition in a way unusual for an ethnomusicologist, with extensive references to the German Lied, but these comparative elements serve as background, not as a focal point, and Qureshi does not scrutinize German and Urdu song practices side-by-side, measure for measure (Qureshi, R.B. (1990), 457–497).

take us, in other words, except to the inescapable conclusion that Turkish music and European music are – just as we suspected – essentially "other"? This suggests a comparison no more satisfying than the cartoonist's absurd revelation that, with all their differences, Davy Crockett and Franz Kafka shared a taste for raw dough.

In the early twentieth century, music historian Daniel Gregory Mason had felt no need to openly reference Herbert Spencer's simple-to-complex historical paradigm, rooted in, or at least resonant with, the evolutionary beliefs of the day. For Mason, Spencer's formula needed no elaboration: Scarlatti was to Beethoven as an oyster was to a horse, and there is little doubt that his comparison between Ahmet's monophonic kaside and Franz's polyphonic Lied would go any differently. Once again, the responsibility for this reductionist view is not entirely Darwin's or Spencer's. Mason's perspective arose out of a set of beliefs as biblical as they were scientific and are traceable to the eighteenth century and even earlier. These ideas were developed further through twentieth-century genetics, and it is only recently that the ideas have begun to fall from their position as a scientific consensus. In the stark contrasts of Spencer's Darwinism, the distinct habitats of the Austrian and Turkish songs stand for distances between them which are geographical and social, but also historical and developmental. As a result, there the two examples sit on the musical worktable: fundamentally other, free-standing and mutually unenlightening, as different as a mollusk is from a mammal.

This is, of course, a highly selective view of these separate songs which I am presenting here. Behind the long list of what makes the two songs different is what would make them intuitively candidates for comparison in the first place: the obvious fact that they are each solo songs with instrumental accompaniment. This would be a reasonable place to start a comparison of what they shared. For our purposes here, commonalities – as much the shared features of members of a family as the backbones shared by all vertebrates – will be as important as differences.[10]

Song-making in the Garden of Eden

If we ask, "where does a song come from?" we are asking no more than what workers in biology, and in natural history and natural philosophy before them, have asked about the origin of life forms. In music, the

[10] For a critical approach to the emphasis on differences in music studies when comparing two global repertoires see: Agawu, K. (2003).

metaphor of the Genesis story directs attention to Adam, while in biology, attention tends more toward the Garden of Eden. The goal here will be to try to keep both Adam and Eden in view. I am entering the final part of this bio-musical experiment expecting that the individual choices of Hafiz Ahmet and Franz Schubert are not exceptions to ecological laws driving Darwin's tangled bank, but examples of them. We can expect that Leibnitz's rule of compossibility will hold true for the kaside and the Lied. There will be no whales in trees or squirrels in the ocean.

True as this last statement may be, we should keep in mind that in its dogmatism it may be a bit misleading. Biologists understand the close-relatedness of the whale and the squirrel as fellow mammals, but they are also more aware than most non-biologists of the presence of startling originality in nature. The sudden invention of new body forms during the Cambrian explosion some 500 million years ago was one such unforeseeable development, perhaps made possible by a combination of one-time-only conditions and the inherent properties of organic matter. If the conditions supporting trees and oceans were to change during the coming centuries, then perhaps adaptations like terrestrial whales or aquatic squirrels could make a place for themselves in Darwin's tangled bank. Keeping in mind the potential for such surprises, we can still trust that Leibnitz's rule of compossibility will continue to stand firm alongside the rule borrowed from Pasteur, that music will continue to come from music.

A final word should be said about comparison itself, both as a procedure and as a state of mind. Each of these songs could be usefully studied in some detail separately. Keeping the frame of comparison narrow – for example, comparing twentieth-century Istanbul with twentieth-century Istanbul and nineteenth-century Vienna with nineteenth-century Vienna – has been the default frame for contextualizing a piece of music. But the frame of comparison is always set by the comparer, and it is has been increasingly the program of music scholarship and teaching to extend the frame forward and backward in time and in cultural context.[11] The fierce rehearsal debates of the Juilliard Quartet about a piece by Beethoven would commonly include references to pieces by Schubert (a contemporary), Mozart (a predecessor) and Brahms (a successor). Here, the frame I have chosen, following the model of the life sciences, is planetary – that is, intercultural, trans-historical

11 For example, in 1957 *The New Oxford History of Music*, Volume I, on *Ancient and Oriental Music*, edited by Egon Wellesz, offered chapters on prehistoric music alongside chapters on the music of Asia, India, ancient Greece, the Bible and medieval Islam. In their broadening of the historical frame in this way, the editors also associate the earliest ages of European music with current musical practices outside of Europe.

and ultimately bio-musical. If it makes sense to compare individual examples of the stickleback family in the Mediterranean and in the Pacific, and of lichens attached to wood, soil and rocks on every continent and climate, why would it also not make sense to compare individual songs as they arise in Istanbul and Vienna? Science has also demonstrated the usefulness of comparing the genomes of the present day stickleback with its many variants of more than thirty thousand years, recorded in its strings of genetic code,[12] and comparing lichens with the traces they have left in the Cenozoic fossil record between 66 million and 23 million years ago.[13] Why would it not also make sense to compare these songs with instrumental accompaniment by Hafız Ahmet and Franz Schubert, separated by a century and a half? And if it turns out that this family of song-making arises and sustains itself over generations through the same processes as a bed of kelp, why shouldn't we compare songs, not only with other songs, but with nature itself?

12 Recent studies have surveyed the history of the stickleback from the first signs of its earliest ancestors in the Pacific 37.3–78.7 thousand years ago. Fang B. et al. (2018).

13 Schmidt, A.R. et al. (2022).

20

Making song in Istanbul and Vienna: Following where variants lead

The goal here is simple: to enter a bio-musical exploration of two songs from different habitats by way of their variability – through the handful of variant versions of each of the songs which have been left to us by the accidents of history. This may not sound like much to rely on, but it is no less than what paleontologists like Stephen J. Gould have had to rely on to sketch out the evolution of thousands of species from a handful of fossils which survived the accidents of history over hundreds of millions of years. We started with the commonplace idea of variation shared by biologists and musicians, so variation is where we will stay.

A musicologist in a mangrove swamp

If we choose to approach a repertoire as an ecosystem, we can enter it at any point in its tangle of features and feel confident that our questions will ultimately lead everywhere within it. For the indigenous people in tropical and subtropical coastal communities who have for centuries built their livelihood around crustacean fishing in the waters of a mangrove forest, those fisheries are the entry point into the whole system. For them, the muddy water of an estuary surrounded by stands of one or more mangrove species is the breeding ground for colonies of larval shrimp and mature crabs (Figure 20.1). Therefore, over the centuries these human communities have gradually had to learn the importance of seeing, and actively preserving, the crustacean habitat in the estuaries as part of a complex multi-species whole, each one of the species offering an entry point into it. For an ornithologist, the primary interest is the birds in the mangrove ecosystem, so that will be his entry point into the system. And

Figure 20.1 A sketch of the mangrove forest ecosystem. Tsernjavski, N. (2001).

yet, following closely the seeds, fruit, leaves, insects, fish and shellfish that those birds feed on in the system, and the habitat supporting the nesting and reproductive cycles of the birds, the ornithologist's interest will soon intersect with those of the shrimp fishermen.[1] The same would be true of the biologists focusing on the microscopic bacteria and phytoplankton in the water and soil of the mangrove estuary which feed on the detritus from animal waste, from fallen leaves and from the decaying bodies of large and small organisms. Bacteria and plankton, in turn, are the food of worms and insects who are the food of crustaceans and fish who are the food of larger fish and birds. The excretions of them all, and their bodies when they die, add to the detritus in the water which settle to the floor of the stream. All these links in the food chain enrich the muddy water with nutrients which feed the whole system. Each species of tree, insect, bird, or microbe is engaged in generating more trees, insects, birds and microbes just like itself, and each species also enables adaptations which help to guarantee that it will continue. Their distinct genetic codes are recipes for the construction of bodies like their own. But genes are not computer programs or a military chain of command. There are no straight lines in the mangrove ecosystem.[2]

1 Cowan, C. (2025).
2 Saifullah, A.S.M. et al. (2015), 104–113.

As a human observer of these two songs, I am choosing to adopt the role of shrimp fisherman. I am choosing to begin my exploration of the kaside and Lied ecosystems narrowly, by scrutinizing individual variant versions of the two songs, just as I would explore the changing forms of each organism in the habitat. The variations in the lichen or stickleback in different regions of the planet are clues to lichen-making and stickleback-making in those regions. We will enter these systems narrowly, but the wholeness which created and sustained the multiple versions which those systems enabled is where we want to end up.

The musicologist Rob Wegman was also speaking in 2014 of repertoires as ecosystems when he described the highly trained fifteenth-century European choirboys for whom the invention of two and three-part counterpoint in performance was a "musical lingua franca spoken everywhere in Western Europe." It was a language so internalized and unconscious that boys "could handle it as effortlessly as you or I can conduct a conversation on the phone."[3] Wegman's deceptively effortless choirboy art is akin to the deceptively effortless art of Hafız Ahmet and Franz Schubert. Melodies, harmonies and words seem to emerge together like a rabbit from an empty hat, a result of what biologists would call self-organization.

Here is the challenge for music specialists: if a repertoire behaves like an ecosystem, the answer to the question "where does a kaside or a Lied come from?" cannot simply be, "the composer." If we find that ecosystem behavior is observable in musical repertoire, then we can perhaps begin to acknowledge the continuity between song-making, mangrove-making and shrimp-making. From there, we can begin to loosen our grip on the long-established belief that human music-making is an orphan of nature and cannot be anything like the actions of a marine arthropod or a tree.

Where does a kaside come from?

With Ahmet's kaside, as with a mangrove forest or Yellowstone Park, the ecosystem does not reveal itself immediately, but only by repeated passes starting from different points through the ecosystem as a whole. This first pass through the song will begin with just two of the most obvious of the variants left to us three decades after the concert series in 1994: first, another performance of the kaside in Middlebury, Vermont a week after the Boston performance; and second, a notated metrical setting of the same text created by Hafız Ahmet in the same time period.

3 Wegman, R.C. (2014), 15.

Looking further afield among all of Hafız Ahmet's U.S. performances in November of 1994, we find that the performance of his kaside "Derd-i mendim ya resulallah" in Boston was actually only one event in a cluster of events occurring during the three weeks of his visit, each of them built around similar (in most cases, identical) musical programs.[4] Listening to the recording of the Boston performance, and then to the performances in Providence, Rhode Island and in Middlebury, Vermont and in a lecture performance in Cambridge, Massachusetts we are left with a mixed impression of sameness and difference. For reasons of space and clarity on the printed page, I have had to choose just two of those kaside versions in Boston and Middlebury for close comparison on the synoptic score of those two performances in Music Example 20.1. Since the poem by Ali Ulvi Kurucu is the point of departure for both, the two versions are aligned by text. A preliminary inspection of Music Example 20.1 reveals what is in all four versions: each performance differs from the others in duration, in musical detail, in word treatment, and even in aspects of overall structure, while at the same time exhibiting strong family resemblances.

We cannot help noticing that although one performance of the text (Middlebury) is by far the longest of the two, both examples begin in the upper register and follow a similar pathway along a shared diatonic gamut of pitches which falls from an A in the upper register, working through the same series of platforms above A (A-B♭-C#-D) and below A (A-G-F#-E-D), subtracting and adding the same set of accidentals (F#-F, B♭-B𝄳, C#-C), and finally arriving to a common final on the A an octave below. The two versions seem to be moving through a similar pathway, but at different rates. The dramatic opening gesture of the Middlebury version on the syllable "Ah!" features a long-sustained note on E which develops into a long trill before descending to C#. The singer then begins the text on A and remains in the upper register exploring the upper tetrachord A-B♭-C#-D with more trills and shakes on sustained pitches as he passes through two lines of text. In the Boston performance, there are fewer moments of sustained pitch, more syllabic chant passages, and only brief references to the upper tetrachord.

We can also see that the Middlebury kaside's greater length is partly due to its frequent repetition of text, as when he takes the opening line of

4 The dates of these different performances were: a lecture demonstration in Cambridge, Massachusetts on 12 November; Providence, Rhode Island on 18 November; Boston on 22 November; Middlebury, Vermont on 28 November. In the interest of conserving space on these pages, only two of the four performances of the kaside in November 1994 are included in this discussion.

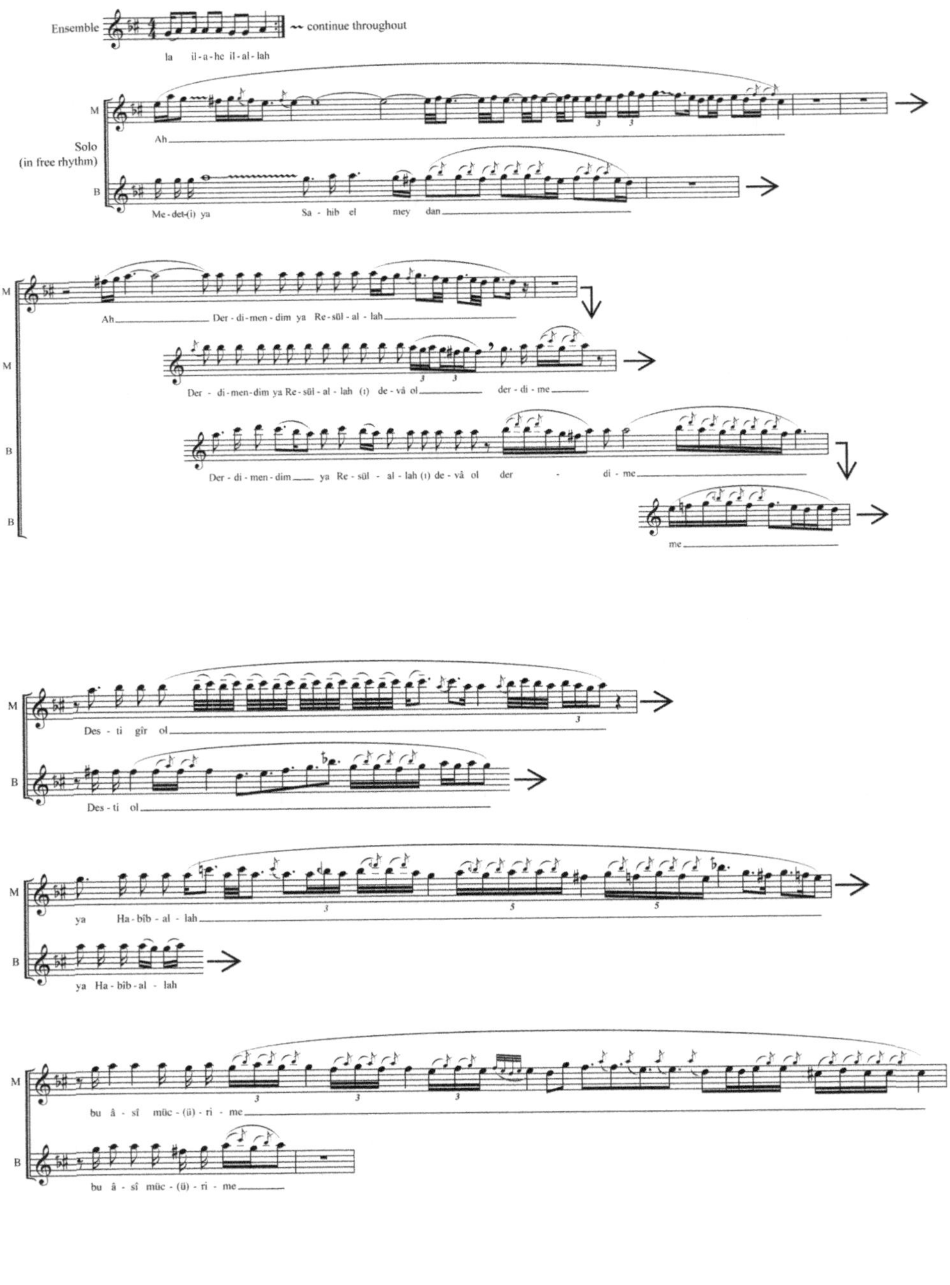

Music Example 20.1 The Middlebury (M) and Boston (B) kasides in synoptic score (first three lines only). For an explanation of the three flats and three sharps in this and other examples, see footnote 8 in Chapter 19. Transcribed by R. Labaree.

Music Example 20.1 (continued).

the poem, "Derdi-i mendim ya Resulallah" beginning on the upper A and immediately repeats it on Bb. These compositional choices resulting in the relative conciseness of the Boston performance turns out to have been an act of economy, a direct response to conditions in the song's environment: the singer had been suffering from a high fever on that evening and he had consciously adapted his kaside to a more limited duration and range.

But there is also the second variant to consider. At the beginning of his U.S. visit, before any of the four performances of the kaside had taken place, Hafız Ahmet revealed another development in the life of this song, a notated version of the same text in a sacred strophic form known as *ilahi,* set to a seven beat rhythmic cycle called *devri hindi,*[5] shown in Music Example 20.2. This seventeen-measure setting was actually composed by Hafız Ahmet himself only days before arriving in the United States for the performances above. Typical of Turkish practice, it was only written down before rehearsals began in Boston to facilitate sharing it with the ensemble.

Music Example 20.2 "Derd-i mendim ya resulallah," the *ilahi* versión in 7/8. Hafiz Ahmet Çalışır, November, 1994 (the circled items are edits added during rehearsal). These edits to the original ilahi notation in the composer's hand were added by a member of the ensemble, the yaylı tanbur player Beth Bahia Cohen, following the directions of Hafiz Ahmet.

Its place in the program in each venue was the same, about five minutes before the performance of the kaside on the same text, separated from the kaside by another ilahi in a duple meter. If we were persuaded of Spencer's simple-to-complex doctrine, the kaside's extended melismas could easily be taken as elaborations of the ilahi's concise syllabic seven

5 *devri hindi:* 3+2+2. The usul would be recited using the syllables for equal drum strokes: *Dum-tek-tek Dum tek.*

beat measures – as if the singer was "riffing" on a well-known tune, in keeping with the conventional view of what a jazz musician or other improviser would do. However, as will be apparent in the discussion ahead, the intricacies of the kaside are not so much a development of the simpler ilahi, as both ilahi and kaside are variations on a shared structure.

Comparing the recordings of all four of the performances of the ilahi on the tour, the singer's rendering of the three verses remained strikingly consistent – what Paul Zumthor, comparing different versions of medieval lyric poetry would have called a tendency toward zero mouvance. This notation of the ilahi became the standard of correct performance for the group, so that the composer's edits – all of them concerned with particular treatment of the pitches F, B and C circled in Music Example 20.2 – were written by the musicians into the score as a uniform standard, at the composer's insistence.

And yet, looking still further beyond a single notation or the occasion of one particular performance, the metrical ilahi setting of the "Derd-i mendim" text is not so simply represented as we might be led to believe by the prescriptive notation in Music Example 20.2. In the case of the ilahi, mouvance is more noticeable in each of the instrumental lines in the accompaniment which show multiple small but distinctive variations to the tune in each strophic repetition, and in each performance. Music Example 20.3 compares the vocal line as sung by Hafız Ahmet with the opening measure of just one verse by the ensemble. The distinct simultaneous variations on the melody by voice and instruments is considered a normal feature of Turkish monophonic performance, the locus classicus of *heterophony* in music.

Music Example 20.3 The ilahi "Derd-i mendim ya resulallah," opening measure, showing varied treatments by three of the accompanying instruments. Transcribed by R. Labaree.

Looking at the sample of instrumental variations in Music Example 20.3, it is as if the multiple variations of one line of "Going to mass last Sunday" which the Irish singer Joe Heaney had demonstrated to James

Cowdery were all being sung by multiple voices at the same time. The impulse for these variants, according to Heaney, is a love for the song expressed in special attention to its details – an impulse which would also describe the behavior of each musician here at this moment in the performance of Ahmet's ilahi. Further comparison of the kaside and ilahi will wait for a second pass through the kaside ecosystem ahead.

Where does a Lied come from?

When we turn from Istanbul to Vienna, we can expect that two regions with different transmission histories – one musically literate, but emphasizing orality, the other de-emphasizing orality and emphasizing music literacy – will provide us with different materials for comparison, even as we place variability at the center of attention in both environments. To answer the question, "Where does a Lied come from?" we cannot expect to follow the same threads as those in the kaside ecosystem by turning to the kind of real-time performing which I was personally able to observe and record as a participant in Hafiz Ahmet's tour thirty years ago. No cassette tape recorder was capturing Franz Schubert's voice and piano at his presumed premiere of "Der Jüngling am Bache" at the home of Franz Schober. What we do have in the Vienna habitat, and which is far less prominent in the kaside ecosystem, is a carefully-tended archive of Schubert's different visions of Schiller's poem notated between 1812 and 1827 which were published two generations after the composer's death as part of the *Schubert Gesamtausgabe*.[6]

As our exploration of both ecosystems deepens, it needs to be stressed that the comparison of written variants of musical objects is also possible in Istanbul, though these artifacts are less obvious to the outside observer of Istanbul musical life. A commitment to the preservation of fixed and authored works and the use of musical notation is common to both regions going back centuries, a fact about Ottoman musical tradition which is less well-known in western circles. The two regions approach notation in different ways and the sheer volume of notated evidence is greater on the European side of the line, but musical literacy is still a shared quality of both systems.[7]

6 Also referred to as *Die Alte Gesamtausgabe*, "the old edition," in light of the work still being done on *Der Neue Schubert-Ausgabe*, "the new Schubert edition," since 1965.

7 Ottoman notation collections are preserved in three indigenous Mediterranean notation systems: *Ebced* (employing Arabic letters), *Hamparsum* (based on Armenian church notation) and the notation system used in Greek Orthodox church music,

To take a first pass through the Lied ecosystem as we did through the kaside ecosystem, I offer here two relatively obvious and accessible examples: first, an anecdote about the commissioning of a song told by a personal acquaintance of Schubert, and second, a set of published variants in the Schubert archive.

In a well-known memoir of Schubert's friend Anna Fröhlich, we learn something of the circumstances surrounding the composition of the song known as "Ständchen" ("Serenade"). Fräulein Fröhlich tells how, in 1827, on the birthday of a friend, she commissioned a new poem from the poet Franz Grillparzer (1791–1872), which she then quickly presented to Schubert to set to music:

> I said to him, 'Look, Schubert, you must set this to music for me.' He: 'Well, let's have a look at it.' Leaning against the piano he read it through repeatedly, exclaiming over and over again 'But how beautiful it is – it's so beautiful!' He looked thus at the sheet of paper for a while and finally said 'There, it is finished now, I've got it already.' And only three days later he really did bring it to me, finished, set for mezzo-soprano (that is, for my sister Pepi) and four men's voices. At this I said to him 'No, Schubert, I can't use it like this, it's meant to be a tribute from Fräulein Gosmar's women friends only. You must write me the chorus for women's voices.'...And soon afterwards he brought it to me, set for Pepi's voice and women's chorus, as it is now.[8]

This is a story retold in western art music circles as a testimony to a general faith in the exceptional qualities of individual artistic genius. As it and other stories like it are commonly interpreted, Schubert possessed a rare creative virtuosity which enabled him to compose rapidly and profusely. Other anecdotes about the hours he was known to spend at the keyboard entertaining his friends with improvised waltzes and Ländler are often offered as a confirmation of his uncommon gifts. (Both of these qualities of effortless creativity he is known to have shared with his more celebrated contemporary, Ludwig van Beethoven.) Reading these stories from a bio-musical perspective, I interpret them as belonging in the tradition of a much older tale of faith, the one in which the Holy Spirit in the form of a white dove sat on the shoulder of Pope Gregory I in the sixth century CE and dictated to him the thousands of Gregorian chant melodies. I am choosing to retell the Schubert story here and to place it in a line

none of which employ the western staff. The application of western staff notation was introduced to the Ottoman region in the nineteenth century, but was not adopted as a universal standard until after the end of the Ottoman dynasty in 1923.

8 Flothuis, M. (1974), 61.

with the tale of Pope Gregory, not to confirm or deny the truth of either story. Instead, I believe both stories testify to a largely unacknowledged faith within European society in the presence of self-organization and emergence in the making of music. This interpretation is an alternative to the default assumption that the answers to any questions about how music is made will be found by a going deeper into the personality and personal history of an individual human creator.

Unlike the tale of the pope and the dove, Anna Fröhlich's story is personal, but at the same time, it also confirms the ecological nature of the Lied composition process for Schubert. The song depicted in the anecdote originates where many of Schubert's commissions began, in the ecology from which the song sprang, in this case, an occasion arising from the social connections of friendship. This particular commission was weighted by the cultural importance of German poetry to middle class Vienna at the time (Fröhlich had also commissioned the new poem from Grillparzer). It was further shaped by the frame of vocal and instrumental resources within which the composer was obliged to work. In this case, the frame happened to not be a soloist with piano accompaniment, an established orchestra or a church choir, but a group of women friends. The story also makes clear that, no matter how much the final product looked final on paper, it was only final for as long as the circumstances stood still. As soon as the circumstances changed – a range considered too low, certain vocal or instrumental forces made available, or simply a different day in the composer's life – the song changed, too. It is significant that the impulse for creating clear copies in Schubert's hand – copies which often contained variants of the original – is most often a publication deadline, an unusual moment when zero mouvance had to take precedence.

Although it is common in music study and teaching to ruminate at length over why and how the final choices recorded in Schubert's manuscripts improved or diminished a song,[9] I believe that from a bio-musical perspective, it is fair to say that each clean copy or scribbled draft represents an adequate solution to the problems posed by the text. Each new revision may possibly express Schubert's discontentment with what he had done before, but each of them is also a record of what he apparently considered to be sufficient at a given moment. In a comparable way, Hafiz Ahmet had little to say about each kaside performance – they were sufficient. However, I do remember him shaking his head while

9 For example, see: Flothuis, M. (1982) and Hallmark, R. (1982) in Badura-Skoda, E. and Branscom, P. (eds) (1982).

listening to the recording of the Boston performance, recalling the high fever he was experiencing at the time. If presented with a choice to publish as Schubert regularly was, this recording would probably not be the one which he would chose to represent his art. And yet, it was sufficient to the moment.

Four items in the catalogue

When we speak of repertoire, we commonly mean a group of pieces that share certain qualities, as when we refer to baroque repertoire or big band repertoire. Repertoire in this sense is a catalogue, an archive of pieces sharing certain family resemblances in form (sonata, minuet, aria) and/or function (song, dance, worship). To the audience at the performance which took place at Schober's home on that evening in 1821 (or, for that matter, at any one of its subsequent performances over the following centuries), there would have been no reason to think that the 27 measure song in D minor written down in the early spring of 1819 to which they were listening was not a unique, stand-alone work. The general expectation of the audience would have been that the performer, who was also the composer, if asked to, would have taken the notation in hand and repeated the song, creating not a new song but a new performance, perhaps one with small, but noticeable differences. These differences might even have been so striking as to lead some listeners to say, like piano afficionados comparing Cortot's Chopin with Rubenstein's, that they preferred one rendition over the other. These are the kinds of variants which we will never have access to, as we did in the case of the Boston kasides of 1994.

However, some in Schubert's inner circle, perhaps his old schoolmate Joseph von Spaun, who was also in attendance that night in Schober's living room, might have been aware that other versions of the song already existed: perhaps the 88 measure version in F major written down on September 24, 1812 when Schubert was 15 years old; or the 40 measure version in F minor completed on May 15, 1815. Here, I am presuming, for the sake of argument, that it was the D minor version most recently notated in 1819 which was premiered at the Schobers' in January of 1821. This is the third but not the final rendering which has survived. A fourth version seems to have been written down at some point later in 1819, in the form of a transposition to C minor of the third (D minor) version notated earlier that year, but including certain other changes as well. It is this fourth version which Schubert himself submitted for publication by August Pennauer in 1827 and, as the most recent version, it was given pride of place in Otto Erich Deutsch's *Das Neue Schubert Ausgabe* in 1951, with all the others relegated to a separate volume as "alternates." Taken together, these four versions – 1812, 1815 and two from 1819 – present a complicated identity for this modest song. All four versions in the *Schubert Ausgabe* sit on the page as separate and distinct works numbered D30, D192, D638a and D638b respectively, thus offering themselves to performers as mutually exclusive performance choices. In the synoptic array in Music Example 20.4 below, all four versions are arranged in chronological order and aligned by text. The piano preludes and accompaniment patterns of each are briefly shown, along with the opening line of each.

Music Example 20.4 "Der Jüngling am Bache," four published versions (opening measures).

As was the case with Hafiz Ahmet's kaside, certain differences among these four renderings of Schiller's poem jump off the page. The summary below highlights these differences in four areas: the presence of a strophic or a through-composed settings of the text; the key of the setting; the overall length of the setting; and the presence of a prelude and/or postlude in the setting.

	1812 (D30)	1815 (D192)	1819 (D638a)	1827 (D 638b)
Strophic/mixed strophic	Same music in v.1 and 3. Different music in v.2 and 4	Strophic: same music in v.1–4	Strophic: same music in v.1–4	strophic (except 3 m. in v. 3)
length	88 m	22 m	27 m.	33 m.
key	F major	F minor	D minor	C minor
prelude/ postlude	no prelude or postlude	4 m. prelude	4 m. prelude	4 m. prelude repeated as postlude
other characteristics		Opening line is minor key version of D30, the rest differs		Same as D638a except some changes in piano part

Leo Treitler's words about medieval song are also relevant here. The opening phrase of the 1815 version impresses us as a minor key version of the first phrase of the 1812 song – as Treitler put it, "the same work, with variations" – but beyond that first phrase, the two songs seem like altogether "different works," diverging significantly in melody, length and form. Among these four versions, the variability is so wide that it would be difficult to claim that any one of them is "a copy with errors."[10] Scanning the comparative score of the four versions in Music Example 20.4 and the summary above, why would we not begin by assuming that, at the end of each work session, the composer considered the form he had just given the song sufficient? On the other hand, why would there not also be equal reason to assume the opposite: that with the completion of each draft, the composer felt a sense of gnawing dissatisfaction, as if the statue were still only half-formed out of the marble?

Scholarly scrutiny of sources might support arguments showing how the composer leaned in one or the other of these two directions, but I am choosing here to leave these questions in favor of either point of view unresolved. Finding ambiguity in the transmission record of Jüngling, we will not attempt to remove the ambiguity in order to study it. Instead, I suggest that we entertain the possibility that the song just might be, in some sense, *essentially multiform*, a metastable entity that we as scholars and students of music will need to adapt to rather than dismiss. What is more, if metastability is a quality shared by both the Turkish and Austrian

[10] There is no suggestion in the scholarly literature that any of these "Jüngling am Bache" variants are the result of copyist error.

examples, then it can offer us another potential avenue for admitting them to the same universe and comparing them.

If we allowed ourselves to accept that Schubert's song is as multiform, indeterminate and dynamic as a folk song, a starling murmuration, a stickleback or a lichen, we would be acknowledging that polymorphousness is as characteristic of art song families as of natural ones. In both the natural and musical families, mouvance prevails, but so do patterns. In both families, the Joneses remain recognizable, even though they don't all have red hair.

In 1974, the musicologist Marius Flothuis examined the versions of Schubert's "Am Flusse" and offered two apparently contradictory, but perhaps complementary conclusions about the making of song. On the one hand, he observed that "[w]henever Schubert revised or altered a song, the later composition is superior to the earlier one,"[11] confirming that Schubert's genius conformed with the irresistible logic of Spencer's doctrine of all progress. When genius is applied to a song version, according to this view, the result is inevitably improved. On the other hand, on the very next page, Flothuis suggested the possibility that "Schubert could not express the two aspects of a poem in one song,"[12] which recognized the multiform potential of any setting of a text.

Both observations are professions of faith in the power of the creative executive which can be applied, if we wish, to the mythic accounts of Anna Frolich and Pope Gregory. The different assumptions at work in each statement lead to different conclusions about the same evidence of variability, but I believe a bio-musical approach encourages us to resist resolving their apparent contradictions and to let them express the fuller song potential contained in both stories. At the very least, simply the acknowledgement that variants in musical objects exist places music in a common ground with biology. As long as a bacterial colony is alive, a sample taken from it on one day cannot be expected to be identical to a sample taken on another day. Neither can two enactments of a song on the same text by one musician – whether on paper or in performance – be expected to be identical on another day. They would only be unchanging if they were no longer in motion in the throat and hands and mind of at least one human being.

11 Flothuis, M. (1974), 83
12 Flothuis, M. (1974), 84.

A "Darwinian" view of the Schubert catalogue

This grouping of four versions of a song on Schiller's poem "Der Jüngling am Bache" also displays a particular type of variability which is characteristic of many of Schubert's song variants. These are characteristics which jump off the page in the summary: the set of variations stake out different positions between the poles of strophic and through-composed settings. As he revisited compositions completed the day, month or year before, Schubert often refashioned them to suit some new vision he had of them. Sometimes that new vision entailed writing in musical variations for the same lines in different strophes where before he had simply called for repetitions of the same music. As in Flothuis's statement favoring the improvement that happens as the composer revises, this is another place where Herbert Spencer's simple-to-complex "Darwinian" evolutionary path can slip unnoticed into an analysis.

Looking at all four variants of this song, it is possible to imagine them all together representing a common pool of solutions to problems of text-setting and performance, solutions which were readily accessible to the composer sitting at the piano with song notation before him. Even when the composer was prepared to treat the song as essentially strophic, it is easy to imagine that he would have allowed himself micro-variations like those in any of the other published versions. The possibilities of variation were virtually infinite, like the modifications of the finch's beak which were different in each individual bird, but which could also be summarized as a limited set of solutions adapted to insects, fruit, seeds or nest-building. In both cases, a stored repertoire of proven solutions awaits the musician and the finch, deposited in the genome and in the memory, waiting for the moment-by-moment input of ecology to favor a solution for now.

In the case of "Jüngling," simple-to-complex is not the paradigm which is evident in the summary of the four versions: in the 1812 setting four lines of poetry were carried by six musical phrases with additional text repetition, compared to the four musical phrases with no text repetition in the 1815 setting, contributing to an overall reduction in three years from 88 measures of music to 27. While two of the four versions may be usefully referred to as strophic and through-composed (or some hybrid of the two), this does not rule out the equally valid perception that these categories simply mark general locations on a continuum of a song's compositional variability, the overall identity of which is, in the end, ambiguous because it is polymorphous. To use the biologist's term, the song's identity is metastable and subject to inflection through epigenetic influences.

And there is more to take in with this first pass through the Lied ecosystem. Knowing that it was Schubert at the piano, singing and playing many songs that evening at Schober's Liederabend, can we be certain what he played and sang once the manuscript pages were before him at the piano? Between the 1812 version of 88 measures and the 1819 version of 27 measures that we are assuming he played from, lie examples of shorter and longer versions, through-composed and strophic versions, major and minor key versions, and versions with and without preludes and postludes. All of them are potential solutions to the requirements of each verse of text. They are also responses to the environment around him at the moment. Were people listening closely or were there other distractions? Did he feel that a dramatic shift from major to minor or a more vigorous figure in the piano in this or that verse was called for to heighten the contrast in all the hubbub? (Compare the 1822 and 1815 versions, for example.) This we cannot know.

We do know that these are the kind of adaptations which a biologist would recognize and perhaps even consider epigenetic – standing for changes which come from outside or other than the genes – the way the fungal and algal symbionts of the lichen shift their chemistry in response to the temperature, moisture and mineral surface on which they land. When we are oriented to ecosystem, complexity and emergence are no surprise. A systemic point of view will also not permit us to exclude from consideration what else we know about Schubert: living up to his reputation as a pianist that evening, the composer may well have interspersed his songs with spontaneously created preludes, interludes and dances. Nothing in this whole array of potential variants and potential pianisms leaves us certain that he played what was eventually printed on the pages of the four published versions from which singers today select for their recitals.

David Sudnow's discoveries about the ways of the hand which precede conscious thought remind us that shifting the melodic line from F major to F minor for a verse or part of a verse would have required changing just one note in the voice and fingers. This is a shift so instinctive for a musician of Schubert's background as to be almost unconscious. As I hear these different versions, I am reminded that shifts of mode like this are a signature quality of Schubert's music, as in the last line of his "Gute nacht" from *Winterreise* when the words "an dich hab' ich gedacht" ("I have thought about you") are repeated, first in major and then, altering just one pitch, in minor. The artistic effect is of emotional ambiguity, an ambiguity which is not resolved – in the same way that the ambiguity of major and relative minor was unresolved in the previous century in Handel's "V'adoro pupille." Taken together, we can be certain that the four printed versions provide more than alternative compositions to choose from. They also offer a crude sketch of Schubert's compositional inclinations. The option to repeat or to alter the music for individual stanzas was before him at every moment, and it is irrelevant whether Schubert made a choice as a composer with pen in hand or as an I-person with hand on keyboard.

Putting this another way, we can say that Treitler's description of the problem of variants applies equally to concise and prolix settings of the same text, regardless of our opinion that they are mature or immature products of the composer's oeuvre. and regardless of our view that the variants are composed and intentional rather than, say, circumstantial and incidental. For our purposes here, all these versions are signs of variability and demonstrations of Schubert's polymorphous practice. This is what his transmission record shows. The question for all of us as observers, teachers and performers is: how does our awareness of this polymorphousness – of the metastability exhibited in the historical record of these songs – impact how we think about Schubert's music and even the creation of music in general? Wherever these questions lead us, they do not appear to lead to Spencer's default progress from simple to complex. These initial flashes of possibility in the four published variants suggest that another round of pulling on the threads of musical practice in the Lied ecosystem will be worth the effort.

21

Where does song come from?

In a second pass through a conjectural kaside ecosystem, I am choosing to enter by a different portal. This time, the starting point will not be the two versions of Hafız Ahmet's kaside performances in Boston and Middlebury. Instead, I will be following four specific constituents of the ecosystem which I observed in another of the performances of Hafiz Ahmet's American tour, his classroom-like lecture-demonstration at Harvard University.

As a teacher, Ahmet was also doing what Ernest Ferand exposed as a historical European practice, a practice which Ferand labelled improvisation. The pages of Ferand's book include numerous examples of published variants of the same piece, as we have already done here in a discussion of Schubert's song-making. However, many of his examples are pedagogical in origin. They are notations published by master composer-teachers in different countries and different centuries for the express purpose of showing students how they could vary the notation presented on the printed page. To these authorities, the metastability embodied in the multiple alternatives they presented to students was a necessary correction to the fixity of the printed page. Repertoire as a library of related musical objects merges on those pages with repertoire as an infinitely variable practice. Each one of Ferand's musical examples shows not just what a particular madrigal or sonata *is*, but *what it can also be*. It was not a stretch for Ferand's exemplary musicians to see beyond the definition of a repertoire as a library. In his Cambridge lecture-recital Ahmet was also trying to communicate the metastable nature of the kaside repertoire in particular and of the Turkish musical tradition in general.

To facilitate a deeper dive into the kasıde ecosystem, I have chosen to follow just four concepts native to Hafız Ahmet's makam practice. For the sake of brevity, I will be calling them simply *seyir, çeşni, aruz/usul,* and *tecvid.*

Makam and seyir: theory in real time and the making of transitions

As part of his general exposition of Turkish sacred vocal performance, Hafiz Ahmet chose to sing his newly-minted ilahi. Before singing it, he announced that he would first sing a short vocal improvisation in the mode of the ilahi "to give you a sense of Hicaz makam," as he put it. By referring to *makam,* and specifically to *Hicaz makam,* the primary makam of his ilahi and kaside, he was inviting his audience into a layer of the kaside ecosystem which he felt they needed to hear if they were to understand the practice which produced the two songs. As a general concept, makam belongs to a domain of global melodic practice that is usually referred to in western music as *mode,* a term commonly reserved for monophonic traditions such as the eight mode system of Gregorian chant and the various *raga* systems of India, but it is also applied to polyphonic traditions such as the Javanese *patet* system. In each of these contexts, a particular mode is understood as a family of distinctive melodic behavior.[1] The most basic representation of mode in all these instances is a *scale,* the set of usually five to seven pitches which helps distinguish it from other melodic families.

The scale of *Beyati makam* is also shown here since in his New England performances the singer had chosen to make the kaside itself serve as a bridge between portions of the program in those two makams. In the sequence of three short devotional songs *(ilahi)* – two in Hicaz in 7/8 and 4/4, followed by one in Beyati in 2/4. Hicaz and Beyati are considered to share a common final or "decision" note (*karar)* on A, and a common "strong" note *(güçlü)* on D, represented as **A** B♭ C# **D** E F# G A and **A** B𝄳 C **D** E F G A, but the modulation requires a modification of three accidentals, B♭, C# and F#. The kaside serves as a lengthy modulating bridge between these two makams:

ilahi 1	ilahi 2	KASIDE (free rhythm)	ilahi 3
Hicaz (7/8)	*Hicaz (4/4)*	*Hicaz* ⟶ *Beyati*	*Beyati (2/4)*
		[*Zikir (4/4)* begins under kaside,to ilahi 3...]	

The set consisted of three sacred songs (ilahi 1, 2, 3) and a kaside accompanied by repetitions of a zikir pattern in the ensemble. Beginning with ilahi 2, the tempo gradually increases until the conclusion of the kaside.

1 See the discussion of mode as a trans-cultural phenomenon in Powers, H.S. et al. (2001).

The making of smooth connections between distinct and independent entities or events was also important in the psalm-singing practice in medieval European monasteries, as in the case of an antiphon which needed to be gracefully elided with a chanted text. Such transitions are also not unique to music. The individual cells in the leaf of a plant are clearly bounded by a membrane of cellulose which separates them from surrounding cells and contains the organelles and the fluid medium *(cytoplasm)* of the cell. Without that outer membrane, the energy of the cell would be dissipated rather than contained. But the passing of nutrients and electrons across that membrane (and in bacteria, genetic information), is also necessary. Recognizing this, in recent biology the membrane has come to be viewed as a point of exchange, a region of transition between distinct entities. The communication across that barrier must be nimble and hypersensitive, a microscopic and stationary version of the communications among starlings in flight. An individual bird's position in the moving flock constantly shifts, but it must also maintain its close relationships with its flock mates if it is to benefit from the protection the flock provides. Transitions in music are also linkages, relationships which are in motion. The linkages and the transitional regions between the birds are no less real for the fact that they take place across a space in the sky between wingtips rather than a distance measured in cellulose and cytoplasm or in a patch of melody in the voice of a singer.

Even though Beyati makam also makes a brief appearance in his ilahi and even though his kaside comes to rest in Beyati, Hafiz Ahmet clearly viewed Hicaz as the principal makam of these two parts of the program. In order to give a sense of Hicaz makam, he sang for about a minute and a half, unaccompanied, a free-rhythm melody which was kaside-like in style, only with no text (Music Example 21.1). In place of a text, he sang solfège syllables, one syllable per note. Singing composed pieces from staff notation using the syllables *do, re, mi, fa, so, la, si, do* is a pedagogical method common in Europe since the middle ages. The practice was first introduced to the Ottoman region in the 1830s. It is a practice still in use in Turkey, where it was part of Ahmet's musical training (although not strictly a part of his koranic regimen). He first referred to what he was about to sing as a *taksim,* the word for a free-rhythm improvised instrumental melody, but he later called his unnotated vocalization a *seyir,* a word not limited to music meaning "course," "process" or "path."

In Turkish music theory for the past century, the term *seyir has* been applied in two ways: first, as a theoretical description of the general pathway

Music Example 21.1 Seyir in Hicaz makam, with solfege, showing two çeşni examples (1 and 2), as sung by Hafiz Ahmet Çalışır in his lecture in Cambridge, Massachussets, 1994. Transcribed by P. Labaree.

of the makam, including its overall direction,[2] and second, as a melody embodying the essential melodic features of a makam which teachers of makam would play or sing as models for their students.[3] In a similar way, the model antiphons were a tool for teaching Gregorian melody-making

2 Twentieth-century Turkish theory considers seyir to be of three general types: *ascending, descending,* and *ascending-descending (çikici, inici and çikici-inici).*

3 This seyir practice was also extended to writing. For example, in his book *Türk Musikisi* in 1922 the Turkish theorist Rauf Yekta Bey provided written seyirs for a selected set of Makams, all of them in triple meter.

in the European Middle Ages. Seyirs by renowned musicians, most of them recorded performances, are prized by students of Turkish makam and are copied, shared and memorized, and in recent years, posted on YouTube.[4] Even in a cursory comparison of Hafiz Ahmet's seyir with his kaside performances and with his ilahi notation in Chapter 19, we can see family resemblances shining through the many obvious differences. The most basic similarity among them is their shared usage of the Hicaz scale, but both seyirs begin around the D or A in the upper octave, then showing the tetrachrd above A and falling to the lower A, the final (karar) once again. In both we can see the upper accidentals, F#, B𝄳 and C# being changed, more often than not in the general pattern of raising the pitch on ascent and lowering the pitch on descent toward the final.[5]

On the recording of the Cambridge lecture recital, Hafız Ahmet can be heard singing the seyir in Music Example 21.1 deliberately. His eyes were closed as, when he sang his kaside in concert. He seemed to be paying the same careful attention to vocal production, melodic nuance and dynamics, even allowing occasional flashes of the same emotional intensity (not preservable in notation) as his kaside and ilahi performances. In the same way, a seyir performed on an instrument will, in general, be indistinguishable from any other performance of a taksim.[6] In his role as teacher in the lecture-recital setting, the singer seems to have felt the need to bring an uninitiated audience into the family of melodic behavior from which his new ilahi emerged by first presenting this concise statement of the ilahi's family identity, something he very likely would not have felt necessary before an Istanbul audience. As a musical experience, there is little in the seyir's execution which would lead us to conclude that it is simply an abstract auditory snapshot of a more complex set of

4 Another example available on YouTube is a seyir in Hicaz makam by Niyazi Sayin, a renowned player of the end-blown cane flute *(ney).* This was one of 17 seyirs for 17 different makams recorded in 1985 by Niyazi Sayin in Boston, Massachusetts during his first teaching residency: https://www.youtube.com/watch?v=dmUZJMyb3_0 (accessed 7/10/26). In the early 1990s Feridun Özgören recorded Sayin's student and colleague İhsan Özgen performing the same sequence of makam seyirs. These also are available now on YouTube: https://www.youtube.com/watch?v=DPUA--APTWQ (accessed 7/15/26).

5 Note that the general direction of the kaside melody in Chapter 19 begins around the upper A, not the lower A or D, which are commonly the entry point of the general "ascending" seyir of the Hicaz family. This is explainable by the fact that in kaside, the singer normally enters high in his range, a practice reflecting its natural habitat, the noisy environment of a Sufi ceremony, which requires greater projection.

6 See the examples in footnote 4 above.

phenomena. It resembles, to some extent, Robert Craft's little sketches of historic repertoires, a kind of crude look at the whole, but its spirited performance and its specific pedagogical purpose add layers of detail which the Craft sketches lack. It would be more proper to say that, rather than a crude look, his exemplar simply adds another variant to the melodic pool from which both the ilahi and kaside versions of "Derd-i mendim" emerged. The use of solfège syllables rather than text marks this melody as having a special pedagogical purpose, since the *do-re-mi* pitch names provide for the student an aural notation of what is being sung. Ahmet the teacher shows no sign of trying to establish a hierarchy of "Derd-i mendim" variants, as if to say "here is the song and now here are variations of it." Instead, he affirms the essential multiform nature of his musical language, as if to say "*this* (his solfeged seyir) is Hicaz makam, *this* is my newly-composed ilahi, and *this* is my kaside on the same text. If you don't hear how they are versions of each other, then you are missing something important." Each instance of Hicaz, each new variant, each newly created family member, becomes a potential heuristic of the family as a whole.[7]

Seyir is an exemplar and teaching tool, but Hafız Ahmet's Hicaz seyir would be different in its details on different occasions, as was true of each of his four kaside performances. The fact that each member of the instrumental ensemble acted out different versions of the ilahi melody as they accompanied the singer is further proof of the complexity and the limitless potential of the kaside ecosystem. We can say that Hafiz Ahmet's seyir is a theory in real time which demonstrates the connection between *the song that is* and *the song that could be (or could have been).* No less so for J.S. Bach's thirty (Goldberg) variations in 1741 and his son Carl's examples of how to vary music at a repeat sign in 1760. Here is the pedagogical message of the Turkish singer's theory: listeners and performers must be aware of variability because the fullest meaning of any one of the separate songs in this group of Hicaz melodies is communicated to the listener who hears the fullest range of the songs' metastability. *You can't just listen to a melody,* he is saying, *you must listen to the family.*

7 In recent generations, the transcription and memorization of recorded improvisations or taksims by respected early twentieth-century performers has become standard practice for Turkish students of makams. This is a practice which mirrors the study of solos by revered jazz musicians through transcription and imitation by students in American conservatories. In both the Turkish and American cases, individual performances are acting as exemplars.

Modules of pitch, taste and terroire

The term çeşni did not come up in conversation with Hafiz Ahmet. However, over the past century, it has been a widely-used term, a remnant of the period when Ottoman music was emerging from Persian influence during the Selcuk period (c. 1037–1194). When a Turkish cook samples a spoonful of broth, it is also a çeşni – a small taste which, to an experienced cook, will give an adequate sense of the whole dish. In the same way, an experienced sommelier can sniff a glass of wine and detect enough of the hundreds of different chemical compounds in it to identify their geography and climate, and even their soil ecology – the *terroir* of the grapes from which the wine was made. What is "nose" to a French sommelier is çeşni to a Turkish chef.

To a Turkish musician, çeşni is also a taste and a nose, a recurring melodic gesture which can, to the sensitized ear, communicate the presence of the melodic family of a particular makam. The short segments labelled 1 and 2 in Hafiz Ahmet's seyir performance, are examples of çeşni – melodic gestures, infinitely variable, which are also detectable by the ear and eye at several points in the course of the kaside and in the ilahi.

Çeşni can be taught to makam students as markers of the Hicaz family's typical behavior, but a çeşni is also a potential building block of composition. The chemical compounds in the wine that the sommelier's sensitive nose detects would also serve to simply identify a family of wines for a potential buyer. But those compounds would also be part of the necessary data for a vintner who wants to recreate the essential quality of a wine family when planting a new vineyard. As instances of typical physiological development, the wing-leg-arm-flipper structures are modules whose usefulness has been repeatedly proven over millions of years of evolution. In embryonic vertebrate development, in the practice of the medieval master singer and in the practice of a singer like Hafiz Ahmet, assemblages of features can be pulled from storage, so to speak, and adapted to context the way the alterations in the genome of a bacteria in the gut of a mosquito are enabled, in the time it takes to complete lateral gene transfer.

Once again, in the case of Wegman's fifteenth-century European choirboy, strict training in counterpoint and years of imitation of authoritative models left him with a storage room stocked with ready-made and accessible combinations of notes with proven adaptability. All of it was accessible to the singer in the time that neuropsychology tells us it takes to blink an eye.

Aruz: modules of sound and time

Fa'ilâtün is a word which stands for another kind of modularity in Turkish music. For anyone familiar with *aruz*, the syllable-counting prosodic system of Ottoman Turkish secular poetry, the non-lexical word *fa'ilâtün* will evoke a series of similar phrases with distinctive rhythmic qualities which themselves sound like poetry. For example, the sequence *fa'ilâtün fa'ilâtün fa'ilâtün fa'ilün* is the aruz pattern of long (—) and short (U) syllables of Ali Ulvi Kurucu's poem "Derd-i mendim ya Resulallah."

Der-di men-dim *ya Re-sul-al-lah de-vâ ol der-di-me*
Fa**-i - **lâ-tün** **fa** – i - **lâ-tün** **fa**- i-**lâ-tün** **fa**-i-**lün
— U — — / — U — — / — U — — / — U —

This is an example of the kind of stringent constraints which operate in the ecosystem of traditional Ottoman Turkish poetry, shaping the actions that take place within its sphere. If your thoughts cannot fit into an aruz pattern, in this environment they fall outside of what is considered poetry. If an organism can't feed or breed in hot, dry environments, then it will not last long here is this open grassland. *Fa'ilâtün* (— U — —) imprints on the ear a rhythmic pattern of four long and short syllables, *fa* standing for a long vowel, while in *fe'ilâtün* (U U — —) *fe* stands for a short vowel. The enormous variety of aruz patterns of different lengths[8] were refined through generations of poetic practice, first in Persian and then in Ottoman Turkish, a process which biologists would describe as hardening through natural selection.[9]

The aruz patterns have a corresponding structure in the Ottoman system of musical meters inherited from Arabic music, known in modern Turkish usage as *usul,* expressed in syllables representing drum strokes. *Devri hindi,* the seven beat usul of Hafiz Ahmet's "Derd-i mendim" ilahi, is represented as:

(7/8) *dum tek tek dum tek*

(1 1 1 2 2)

8 The total number of generally accepted *bahr or* aruz patterns is nineteen.

9 The language reforms beginning in 1928 converted the nation to the Roman alphabet and purged many of the Persian and Arabic words and constructions, creating modern Turkish. The poem "Derd-i mendim ya Resulallah" strikes the modern Turkish ear as antique because of the Ottomanism identifiable in its aruz form and its vocabulary.

In this scheme, *dum* (the deep sound at the center of a drum) receives greater stress than the lighter strokes at the drum's edge. Although both the aruz and usul patterns suggest how the poetry might be recited and the music played, they are more like compositional paradigms or even gestalts than instructions for performance. While some usuls would be an easier fit for a particular aruz pattern, in practice, there is no one-to-one fit of aruz to usul. In the kaside, the same Turkish text is stretched out over a much longer time timespan the seven beat usul of the ilahi, a time span not marked out by drum pulses, where each syllable of the pattern can be lengthened with extensive melismas and ornaments.

Tecvid and the symbiosis of sacred and secular in the ecosystem

In all that he chose to sing and say in the Harvard lecture-demonstration, Hafiz Ahmet demonstrated that the constraints of secular poetry and music were closely woven into his sacred music practice. As we follow the strand of the kaside ecosystem devoted to his training in koranic recitation, we quickly arrive at *tecvid* (Arabic: *tajwid*), the traditional set of rules of pronunciation, articulation and emphasis in Arabic which were the spine of his koranic recitation curriculum as a teenager. Within the global Muslim community, tecvid is cited as the primary method used since the seventh century CE to preserve the oral purity of the Arabic words first dictated to the Prophet Mohammed in a cave outside of Mecca by the angel Jibril. The original revelation was not written on tablets of stone or on parchment, but was transmitted in sound waves – the angel is said to have commanded the prophet to "recite!" (Note that the Prophet Mohammed is traditionally described as "unlettered.") Therefore, recitation is still considered the original and most authentic form of the Koran, which was only converted to written text in the century after The Prophet's death by transcribing the recitations of many reciters who had learned it from other reciters, and as it is believed, from the Prophet himself. This is the importance of the oral treatment of text which shaped the young hafız's intonation of Arabic, which he, like many before him, transferred over time to the intonation of Turkish.

In her study of Koranic recitation in Egypt, Kristina Nelson observed that the goal of tecvid is to preserve the oral nature of the Revelation by protecting the divine utterances which Mohammed was the only human being to ever hear with "a comprehensive set of regulations which govern many of the parameters of sound production such as duration of syllable, vocal timbre and pronunciation." In the end, she says, tecvid "preserves the nature of a revelation whose meaning is expressed as much by its sound as by its content."[10] The teachers of aspiring the young Ahmet Çalışır, an aspiring koranic reciter, holding closely to the rules of tecvid, corrected his pronunciation of each intoned, prolonged or elided syllable, each vowel and consonant as they were embedded in the singer's musical gestures. This is a discipline which was already merged in the singer's ear and voice with the melodic families of makam. Makam itself, rooted in secular musical practice, had over the centuries shaped the distinctive melodic style of koranic chant heard in Muslim habitats from Malaysia to Bosnia, even though koranic chant is never considered *song* and the word music is never applied to it.[11]

Kaside may be understood as an adaptation of *tilavet*, the Koranic chant practice he had learned as a teenager. If he had chosen the text of a secular love poem in Turkish, the result would have been considered an example of *gazel*. The tilavet-kaside-gazel relationship would have been obvious to any listener familiar with these forms, and the melismatic, free-rhythm melodic style and intense emotion that is common to them all would have been audible, coloring the words being delivered. A more familiar example of a sacred musical style applied to secular music is modern American gospel music. It is inseparable from Sunday morning worship in an African American church, but it is still brimming with the idioms of American popular music and jazz.

To initiated listeners to kaside it would also have been impossible to miss another quality, a scent of the ecstatic qualities of Islamic mysticism (*tasavvuf* or Sufism) in Hafız Ahmet's performances. The gentle rhythmic ostinato of voices on the phrase "la ilahe ilallah" ("there is no God by God") in the ensemble, supported by instruments, is taken directly from the so-called *zikir* ceremonies of many Sufi brotherhoods: the rhythmic repetition of that phrase in a duple meter is shaped by the regular expulsion of air on the syllable "ah," becoming a devotional breath discipline. In many Sufi brotherhoods, the breath discipline of the zikir ceremony would also be accompanied by body movement, either a rhythmic bowing or a rhythmic movement in a circle, arms across each other's shoulders.

(4/4) **LA** i-**LA**-he **IL**-al-**LAH**

10 Nelson, K. (1986), 14.

11 It has long been the custom in Islamic discourse to exclude tilavet from the category of "music," a term which is derived from Greek *(musiki)* and considered alien because it was foreign and because it also included secular music associated with alcohol and sex.

Ahmet's view of the orthodox rules for vocalizing the koranic text functioned alongside the secular orthodoxy of aruz which he had acquired from Persian-Ottoman court music and poetry. The membrane separating secular from sacred which he had inherited turns out to have been porous, with transmission in both directions carefully regulated. The self vs. non-self model, the border wall bristling with armed defenses against the invasion of unwanted influences, does not describe the environment of the young hafiz's training. Rather, they would have something closer to the immunology paradigm of recent decades, something more like Gilbert and Tauber's description of immunitas (Chapter 6), a carefully regulated negotiation in both directions across a membrane which produces a seamless merging of the two streams of text-setting. The independent, and in many ways strictly separated, streams of secular and sacred singing were the blue-green algae and fungi populations of Ahmet's practice. His own voice and his hands on a tanbur enacted at every moment the symbiotic union of the two streams, but their separateness was always present and acknowledged, the tension never wholly resolved.[12]

Following the radical secularizing and westernizing reforms of Mustafa Kemal Ataturk beginning in the 1920s, the sacred and secular orientations in Turkey remain today as unresolved in Turkish society as in Ahmet's own musical performance practice. The habitat of tilavet and kaside has evolved against an increasingly secular background of popular culture, but Hafiz Ahmet's kaside is still compossible within that habitat, in the way a squirrel originally from a forest habit is still compossible today in a city.

I have highlighted just a few of the ecological dynamics driving kaside. Makam, seyir, çeşni, aruz, usul and tecvid are ultimately adaptable, though in different ways, but they provide ready-made tools for both creation and comprehension. To a professional like Hafız Ahmet Çalışır, there is no room for mistakes of pitch, mode, diction or rhythm. The expectations of his profession as bearers of the miraculous message of The Revelation to all of humanity sharpen the real choices he makes as he puts together words and melody out loud before an audience, and even silently in his own head.

12 Many of the most celebrated popular Turkish singers of secular love songs in the first half of the twentieth century were trained, like Ahmet Çalışır, as a *hafiz*, showing that the not-entirely-separate cultural standard was already deeply rooted in the Ottoman period. When the recording industry first established itself in formerly Ottoman cities between the two world wars, names like Hafız Sami (1874–1943), Hafız Kemal (1884–1939) and Hafız Burhan (1897–1943) became national celebrities on 78 RPM discs of secular love songs, sung in a style saturated with tecvid and tasavvuf.

The apparent effortlessness and seemingly infinite flexibility of Hafiz Ahmet's fluent compositional craft in real time obscures the systemic interactivity that produced it. The singer, at the moment of performance is dwarfed by all that is passing through him from the ecosystem surrounding him at that moment. His song is not simply the gorgeous egret wading in the mangrove estuary, nor is it the delicate shrimp larva in the mud below. At the moment of composition, the singer's song is the egret and the shrimp, the mangrove and the mud and the microbes. The living thing to which he is giving voice is a momentary instance of the whole ecosystem, the song of the estuary in its entirety, a song which could not have happened without him, but it is not so much *his* as it is something which he has enabled.

Surveying the components of the Lied habitat

Reaching deeper into the Lied ecosystem for a second pass it will be hard to find the same kinds of tools which were articulated so clearly in the kaside ecosystem. The two cultures differ in the ways they enable their musicians to create music, but they also differ in the ways they enable those musicians to talk about and teach what they do. The less central role for musical notation in the Istanbul habitat and the greater emphasis there on transmission through live performance helps to explain why a practice of theorizing in real time like seyir floats to the surface of the ecosystem of kaside. This will be less likely in the notation-dominated ecosystem of Lied. Given Ahmet's background of explicit pedagogy in real-time composition, it is no wonder that he would turn to seyir and that he would feel the need to give his audience "an idea of Hicaz makam." Such moments are fewer and farther between in the record of song-making in Vienna. Anna Frölich's anecdote tells us nothing about the making of "Jüngling am Bache" but it does direct our attention to Schubert's personal way of working. True as the artistic exceptionalism in the story may be, Schubert's extraordinary gifts are not the last word to be spoken about his song-making. Nor do I believe that they are even the most important or interesting word. If we are viewing the Lied repertoire as an ecosystem, then the exceptional behavior of individuals is plain to see in the "Ständchen" story, but that exceptionalism does not help us move deeper into the question of where does a Lied come from?

Surveying the components of the Vienna habitat revealed in the historical record of Lied, I have selected a handful of phenomena I find there which are a mix of specific practices, widely shared tendencies and pedagogical tools of Schubert's day that are potentially as revealing as makam, aruz, usul and tecvid. I am attaching to them short captions, to be expanded in the discussion which follows. For convenience, I will simply refer to them as: *singers' embellishments, improvised song, counterpoint pedagogies* and *Volkston.*

Patterns of embellishment: Vogl's variants

Franz Schubert may have premiered his song on a text by Schiller at a house party on a particular night in 1821, but the four versions of "Jüngling" over seven years sketch a plausible spectrum of different renderings of it which encompass strophic and through-composed options, including additional variables of key, length and piano accompaniment. These variants were certainly embedded in the hands, voice and memory of Schubert himself and, as such, they might even have shown themselves as variants of the written score during the presumed premiere. Two centuries later, these variants sketch for us the range of the song's essential polymorphousness. Taking this view of the variants, the question of which is the final version ("Die Fassung auf letzter Hand" as it has been talked about in twentieth-century musicology) is of less interest than the fact that all variants emerged from the same ecosystem and therefore may represent clues to that ecosystem.

We do know with some certainty that the most famous singer of Schubert's songs during the composer's short lifetime was the baritone Johann Michael Vogl (1768–1840), thirty years Schubert's senior and retired from a distinguished career at the Vienna Opera. Between 1817 and the composer's death in 1828, Vogl became an important advocate for Schubert's music in recitals with the composer at the keyboard. Vogl's personal handwritten copies of the scores of the songs were meticulously notated to show how he executed them. In his written annotations and edits of Schubert's songs, the added ornaments, pauses and other alterations occupy the micro-variation end of the spectrum in contrast to the macro-variations of the four song versions published after the composer's death. Since Schubert so often accompanied Vogl himself, we can conclude that both ends of the variant spectrum from micro to macro were approved by the composer. We know that the composer at times took issue with some details of Vogl's execution, but that overall, the versions in Vogl's notebooks most likely represent how the two performed the songs. Vogl's variants of different verses of strophic settings is especially revealing. They reflect the kinds of alterations which a performer, including Schubert himself, might have applied to the different moods and meanings of each verse of text, even though the notated music gives the impression that each verse was sung identically (Music Example 21.2).

Music Example 21.2 Schubert, "Der Fischer" (D. 225). Michael Vogl's execution of a strophic line of music compared with Schubert's written notation. Dürr, W. (1979/1982), 131. The Vogl variants have been transposed to facilitate comparison.

In his study of Vogl's song books, Walter Durr found that Vogl's edits mostly fall within the standard practices of eighteenth-century opera performance which originated in Italy a century earlier. By the time of Vogl's collaborations with Schubert, these conventions were already passé, according to Durr. But passé or not, they were familiar performance idioms of the time, and as such, they were stereotypic melodic behavior which was expected, anticipated, and in their absence, perhaps even missed. In the ears of the audience, they mark the performances as typical. They correspond with the kinds of standardized gestures and melodic behaviors in Hafiz Ahmet's kaside versions. Çesni and seyir are factors in the creation of typical musical behavior which, like the conventions of musical behavior of eighteenth-century opera in Schubert's Vienna, mark them as *belonging*, as true to their habitat. Their naturalness within the habitat of Viennese salon and stage also suggests the presence of a musical system where self-organization can happen.

Pulling off improvised song outside of Vienna

I believe we can feel confident in assuming that any variant we identify in the record of "Jüngling" between 1812 and 1827 is an example of Schubert's native vocabulary and could therefore have appeared at any moment, in any performance before any audience, or in any composing session. The Vogl notebooks provide another layer of potential variations to think about. They testify to a strand of idiomatic music-making which is equivalent to the idiomatic layer of Hafiz Ahmet's craft shaped by the array of secular and sacred traditions of music and text-setting of his day. Whether the variations were expressed with voice and hand on keyboard or with hand and pen on paper is secondary to their role as emergent properties of the Lied ecosystem.

Looking outside of Vienna in northern Germany, we find that during Schubert's lifetime there was another song tradition which ran parallel to that which Viennese called simply *Lied* ("song") and which in the English-speaking world gradually acquired the label *art song*. This parallel practice of song-making was one with which Schubert seems to have had little direct involvement, but it provides a link to important strands in the web of Schubert's own musical habitat which we might otherwise miss. During Schubert's lifetime, in Berlin and Mannheim substantial reputations were being made by singer-pianists who would sit at a keyboard on a concert stage and create whole songs before a live audience with nothing before them but a poetic text, sometimes even a text submitted by someone in the audience. The most famous of these figures was the singer, pianist and conductor Carl Loewe (1796–1869), known after Schubert's death as "The Schubert of the north."[13]

As the pianist and musicologist Dana Gooley described it in 2018, Loewe's performances were more than displays of musical acrobatics. The fact that Schubert himself did not identify openly with this parallel tradition of improvised song does not rule out the intersection and continued co-development of the two traditions in the habitat of German music-making in the first half of the nineteenth century. Gooley acknowledges that "by any standard, the task of improvising both the melody and an accompaniment simultaneously, while singing and playing, on a poem just offered, is extremely difficult." But he then goes on to ask a question which by now we can recognize as ecological in its implication: "What skills and forms of training gave Loewe the ability to pull it off?"[14]

Gooley's account of improvisation in Germany up to the 1880s traces the movement, not through figures like Beethoven, whose improvising prowess is now a footnote to his monumental sonatas and symphonies, but through the many musicians who were famous as improvisors in their own day. In comparison with Beethoven's written legacy of works today, most of the preserved works of these improvisers are today considered less worthy of notice. Certainly, any outline of the components of Loewe's

13 Gooley, D.A. (2018), 123.
14 Gooley, D.A. (2018), 120.

song-making ecology would have to include the centuries-old tradition of the *Kapellmeister*. J.S. Bach is only the most famous of these traditional keyboardist-teacher-composer figures who were masters of the daily religious rituals of churches in eighteenth and early nineteenth-century German-speaking Europe. For Gooley, the iconic Kapellmeister in the Romantic period was the famous pianist and organist Georg Josef (Abbé) Vogler (1749–1814). Loewe's early training was with Daniel Gottlieb Türk (1750–1813), another famous Kapellmeister. Türk and Vogler and the many Kappelmeisters like them throughout Germany would also be counted among the most important purveyors of the keyboard-based practice of basso continuo and the pedagogy of partimento.

Loewe's published songs give a sense of what his performances before an audience might have sounded like. In his setting of Goethe's *Der Zauberlehrling* (The Sorcerer's Apprentice), for example, a variety of repeated patterns in the hands in sixteenth notes animate the entirely syllabic vocal line moving scale-wise, which is frequently doubled or shadowed at the interval of a third on the piano. Harmonies remain basically diatonic in C major, with a dramatic half step up to D♭ at the refrain. This is an example of Loewe's strophic treatment, so the melody and accompaniment of the first strophe could be expected to repeat on the subsequent six verses, most likely with variations.

But in his career, Loewe tended to favor a through-composed approach to story poems like this one, which he referred to as *Balladen*. They often featured distinctive treatments for the different speakers in the tale, as in Schubert's famous settings of two such story poems by Goethe, "Gretchen am Spinnrade" (1814) and "Der Erkönig" (1815). Early in his song-writing career, Schubert also seemed to favor through-composed settings, as in his first version of "Der Jungling am Bache" in 1812, when he was still a teenager. In the four settings of Schiller's poem we have seen how the 1812 model gave way to settings with more repeated music. And yet, his famous through-composed setting of Goethe's "Prometheus," full of dramatic declamation and supported by recitative-like chords, was written down in 1819, the same year as the last two hybrid strophic versions of Jüngling. Once again, these were options for the composer at the writing desk or at the piano the way they were also options for the public improvisor.

Wherever else Gooley's ecological question leads, from a bio-musical point of view, the question must ultimately land here, in the habits of hand and voice and imagination which have been honed for a century before Schubert in the teaching of keyboard counterpoint. Loewe was able to pull off the feat of improvised song for the same reason that Hafiz Ahmet was able to pull off the strenuous difficulties of kaside: because of years of rigorous hands-on education and imitation of masters. In both cases, the pedagogical goal was to enable the musician to harness the tools evolved within their respective traditions for what biologists call self-organization.

Volkston: the peasant village in a Viennese habitat

Volkston might be translated to modern English as "in folk style," or even (in American English) "folky," meaning that to a lover of poetry and music in Vienna in the 1820s, both the words and the music of a song "in Volkston" embodied the traits of simplicity and authentic emotion which were thought to be the qualities of rural life outside of cities. It was a life lived closer to nature. As an active shaping ingredient in the Lied ecosystem, Volkston is closely linked to the strophic form which Schubert relied on for some thirty percent of his song output. In another thirty percent of songs the strophic form is there in modified form, a strophic/through-composed hybrid like the 1815 and 1819 versions of "Jungling am Bache."[15] Concretely, the shaping power of Volkston was as much in the form, rhythm and rhyme scheme of the poetry, which then acted as a constraint on the music made for it. Strophic form, in fact, was thought to be the essence of Volkston in music since a single melody had to serve for all stanzas – the essence of simplicity.

Walter Frisch describes the strophic song as incapable of reflecting much subtlety from verse to verse, and therefore "introspection and deeper reading of a poem are not part of its world."[16] His judgement captures something of the flavor of the folk ideal among early nineteenth-century educated audiences of poetry and song in Vienna. Similarly, when Hafiz Ahmet in 1994, like many other Turkish musicians of his time, recreated a Turkish Sufi ceremony in the secular setting of a concert, he was choosing to evoke the flavor of a former time and a setting which was clearly not the modern concert hall. The allusion to tasavvuf evoked an evening ceremony in a room hidden from the disapproving gaze of mid-twentieth-century secular Turkey. The repeated phrase "la ilahi illallah" conjures up an idealized event in which men shuffle in a circle to the sound of rhythmic breathing, the pulse of the drums and the wail of the ney. In the formality and vocabulary of Ali Ulvi Kurucu's poem shaped by aruz,

15 For the years 1826–27 – the final productive years before Schubert's early death in 1828 – one study has estimated that he composed some fifteen strophic songs. Kuljurgis, C. (2010).

16 Frisch, W. (1986), 179.

a Turkish audience would also hear an evocation of a former time,[17] just as Schiller's poem also evokes an idealized former time: "Come down, gracious beauty / and leave your proud castle! / There is room in the tiniest cottage / for a happy, living couple." Short, tightly rhymed poetic lines in German would have triggered in a Viennese audience in Schubert's day images of an idealized village setting, one perhaps closer to the experience of their grandparents than their own.

But Frisch's judgement that the strophic song could not reflect a deeper reading of a poem does not take into account the inherent variability generated by the Lied's ecology. With the four published variants of "Jüngling" before us, created over seven years, and imagining the composer as singer at the keyboard, or as accompanist to a singer steeped in the vocal reflexes of Italian opera, variant readings would be likely to emerge "en parlant," as von Kleist put it. At every moment, the music would be subject to Sudnow's "ways of the hand" and the habits of keyboard variation honed by the partimento curriculum.

I believe that the concept of Volkston functions as one useful path into a presumptive Lied ecosystem in Schubert's Vienna, perhaps surprisingly, Volkston functions in the Lied ecosystem on a par with the Sufi elements in the kaside ecosystem. Volkston and Sufism are both words which stand for intangibles, customary actions and social assumption, but both words are also embodied in concrete practices which leave a tangible footprint behind for us to interpret in retrospect.

Paul Weiss's example of the growth rings of trees and patterns on snail shells provides one model of how the unseen creative processes of nature leave behind footprints, registering themselves on the solid materials of cellulose and calcium. In music, the footprints put a dent in the human senses as sound waves in the ear, as neural excitements, as subtle articulations of the vocal cords and as fine movement of the fingers. These imprints are also stored in memory which last long after the sensory footprints have dissipated. The memories themselves are active agents in the haside and Lied ecosystems, adding another ingredient in the creation of more kasides and more Lieder. Each replication of an ecosystem seeks to sustain itself, but each replication also "sings its own," as Herder said of folk song. Each Lied or kaside, each instance of "Jüngling am Bache"

17 In structuring one of the sets in his U.S program around a form of the Sufi zikir ceremony Hafiz Ahmet was following a concert convention of the time in Turkey. In the 1980s and 90s, allusions to Sufism were commonly woven into secular concerts of compositions from the Ottoman period billed as *Türk sanat müziği* (Turkish art music) or *Türk klasik müziği* (classical Turkish music).

or "Derd-i mendim," resembles those that came before, but not exactly, because each new kaside or Lied, like each organism in an ecosystem, must adapt in ways both subtle and grand to its fresh circumstances. When we add the endemic variability of the Lied ecosystem to the performance of "Jüngling am Bache" at Schober's Liederabend in 1821, I believe we are approaching a fuller understanding of the song recorded on the carefully edited pages of the *Schubert Ausgabe.*

The ten-piece rule

Kudsi Ergüner (b. 1952), the famous player of the end-blown cane flute, the ney, offered to his students of Turkish makam at the Istanbul Conservatory a maxim which he claimed to have been passed down to him from his grandfather and father, also revered neyzens: *if you want to learn how to play taksim in a particular makam* (or, translated to the world of the I-word, if you want to learn to improvise in that makam) *you must first learn ten compositions in that makam. You cannot play taksim without knowing repertoire.* The maxim also came with a corollary, Ergüner said: *you cannot play repertoire without knowing taksim.* Music comes from music – Pasteur's dictum originally applied to bacteria – is a conceptual tent with room in it for the Turkish habitat's ten-piece rule. I cannot think of a comparably succinct maxim for the Lied ecosystem, but *all music comes from music* is a maxim which allows for the huge repository of compositional exemplars, teaching methods and notational traditions of the German habitat that fostered the likes of Franz Peter Schubert. In the Vienna environment and in the Istanbul environment, creation is also re-creation and re-creation is also creative.

I have taken Ergüner's advice literally over the years and passed it on to students. His story has always made clear the continuity of improvisation and composition in Turkish music. This is a point which did not always occur to young musicians of western art music, for whom the line between I-people and not-I-people was hard and dark. For them, it was potentially a challenging story about an intimidatingly porous membrane separating their own creative abilities from the revered instrumental pieces printed on the pages of ur-text scores.

If we are correct in thinking that a musical repertoire is an ecosystem, then each instance of that ecosystem – each kaside or Lied – is an exemplar for the creation of more instances of itself.

Each gesture of a kaside in Hicaz makam registers in the ear and memory. The wholeness of the text-Hicaz combination sinks in slowly as Hafiz Ahmet deepens his experience of its many exemplars, which merge with his rigorous training in tecvid. At some point, the wholeness of kaside art passes through him and into the recording transcribed on these pages. The text-setting art of Lied emerges in the same way, from the song-making art embedded in the songs of others which a musician like Schubert has studied and imitated, and from the many generations of contrapuntal practice embedded in his voice and hands.

In the creative moments generated by these ecosystems (if we were to pause long enough to reflect on them at all) authorship can feel unclear. "Was that me just now? My hands? My voice?" When that happens, human actors are experiencing self-organization close to home, as a musical phenomenon. Like David Sudnow, when his teacher confessed that he was unable to replay what he himself had just played, musicians of all kinds are left to ponder the contingent nature of their own agency. "Ways of the hand" and the "ten-piece rule" may not be concepts which can completely assuage the flash of discomfort that comes with moments when we experience self-organization passing through us. These are moments when we recognize ourselves as musical actors who are, in the end, necessary participants and recipients, moments when we are all – musician and non-musician – lichens.

So much of what we hear and admire in music is its aliveness, the mouvance arising from self-organization. I have tended to stress the mouvance here, but only as a corrective, an alternate view to the objectifying habits of our speech about music and a reminder of our continuity with the mouvance of all living things. But this is not to deny the real human impulse toward replication and zero mouvance in musical thought. These are states which musicians have always aspired to. They give us moments when we feel we are autonomous agents engaged in creating contrivances which, in their appearance of fixity, are exceptions to nature. They reassure us that our musical experience is not ephemeral but enduring. This reflective and objective quality of creation allows us to, so to speak, walk around a piece of music and look at it from all sides as if it were a marble statue, editing it here or there with our chisel.

This rational and non-dynamic face of composition is, I believe, an important reason that music appears to stand so decisively apart from nature. The moments when our sense of music's enduring quality is strongest are when the conviction that "I did it" described by Gregory Bateson is also the strongest. The sense that music consists of a series of objects and that we are the exclusive creators of those objects, go hand-in-hand. Together, these experiences argue that all of our works, musical and non-musical, stand apart from nature, as different from nature as a new car coming off the assembly line is from a polar bear or a computer on our desk is from a mushroom. If these works stand apart from nature, surely we – humans, and all of human civilization – also stand apart from nature.

If we find it hard to wrap our minds around the apparent paradox of a self-organization that lives in and through individual human imaginations, we must forgive ourselves. Our path to such a realization has not been made easy for us by our civilization's tendency to favor homo sapiens' exclusive authority over everything, including nature and art. If I was given by God "dominion over every living thing that moves upon the earth," as is written in the Book of Genesis, how can I imagine surrendering even a small part of my dominion over the sounds just now emitting from my saxophone or the corrections I have just made to a musical passage I wrote only yesterday?

Epilogue

A Copernican moment for music?

Eppur si muove.
[And yet, it moves.]
(attributed to Galileo Galilei after his public renunciation of his findings that it is the earth that moves around the sun)

What remains to be said now can hardly be called a conclusion. I have thought that *Epilogue* or even *Afterword* is a better fit, especially when I remember that the threat of Amitav Ghosh's great derangement, which I introduced in the opening pages, is still hanging over the bio-musical landscape of this book, barely addressed. As I said at the beginning, this threat has been playing an active, if mostly invisible, role at every stage in the development of the thinking in all of these chapters. For those who care about both music and the planet, a paradoxical benefit of our worsening planetary crisis is that it has added an increased sense of urgency to these ruminations about the continuity of nature and music which might otherwise seem academic. The great derangement might just end up being the force which actually accomplishes what the sequestering habits of music scholarship and education have so far resisted.

Musicians, like all other human beings, wake every morning to eat, earn a living, raise a family and travel, but it is only in recent years that disruptive planet-wide natural events are acting out for us, as never before, the continuity of all things human with green, growing, replicating, inventing and decaying life forms. Arriving at this moment leaves a bitter taste in the mouth. We are not coming to this realization willingly. We are, at long last, being driven to it. As we see the natural order collapsing and realize belatedly the outsized role of human civilization in the collapse, it makes more sense to heed the signs of natural processes in the beautiful objects and events we value so highly. What if all those exquisite man-made moments were a demonstration of our connection with nature? What if we were able to hear each dance tune or symphony as testimony to a continuity with nature that we have for many centuries chosen to either minimize or ignore?

In 1927 the musicologist Heinrich Besseler described the beginning stages of what we can now see was a decentering of music thought, just as the twentieth-century expansion in global music-making through communication, travel and recording technology was getting underway in earnest. New categories of musical awareness and marketing – what by now have been normalized as "early music," "folk music," "popular music" and "world music" – were beginning to find their way into music studies. It was still several generations before the current period of awakening environmental awareness (using the publication of Rachel Carson's *Silent Spring* in 1962 as a benchmark), but I read Besseler's musical reflections as a hunch with bio-musical implications. "The plurality of equal historical periods," Besseler wrote, and "our current engagement with foreign musical cultures" has made it impossible to accept "the blissfully self-evident framework of a closed musical tradition." His word "closed" was directed at the culturally limited range of assumptions of music science as defined by his teacher Hugo Reimann (1849–1919), but he was hinting at what can happen in any field when what has been closed is opened.[1] In the case of music, that opening now extends to the extramusical domain of nature itself.

My goal here has been to lay out a sample of bio-musical explorations I have stumbled upon over decades which have been enough for me to at least initiate the kind of open conversations foreshadowed by Besseler. My explorations have been modelled by others in music, and I have tried to include some of them in this story. In the end, the goal has been to allow musicians to feel comfortable seeing their musical sensibilities reflected in nature and to then look into music for signs of natural processes. I see nothing out of the ordinary in this expansion of musical thought. It is nothing more than human curiosity, directed to where we are unaccustomed to applying it, in response to changing circumstances. Our curiosity is often stimulated at first by metaphors which occupy the borderland between what we think we know and what our hunches are whispering to us. *A repertoire is like an ecosystem* is a metaphor which can, with time, become a rule of thumb – *Let's say a repertoire is an ecosystem* – and from

1 Wegman, R.C. (1998), 435.

there, it can become a tool for study and even teaching: *if a repertoire is an ecosystem, then where does song come from?*

A Copernican moment?

The struggles of Galileo Galilei with the Roman Catholic Church are well known. They began in 1610 and culminated in 1633 when he was summoned to appear before the Inquisition in Rome. His crime was his support of the heliocentric theory of the solar system, first articulated by the mathematician and astronomer Nicolaus Copernicus in 1543, which was judged to be heretical, a contradiction of the theories of Aristotle and Ptolemy on which the Church had staked its interpretation of the creation. The Inquisition wanted Galileo to renounce what he had been teaching, that the earth rotated around the sun. Faced with torture, in 1638 Galileo publicly recanted his claims and submitted to house arrest and silence on this issue to the end of his life. As Bertolt Brecht portrayed him in his play *The Life of Galileo* in 1943, written in the shadow of both German National Socialism and Soviet Communism, Galileo submitted to these conditions for a simple reason. "I recanted because I was afraid of physical pain...They showed me the instruments."[2] But Galileo had seen with his own eyes through his improved telescope what Copernicus and his own mathematical calculations of planetary motions had first alerted him to, that the planet Venus could be seen moving through phases like the phases of the moon, suggesting that it was on an orbit, not around the earth, but around the sun. He saw that Jupiter, too had its own set of orbiting moons, and that the Milky Way was not a blurry cloud but a band of more stars, more than could be counted. How many of those stars were also suns, with planets circling them? The universe, it seems, was fuller of worlds than fit either in the Bible or in papal bulls, and the worlds all seemed to move in a perfect harmony. They were moving, as Darwin later put it about nature's tangled bank, according to laws. Galileo said no more of this publicly, but he knew that, no matter what he or the inquisitors or the pope said or did, the earth would still go around the sun.

One lesson to be learned from this story is that in scientific discovery, it is minds that change, not nature. The statement attributed (perhaps apocryphally) to Galileo, muttered in private following his public renunciation of his own observations – "and yet, it moves" – recommends to all of us within music, as well, that it is minds, not music, which changes. What

2 Brecht, B. *The Life of Galileo*, scene 14.

I am calling music in vivo is an approach to a conjectural biology-centered view in which human sound art rotates around nature. A bio-musical view does not change nature or musical practices, but it does attempt to understand them differently. I cannot be certain what the outcome of this re-orientation will be. But I can feel sure that, no matter what changes in musical thought come from this shift, works and artists will remain, like the planets, but our view of them will likely be altered. In that change, enormous adjustments, restructuring of thought and practice, are likely. In the process, there will surely be disruption, but I believe that the value of neither musical events nor artists will be diminished. I have no right to be certain about this, but our own time just might be a Copernican moment in music.

The artist disappears

As a musician and member of the liberal arts tribe, what appeals to me most of all about the work of many contemporary biologists is not the nerdy rigor of their empiricism, though I have tried to test the impact of that rigor on a musical conversation. Rather, what I admire in them is their refusal to forsake the material properties of the biome, even while they refuse to reduce its dynamism to the simple dualisms of normal vs. abnormal, fixed vs. open, primitive vs advanced, and ultimately, them (not human) vs. us. The normalcy of the musical object, as presumed in so much musical conversation and moralized in Reverend Walter's eighteenth-century indignation over the prodigal variability of congregational singing, sets us up to think of variability as a problem. It is as if variability is somehow a symptom of deviant influences[3] rather than a clue to the processes by which humans create the musical objects they value.

Compare this to the objects of biological study: each of the components of the biome have weight and dimension and obey physical laws, and those laws are amenable, to a substantial degree, to description through quantification, placing them squarely within the comfort zone of Crosby's tribe of nerds. Within this irreducibly materialist frame, the full range of invariability and variability fill a biologist's day. And yet, like someone viewing a distant hillside through different ends of a telescope, the biologist cannot afford to accept either the invariability or the variability as the truest of the two views.

3 Nicholas Cook (citing Peter Kivy) reminds us of a "composer worship" at work in the habits of art music scholarship, according to which (citing Richard Taruskin) "performers are *essentially* corrupters – deviants, in fact (Cook, N. (2001), 2).

I have already proposed that if we are willing to adopt a bio-musical point of view as we lead our musical lives, we must be prepared to accept the de-centering that comes with it, the de-centering not only of musical objects, as Besseler observed, but of human creators. The matrix of bio-musical metaphors in Denis Noble's *Music of Life* (Chapter 15) suggests exactly this: that music, like life, "is a process, not a thing."[4] Biologists of the past two centuries first had to relinquish the top-down certainty of an all-powerful Creator, and then the certainty of an all-controlling genetic program. In the place of these constructs implying unique command-and-control agency, biologists have steadily substituted a systemic complexity which is at once more explanatory as science and more mysterious as human experience.

If the story of how reptiles eventually led to humans was a compelling model for writing music history in 1850, will the story of a plastic, interactive and metastable genome, of epigenesis, symbiosis and compossibility, of biological relativity and the Gaia hypothesis, be compelling musical models in 2050? And how, on that future date, will we factor in the self-organization we find all around us, if we choose to look – in the tangled autopoiesis of cellular functioning, in the starling's murmuration, in David Sudnow's ways of the hand, in T.S. Eliot's artist as catalytic agent, in the apparently effortless creativity of Franz Schubert and Hafız Ahmet Çalışır, who seemed to create *en parlant*? Noble concludes that, looking from the biology side of his bio-musical metaphors, "the artist disappears." On the musical side, these words may sound to some of us like no metaphor at all, but rather a warning shot, the beginning of the end of all that we know. For my own part, I look forward to that moment as a long-awaited beginning.

It was never my intent to use science to "get beyond" metaphors. I only wanted to test their edges. With Denis Noble's music of life, Mark Bickhard's flame state, James Lovelock's Gaia, and all the roots, branches, blossoms, trees and families sprouting from the pages of more than a century of writing about music, let me add two more metaphors. I can promise that they will be the very last on these pages. The first of them is also a remembrance of my teacher of Turkish makam, the multi-instrumentalist İhsan Özgen, who died at the age of 79 in 2021. The metaphor takes up where William Butler Yeats' image of the water-walking, long-legged fly leaves off.

4 Noble, D. (2006), 143.

Up to our necks in the stream

İhsan Hoca[5] *is standing with me in a rapidly moving stream. Next to us in the water is a fly fisherman. I am in water up to my ankles, the fisherman is in up to his knees and the teacher is in up to his chin. None of us is new to the water. Each of us has sharpened his attention to what we see and feel there in the current through years of sensitization. We are all waiting, watching and listening, though people observing us from a distance will perhaps not understand our passion for the stream and especially the way we occasionally pull something from it.*

To the observer, it all seems effortless, so when something does emerge from the water it seems a bit like a rabbit pulled from an empty hat. But we in the stream know that it is also a matter of serious practice and discipline. We, musicians and fly fisherman, are lost in the internal activity, like Caesar in his tent and Michaelangelo on his scaffold. But to others, so much of the time in the water looks empty and insignificant, followed by brilliant moments when suddenly when we produce something shiny and alive.

Three men in a stream is an image that works, but, like the Russian dolls image recurring on these pages, it is limited. The teacher's stream (and the student's) is a fluid archive of musical memory and reflex which has in common with the fisherman's stream the fact that the stream is there whether or not artist, student or fisherman is attentive to it. We each experience the stream as something we can choose to visit or not. It will be there when we come back, constantly churning, always revealing to us its content in momentary flashes.

But comparing the two streams closely, we will have to admit that the teacher's stream (and therefore, to some limited degree, the student's) differ from the fisherman's in the simple fact that there is more of the teacher himself in his stream than there is fisherman in his. The fisherman's stream is full of a great deal of water and a few fish. İhsan Hoca's stream is an animated archive of memories and muscular reflexes, stocked with melodies, forms, gestures, models and teachings which are shot through with the contingencies of his own life, and which are unique to him. When he contemplates the archive, he gets to discover in fleeting moments that much of what emerges is what he himself has put there: whole compositions by other musicians which he has meticulously learned,

5 In Turkish, the word *hoca* [pronounced hodja] is an honorific, a title meaning teacher.

but also fragments, strategies and relationships he has struggled with and mastered, or which he has simply acquired by osmosis.

If it was pointed out that not all fishermen are the same – "well, the Turkish musician is improvising, not so, the musician playing Beethoven" – we have forgotten the experience of the string quartets playing Beethoven, the versions of Chopin Preludes, Herder's "every province sings its own," and the variants in the songs of Hafız Ahmet and Schubert. No two performances of the same piece in music are ever the same. "You played it just like the score" or "just like Horowitz," will ultimately never ring true in a musical ear. Differences are proof that we are in a unique moment, and difference – variability – is also a symptom of life on this planet.

And here is the second metaphor. It's a parable.

Jack Ponders the Sunset

Jack stood with a friend on a bluff looking west over the ocean, watching the sun go down. They gazed at the changing colors of the sky and the ocean as the sun sank below the horizon. The last sliver of the orange orb disappeared from sight and the sky grew dark and stars became visible. The friend chuckled when he remembered how, as a child, he wondered why he didn't hear a giant hissing sound when the boiling hot sun went into the ocean. They talked about how the sun was now rising in the east in another part of the world. They both joked about how some people still believed in a flat earth. How foolish, they thought.

The next day they stood again on the bluff at sundown and watched the sun sink below the horizon. "Let's try an experiment," said the friend. "Keep your eyes on the sun as it disappears and see that we are standing on a giant beachball that is rolling backwards away from the sun. As the ball rolls, the sun will slowly disappear. The sun isn't going down. We are falling away from it." They concentrated on the experiment.

But Jack's eyes kept telling him the sun was simply going down behind the broad line of blue water. So he closed his eyes. Instead of seeing, he was now imagining. He pictured himself standing on a giant beachball which was turning away from the horizon. For just a moment, he felt the ball moving beneath his feet. "Whoa, careful!" cried the friend as Jack found himself on the ground, his head spinning.

Bio-musical thinking asks us to imagine something that we don't find easy to see or hear – that the forces driving the endless inventions of ecosystems ever since single-celled life emerged on earth more than three billion years ago might also be driving the endless inventions of human sound art today. Hard as it may be to accept such a nature-centered view of music, we also know well that for millennia it was no easier to accept the view that the earth orbited the sun. After all, isn't it what our senses have told us, that the sun clearly moved across our own piece of sky from dawn to dark and that this is what humans have observed wherever they were standing on the earth's surface, at any time in recorded history? Who would question such an obvious truth?

To come to a different understanding eventually required a leap, a leap based on an intuition which was fortified by changing cultural assumptions, recently acquired knowledge, renewed scrutiny and improved tools. We are now offered the opportunity to renew our scrutiny of music by following where recently acquired biological knowledge leads, using the biologist's continually evolving assumptions and improved tools for understanding the making of songs and sonatas. I believe that the benefits of a biology-centered in vivo approach to music study and teaching will be persuasive in this century the way that the idea of the gradual evolution of species was persuasive to music thinkers 150 years ago.

When organisms and repertoires are placed, so to speak, side-by-side on the music stand, I am betting that the benefits will eventually go beyond music itself. To take even a small (musical) step toward rethinking humanity's image of itself as a special race standing apart from nature and above nature is to take a step closer to re-orienting human civilization itself within nature. If we were able to live our musical lives aware that all the beautiful objects we engage with were as much the transcriptions of natural processes as the annual rings of a tree or the swirling patterns of a snail shell, would "the climate crisis" still feel so removed from music-making?

I recall that, for my music colleagues and me, the mutual otherness of music and nature was a visceral conviction that left us feeling powerless to respond – as musicians – to the catastrophic effects of climate warming going on outside the practice room and classroom. Yes, we could always, like everyone else, carry a sign, write a Congressman or play a benefit concert – the obligation of all citizens in a republic. But our conviction that musical processes are inherently different from natural processes, that there is a hard line which divides matter and imagination, leaves us with little to hold onto other than the conviction of our specialness. Humanity has staked its existence on this self-image: that we are the proven masters of nature, uniquely able to do what nature cannot do. Musical art is a poster child of that self-image.

In my own experience, a bio-musical posture will feel strange at first. But it can also feel, if only for fleeting moments, true, the way Jack felt, for just a moment, the truth of the giant ball falling away from the horizon. Jack's truth was no more provable than Darwin's dear child, his law of natural selection. Nor did the hazy images in Galileo's telescope prove that the earth circled the sun. But these discoveries were enough to make it possible for Darwin and Galileo to make the final leap headlong into the heart of their respective metaphors. This nervy plunge into an educated hunch, and the inevitable disorientation that comes with it, may be all that is possible, here, on the third planet from the sun.

BIBLIOGRAPHY

Abzhanov, A. (2010). "Darwin's Galapagos finches in modern biology." *Philosophical Translations B Royal Society London*. 365(1543), 1001–1007.

Afflighemensis, J. (11th c.) *De Musica*. In Palisca, C. (ed.) (1978). *Hucbald, Guido, and John on Music: Three Medieval Treatises* (New Haven: Yale University Press).

Agawu, K. (2003). "Contesting Difference." Ch. 7 in *Representing African Music: Postcolonial Notes, Queries, Positions* (New York: Routledge).

Alexander, C. (1965). "A City is Not a Tree." *Architectural Forum* 122(1), April, 58–62.

Allanbrook, W.J. (1983). *Rhythmic Gesture in Mozart: Le nozze di Figaro & Don Giovanni* (Chicago: University of Chicago Press).

Allen, A.S., Leapaldt, T., Pedelty, M., and Titon, J.T. (2024). "The sound commons and applied ecomusicologies." In Dromey, C. (ed.). *The Routledge Companion to Applied Musicology* (New York: Routledge), 143–159.

Allen, A.S. and Titon, J.T. (eds) (2023). *Sounds, Ecologies, Musics* (New York: Oxford University Press).

Allen, W.D. (1962). *Philosophies of music history.* Ph.D dissertation, Boston University.

Aubrey, E. (1982). *A Study of the Origins, History and Notation of the Troubadour Chansonnier Paris, B.N. f.fr. 22543.* Ph.D. dissertation, University of Maryland.

Aunger, R. (2000). *Darwinizing Culture: the Status of Memetics as a Science* (Oxford: Oxford University Press).

Ayangil, R. (2008). "Western notation in Turkish music." *Journal of the Royal Asiatic Society* 18(4), 401–447.

Badura-Skoda, E. and Branscom, P. (eds) (1982). *Schubert Studies: Problems of Style and Chronology* (New York: Cambridge University Press).

Baedke, J., Fábregas-Tejeda, A. and Delgado, A.N. (2020). "The holobiont concept before Margulis." *Journal of Experimental Zoology—Part B, Molecular and Developmental Evolution*, 150.

Bailey, T. (ed. and trans.) (1978). *Commemoratio brevis de tonis et psalmis modulandis* (Ottawa: University of Ottawa Press).

Bar-On, Y.M., Phillips R. and Milo R. (2018). "The biomass distribution on Earth." *Proceedings of the Natlional Academy of Sciences USA* 115(25), 6506–6511.

Barry, P. (1937). "Notes on the Ways of Folk Singers with Folk Tunes." *Bulletin of the Folk-Song Society of the Northeast* 12, 2–6.

Barringer, D. (2006). "Raining on Evolution's Parade." *I.D. Magazine*, March/April.

Barry, P. (1937). "Notes on the Ways of Folk Singers with Folk Tunes." *Bulletin of the Folk-Song Society of the Northeast* 12, 2–6.

Bateson, G. (1973). "The Logical Categories of Learning and Communication." In *Steps to an Ecology of Mind* (New York: Ballantine).

Bayard, S. (1939). "Aspects of Melodic Kinship and Variation in British-American Folk Tunes." *Papers Read at the International Congress of Musicology*, 124.

— (1950). "Prolegomena to a Study of the Principal Melodic Families of British-American Folk Song." *Journal of American Folklore* 63.

Bearman, C.J. (2000). "Who Were the Folk? The Demography of Cecil Sharp's Somerset Folk Singers." *Historical Journal* 43, 751–757.

Beck, J. (1908). *Die Melodien der Troubadours, nach dem gesamten handschriftlichen Material zum erstenmal bearbeitet und herausgegeben, nebst einer Untersuchung über die Entwickelung der Notenschrift (bis um 1250) und das rhythmisch-metrische Prinzip der mittelalterlich-lyrischen Dichtungen, sowie mit Übertragung in moderne Noten der Melodien der Troubadours und Trouvères* (Strassburg: K.J. Trübner).

Bell, M.A. and Foster, S.A. (eds) (1994). *The evolutionary biology of the threespine stickleback* (Oxford: Oxford University Press).

Bellman, J.D. (2000). "Chopin and His Imitators: Noted Emulations of the 'True Style' of Performance." *Music Faculty Publications* 1 (Oakland: University of California Press).

Berger, A.M. (2005). *Models for composition in the fourteenth and fifteenth centuries* (Berkeley: University of California Press).

Bergeron, K. and Bohlman, P.V. (eds) (1992). *Disciplining Music: Musicology and Its Canons* (Chicago: University of Chicago Press).

Bergeron, K. (1998). *Decadent Enchantments: the revival of Gregorian chant at Solesmes* (Berkeley: University of California Press).

Berliner, P. (1978). *The Soul of Mbira* (Chicago: University of Chicago Press).

— (1994). *Thinking in Jazz: The Infinite Art of Improvisation* (Chicago: University of Chicago Press).

Bickhard, M.H. (2005). "Consciousness and reflective consciousness." *Philosophical Psychology* 18(2), 205–218.

Billingham, R.E., Brent L. and Medawar, P.B. (1953). "Actively acquired tolerance of foreign cells." *Nature* 172(4379), 603–606.

Blackmore, S. (2000). *The Meme Machine* (New York: Oxford University Press).

Blainville, C.H. de. (1776). "The Music of the Turks." In Martin, R. (trans.). Ch. 5 in *A General, Critical, and Philological History of Music* (Paris). *Turkish Music Quarterly*, Autumn-winter 1991–1992, 4.

Bonzi, L.C., Monroe, A.A., Lehmann, R. et al. (2021). "The time course of molecular acclimation to seawater in a euryhaline fish." *Scientific Reports* 11(18127).

Booth, A. and Doolittle, W.F. (2015). "Eukaryogenesis, how special really?" *Proceedings of the National Academy of Sciences of the United States of America* 112(33), 10278–10285.

Born, G. (2010). "For a relational musicology: Music and interdisciplinarity, beyond the practice turn." *Journal of the Royal Musical Association* 135(2), 205–243.

Bowen, J. (1999). "Finding the Music in Musicology: Performance History and Musical Works." In Cook, N. and Everist, M. (eds). *Rethinking Music* (Oxford: Oxford University Press), 424–451.

Boyes, G. (2010). *The Imagined Village: Culture, ideology and the English Folk Revival* (London: No Masters Cooperative Limited).

Brecht, B. (1939/1966). *The Life of Galileo*. Willett, J. and Mannheim, R. (eds). Translated by Willett, J. (London: Methuen Drama/Bloomsbury).

Broadwood, L. and Maitland, J.A.F. (eds) (1893). *English County Songs* (London: The Leadenhall Press).

Bronson, B. (1950). "Some Observations About Melodic Variation in British-American Folk Tunes." *Journal of the American Musicological Society* 3.

Brown, K.G. (2021). *Those great inspirers: The tactile compositional world of Igor Stravinsky.* Ph.D. dissertation, University of California.

Burkhardt, R.W. Jr. (1977). *The Spirit of System: Lamarck and Evolutionary Biology* (Cambridge: Harvard University Press).

Burnet, F.M. (1969). *Self and Not-self* (Melbourne: Melbourne University Press).

Burnham, S. (1992). "Method and Motivation in Hugo Riemann's History of Harmonic Theory." *Music Theory Spectrum* 14(1), Spring, 1–14.

Butler, S. (1879). *Evolution, old and new: Or, the theories of Buffon, Dr. Erasmus Darwin, and Lamarck, as compared with that of Mr. Charles Darwin* (London: D. Bogue). https://archive.org/details/evolutionoldnewo00butl/page/n5/mode/2up (accessed 8/12/26).

Campbell, M.B. (1988). *The Witness and the Other World: Exotic European Travel Writing, 400–1600* (Ithaca: Cornell University Press).

Canguilhem, P. (2011). "Singing upon the book according to Vicente Lusitano." *Early Music History* 30, 55–103.

Caporale, L.H. (1999). *Darwin in the genome: Molecular Strategies in Biological Evolution* (New York: McGraw-Hill).

— (ed.) (2006). *The Implicit Genome* (New York: Oxford University Press).

Capra, F. and Luisi, P.L. (2014). *The Systems View of Life: A Unifying Vision* (Cambridge: Cambridge University Press).

Capy, P. (1998). "Evolutionary biology: A plastic genome." *Nature* 396, 522–523.

Castéret, J.-J. (2013). "In quest of communion: The dialogic of multipart singing." In Ahmedaja, A. (ed.). *Local and global understandings of creativities: Multipart music making and the construction of ideas, contexts and contents* (Newcastle upon Tyne: Cambridge Scholars Publishing), 42–59.

The Catholic Encyclopedia. https://www.newadvent.org/cathen/02436a.html (accessed 9/17/24).

Cattin, G. (1984/2011). *Music of the Middle Ages* (Cambridge: Cambridge University Press).

Cavagna, A., Cimarelli, A., Giardina, I., Parisi, G., Santagati, R., Stefanini, F. and Viale, M. (2010). "Scale-free correlations in starling flocks." *Proceedings of the National Academy of Sciences* 107(26), 11865–11870.

Çetinkaya, Y. (2019). "'Senfonik' müziği doğru anlamak..." ["The correct understanding of 'symphonic' music..."] *Fikircoğrafyası.* https://fikircografyasi.com/makale/senfonik-muzigi-dogru-anlamak (accessed 7/9/26).

Chiu, L. and Gilbert, S.F. (2015). "The birth of the holobiont: Multi-species birthing through mutual scaffolding and niche construction." *Biosemiotics* 8, 191–210.

Clarke, B. (2020). *Gaian Systems: Lynn Margulis, Neocybernetics, and the End of the Anthropocene* (Minneapolis: University of Minnesota Press).

Clarke, B. and Gilbert, S.F. (2022). "Margulis, autopoiesis, and sympoiesis." In Jones, C.A., Bell, N. and Nimrod, S. (eds). *Symbionts: Contemporary artists and the biosphere* (Cambridge: MIT Press), 63–78.

Combe, P. Dom, O.S.B. (1969/2003). *The Restoration of Gregorian Chant: Solesmes and the Vatican Edition.* Marier, T.N. and Skinner, W. (trans.) (Solesmes: Abbaye de Solesmes).

Cook, N. (2001). "Between process and product: Music and/as performance." *Music Theory Online* 7(2), 1–31.

— (2013). *Beyond the score: Music as performance* (New York: Oxford University Press).

Covel, Dr. J. (1670–1679). *Early Voyages and Travels in the Levant: Extracts from the Diary of Dr. John Covel 1670–1679. Turkish Music Quarterly,* Spring–summer 1992, 19.

Cowan, C. (2025). "How to tell if mangrove restoration is working? Listen to the birds." *Mongabay.* https://news.mongabay.com/2025/02/how-to-tell-if-mangrove-restoration-is-working-listen-to-the-birds/ (accessed 7/10/26).

Cowdery, J.R. (1985). *Rethinking the Concept of Tune Family.* Ph.D. dissertation, Wesleyan University.

— (1990). *The melodic tradition of Ireland* (Kent: Kent State University Press).

Criado-Reyes, J., Bizzarri, B.M., García-Ruiz, J.M. et al. (2021). "The role of borosilicate glass in Miller–Urey experiment." *Scientific Reports* 11, 21009.

Crosby, A.W. (1994). *Germs, Seeds and Animals: Studies in Ecological History* (New York: Routlege).

Cross, I. (2006). "The origins of music: Some stipulations on theory." *Music Perception: An Interdisciplinary Journal* 24(1), 79–82.

Cumming, J.E. (2013). "Renaissance improvisation and musicology." *Music Theory Online* 19(2).

Czerny, C. (1829/1983). *A systematic introduction to improvisation on the pianoforte, opus 200* (Vienna: Diabelli & Cappi).

Dallaway, J. (1797). *Constantinople Ancient and Modern, with Excursions to the Shores and Islands of the Archipelago and to the Troad* (London). *Turkish Music Quarterly* fall/winter 1992–1993, 14.

Daly, H.E. (2000). *Ecological Economics and the Ecology of Economics: Essays in Criticism* (Cheltenham: Edward Elgar Publishing).

Darwin, C. (1839/1845/1905). *Journal and Remarks,* published in 1905 as *The Voyage of the "Beagle."* (London: Amalgamated Press).

— (1850/1998). *Journal of researches into the natural history and geology of the countries visited during the voyage of H.M.S. Beagle round the world, under the Command of Capt. Fitz Roy, R.N.* 2d ed. (London: John Murray)

— (1859/1993). *On the origin of species by means of natural selection, or, The preservation of favored races in the struggle for life* (New York: The Modern Library).

— (1871). *The descent of man, and selection in relation to sex* (London: John Murray). Vol. 1. John van Wyhe, J. (ed.) 2002–. *The Complete Work of Charles Darwin Online.*

Dawkins, R. (1976). *The Selfish Gene* (Oxford: Oxford University Press).

Dawson, J. (2021). *BIOC*2580: Introduction to Biochemistry,* Figure 22.13 (Open Library, Press Books). https://ecampusontario.pressbooks.pub/bioc2580/chapter/the-citric-acid-cycle/ (accessed 7/28/26).

De Souza, J. (2014). "Voice and instrument at the origins of music." *Current Musicology* 97, 21–36.

Deacon, T.W. (2012). *Incomplete Nature: How Mind Emerged from Matter* (New York: W.W. Norton).

— (2020). "Self-Organization is Not Enough: On Beyond Complex Systems." *SFU Faculty of Graduate Studies.* https://www.youtube.com/watch?v=WpYzLr4kAdM (accessed 7/13/26).

Deleuze, G. and Guatari, F. (1987). *A Thousand Plateaus: Capitalism and Schizophrenia* (Minneapolis: University of Minneapolis Press).

Della Valle, P. (1614). *Viaggi di Pietro Della Valle Il Pelgrino, 1614. Turkish Music Quarterly.* Martin, R. (trans.) (1989). Autumn. 11.

DeLoss, J. (2024). "Apex predators not a quick fix for restoring ecosystems, 20-year CSU study finds." *Colorada State University, Warner College of Natural Resources*, No. 7, April. https://warnercnr.source.colostate.edu/apex-predators-not-quick-fix-for-restoring-ecosystems/ (accessed 3/12/26).

Dempster, B. (2000). "Sympoietic and autopoietic systems: A new distinction for self-organizing systems." In Allen, J.K and Wilby, J. (eds). *Proceedings of the World Congress of the Systems Sciences and International Society for Systems Studies 2000* (Ashland: International Society for the Systems Sciences (ISSS)), 1–18.

Dennet, D. (1995). *Darwin's Dangerous Idea: Evolution and the Meanings of Life* (New York: Simon and Schuster).

DeNora, T. (2000). *Music in everyday life* (Cambridge: Cambridge University Press).

Deutsch, O.E. (1947). *The Schubert reader: a life of Franz Schubert in letters and documents* (New York: W.W. Norton).

Doolittle, W.F. (1999). "Phylogenetic Classification and the Universal Tree." *Science* 284, 2124–2128.

— (2015). "Eukaryogenesis, how special really?" *Proceedings of the National Academy of Sciences of the United States of America* 112(33), 10278–10285.

Doolittle, W.F. and Booth, A. (2017). "It's the song, not the singer: An exploration of holobiosis and evolutionary theory." *Biology & Philosophy* 32(1), 5–24.

Dupré, J. (2012). *Processes of Life: Essays in the Philosophy of Biology* (New York: Oxford University Press).

Dürr, W. (1979/1982). "Schubert's Songs and Their Poetry." In Badura-Skoda, E. and Branscom, P. (eds). *Schubert Studies: Problems of Style and Chronology* (New York: Cambridge University Press), 1–24.

Eigeldinger, J-J. (compiler) (1986). *Chopin: pianist and teacher as seen by his pupils*. Howat, R. (ed.); Shohet, N. with Osostowicz, K. and Howat, R. (trans.) (Cambridge: Cambridge University Press).

Eliot, T.S. (1919/1975). "Tradition and the Individual Talent." In Kermode, F. (ed.). *Selected Prose of T.S. Eliot* (New York: Harcourt), 36–45.

Ellison, E. et al. (2017). "Trees, forests and water: Cool insights for a hot world." *Global Environmental Change* 43, March, 51–61.

Engels, F. (1873–1886/1960). *Dialectics of Nature* [unfinished] (London: Wellred Books).

Ergur, A. and Doğrusöz, N. (2015). "Resistance and adoption towards written music at the crossroads of modernity: Gradual passage to notation in Turkish makam music." *International Review of the Aesthetics and Sociology of Music* 46(1), 145–174.

Etxeberria, A., Damiano, L. and Ruiz-Mirazo, K. (2021). "Editorial: 'Inter-identities' in Life, Mind and Society.'" *Frontiers of Psychology* 12.

Fang, B., Merila, J., Ribeiro, F., Alexandre, C.M. and Momigliano, P. (2018). "Worldwide phylogeny of three-spined sticklebacks." *Molecular Phylogenetics and Evolution* 127, 613–625.

Fedoroff, N. and Botstein, D. (eds) (1992). *The dynamic genome: Barbara McClintock's ideas in the century of genetics* (Cold Spring Harbor: Cold Spring Harbor Laboratory Press).

Felsenfeld, G. (2014). "A brief history of epigenetics." *Cold Spring Harbor Perspectives in Biology* 6(1), Article a018200.

Ferand, E.T. (1961). *Nine centuries of improvisation in Western music: An anthology with historical introduction* (Cologne: Arno Volk Verlag).

Ferreira Ruiz, M. and Umerez, J. (2021). "Interactionism, post-interactionism, and causal complexity: Lessons from the philosophy of causation." *Frontiers in Psychology* 12, Article 590533.

Fertig, W. (2025). "Ten Things You Might Not Know About Lichens, But Wish You Did." *Arizona State University Biodiversity Knowledge Integration Center*. https://biokic.asu.edu/collections/lichen-herbarium/ten-things-about-lichens (accessed 7/7/26).

Feuillet, R.-A. and Pécourt, G.L. (1704). *Recueil De Dances* (Paris: by the author). Library of Congress. Reproduced from *An American Ballroom Companion*. https://www.loc.gov/item/14002125/ (accessed 8/9/26).

Fitch, W.T. (2006). "On the biology and evolution of music." *Music Perception: An Interdisciplinary Journal* 24(1), 85–88.

Flothuis, M. (1974/1982). "Schubert revises Schubert." In *Schubert Studies: Problems of Style and Chronology* (New York: Cambridge University Press), 61–84.

Foulet, A. and Speer, M.B. (1979). *On Editing Old French Texts* (Lawrence: Edward C. Armstrong Monographs on Medieval Literature).

Frere, W.H. (ed.) (1901). *Antiphonale Sarisburiense: A Reproduction in Facsimile of a Manuscript of the 13th Century, with a Dissertation and Analytical Index.* Catholic Church Diocese of Salisbury (England), Plainsong and Mediaeval Music Society (Great Britain), Salisbury Cathedral, and Cambridge University Library Mm 2.9 (Farnborough, Hants.: Gregg Press Limited).

Frisch, W. (1986). "Schubert's Nähe des Geliebten: Transformations of the Volkston." In Frisch, W. (ed.). *Schubert: Critical and Analytical Studies* (Lincoln: University of Nebraska Press).

Gest, H. (2004). "The Discovery of Microorganisms by Robert Hooke and Antoni van Leeuwenhoek, fellows of the Royal Society." *Notes and Records of the Royal Society of London* 58(2), 187–201.

Ghosh, A. (2016). *The Great Derangement: Climate Change and the Unthinkable* (Chicago: University of Chicago Press).

Ghosh, I. (2021). "All the biomass of Earth, in one graphic." *Visual Capitalist*. https://www.visualcapitalist.com/all-the-biomass-of-earth-in-one-graphic/?utm_source=ActiveCampaign&utm_medium=email&utm_content=%5BRuminations+vol++43%5D+%22It+s+not+lack+of+rain+causing+bare+ground&utm_campaign=Savory+Monthly+Newsletter++-+JULY+2022 (accessed 2/3/26).

Gil, D. (2017). *Melancholic Modalities: Affect, Islam, and Turkish Classical Musicians* (New York: Oxford University Press).

Gilbert, S[arah]. (2018). "A Widening Chromaticism: Learning Perception in the Collective Craftwork of the Encounter." In Bull, K.A. and Gali, A. (ed.). *Material Perceptions. Documents of Contemporary Craft* no. 5 (Stuttgart: Arnoldsche Art Publications).

Gilbert, S.F. (2011). "Symbionts as an Epigenetic Source of Heritable Variation." In Gissis, S.B. and Jablonka, E. (eds). *Transformations of Lamarckism: From subtle fluids to molecular biology* (Cambridge: The MIT Press).

— (2023). "Sympoietic view of life: Gaia as a holobiont community." https://doi/org/10.20944/preprints202309.1072.v1 (not peer-reviewed).

Gilbert, S.F., Sapp, J. and Tauber, A.I. (2012). "A symbiotic view of life We have never been individuals." *The Quarterly Review of Biology* 87(4), 325–341.

Gilbert, S.F. and Bard, J. (2014). "Formalizing theories of development: A fugue on the orderliness of change." In Minelli, A. and Pradeau, T. (eds). *Towards a Theory of Development* (Oxford: Oxford University Press), 129–143.

Gilbert, S.F., Bosch, T. and Ledón-Rettig, C. (2015). "Eco-evo-devo: Developmental symbiosis and developmental plasticity as evolutionary agents." *Nature Review Genetics* 16, 611–622.

Gilbert, S.F. and Tauber, A.I. (2016). "Rethinking individuality: The dialectics of the holobiont." *Biological Philosophy* 31, 839–853.
Gilbert, S.F. and Barresi, M.J. (2016). *Developmental Biology*. Eleventh Edition (Sunderland: Sinauer Associates).
Gissis, S.B. and Jablonka, E. (eds) (2011). *Transformations of Lamarckism: From subtle fluids to molecular biology* (Cambridge: The MIT Press).
Glarean, H. (1547). *Dodecachordon I*, Cap. XV, Basle.
Goertzen, V.W. (1996). "By Way of Introduction: Preluding by 18th- and Early 19th-Century Pianists." *Journal of Musicology* 1(3), 299–337.
Goodenough, A.E., Little, N., Carpenter, W.S. and Hart, A.G. (2017). "Birds of a feather flock together: Insights into starling murmuration behaviour revealed using citizen science." *PLoS ONE* 12(6).
Gooley, D.A. (2018). *Fantasies of improvisation: Free playing in nineteenth-century music* (New York: Oxford University Press).
Gorbman, C. (1989). *Unheard Melodies: Narrative Film Music* (Bloomington: Indiana University Press).
Gould, S.J. (1989). *Wonderful Life: The Burgess Shale and the Nature of History* (New York: W.W. Norton).
— (2007). "More Things in Heaven and Earth." In Rose, S. (ed.). *The Richness of Life: The Essential Stephen Jay Gould* (New York, W.W. Norton), 444–466.
Grauer, V.A. (2006). "Echoes of our forgotten ancestors." *The World of Music* 48(2), 5–58.
— (2009). "Concept, style, and structure in the music of the African Pygmies and Bushmen: A study in cross-cultural analysis." *Ethnomusicology* 53(3), 396–424.
Gray, M.W. (2017). "Lynn Margulis and the endosymbiont hypothesis: 50 years later." *Molecular Biology of the Cell* 28(10), 1285–1287.
Graziano, A.B. and Johnson, J.K. (2006). "The influence of scientific research on nineteenth-century musical thought: the work of Richard Wallaschek." *International Review of the Aesthetics and Sociology of Music* 37(1), 17–32.
Green, D. (1983). "Oral and Written Literature in Medieval Germany." *The Spirit of the Court: International Courtly Literature Society.*
Greig, G. (1963). *Folk-Song in Buchan and Folk-Song of the Northeast* (Hatboro: Folklore Associates).
Griffith, F. (1928). "The Significance of Pneumococcal Types." *The Journal of hygiene* 27(2), 113–159.
Griffiths, P. (2006). "The fearless vampire conservator: Genetic determinism and the informational gene." In Neumann-Held, E.M. and Rehmann-Sutter, C. (eds). *Genes in development: Re-reading the molecular paradigm* (Durham: Duke University Press), 175–198.
Griffiths, P. and Stolz, K. (2013). "The reactive genome." In *Genetics and philosophy: An Introduction* (Cambridge: Cambridge University Press), 66–107.
Gündüz, G. (2023). "Entropy, energy, and instability in music." *Physica A: Statistical Mechanics and its Applications* 609, Article 128365.
Haeckel, E. (1866). *Generelle Morphologie der Orgaismen* (Berlin: G. Reimer).
Haines, J. (2001). "The footnote quarrels of the modal theory: A remarkable episode in the reception of medieval music." *Early Music History* 20, 87–120.
Hallmark, R. (1982). "Schubert's 'Auf dem Strom.'" In Badura-Skoda, E. and Branscom, P. (eds) (1982). *Schubert Studies: Problems of Style and Chronology* (New York: Cambridge University Press), 25–46.

Hancock, Z.B. and Bradburd, G.S. (2021). "Neo-Darwinism still haunts evolutionary theory: A modern perspective on Charlesworth, Lande, and Slatkin (1982)." *Evolution* 75(6), 1244–1255.

Hanslick, E. (1986). *On the Musically Beautiful*. Payzant, G. (ed. and trans.) (Indianapolis: Hackett).

Haraway, D.J. (2008). *When species meet* (Minneapolis: University of Minnesota Press).

— (2016). *Staying with the Trouble. Making Kin in the Chthulucene* (Durham: Duke University Press).

Hawkins, H.J., Cargill, R.I.M., Van Nuland, M.E., Hagen, S.C., Field, K.J., Sheldrake, M., Soudzilovskaia, N.A. and Kiers, E.T. (2023). "Mycorrhizal mycelium as a global carbon pool." *Current Biology* 33(11), PR560–R573.

Head, M. (1997). "Birdsong and the origins of music." *Journal of the Royal Musical Association* 122(1), 1–23.

Heckwolf, M.J., Meyer, B.S., Häsler, R., Höppner, M.P., Eizaguirre, C. and Reusch, T.B.H. (2020). "Two different epigenetic information channels in wild three-spined sticklebacks are involved in salinity adaptation." *Science Advances* 6(12), Article eaaz1138.

Henebry, R. Rev. (1928). *A Handbook of Irish Music* (Dublin: Cork University Press).

Hennion, A. (2015). *The passion for music: A sociology of mediation* (New York: Routledge).

Hiley, D. (1995). *Western Plainchant: A Handbook* (Oxford: Clarendon Press).

Hobbs, N., Thompson, D.B., Johnston, K.N., Marshall, E., Wolf, C. and Cooper, D.J. (2024). "Does Restoring Apex Predators to Food Webs Restore Ecosystems? Large Carnivores in Yellowstone As a Model System." *Ecological Monographs* 94(2), e1598.

Holland, J. (2012). *Signals and boundaries: Building blocks for complex adaptive systems* (Cambridge: The MIT Press).

Hollander, J. (1961). *The Untuning of the Sky: ideas of music in English poetry, 1500–1700* (Princeton: Princeton University Press).

Howard, D.R. (1971). "The World of Mandeville's Travels." *Yearbook of English Studies* I, 1–17.

Howell, F.C. (1968). *Early Man* (New York: Time-Life Books), 41–45.

Hubbs, N. (1991). "Schenker's organicism." *Theory and Practice* 16, 143–162.

Hudson, C.M., Lucek, K., Marques, D.A., Alexander, T.J, Moosmann, M., Spaak, P., Seehausen, O. and Matthews, B. (2021). "Threespine Stickleback in Lake Constance: The Ecology and Genomic Substrate of a Recent Invasion." *Froniers in Ecology and Evolution* 8, Article 611672.

Huglo, M. (2001). "Tonary." *Grove Music Online*. Root, D. (ed.). https://doi.org/10.1093/gmo/9781561592630.article.28104 (accessed 8/7/26).

Hunt, G. (2013). "In Murmurations, Starlings Have a Darwinian Dance Partner." *CornellLab All About Birds*. https://www.allaboutbirds.org/news/in-murmurations-starlings-have-a-darwinian-dance-partner/ (accessed 7/8/26)

Huxley, T.H. (1860). "The origin of Species." *Westminster Review* 17 (n.s.), 541–570.

ICTMD Study Group on Multipart Music. https://ictmusic.org/studygroup/multipart (accessed 8/7/26).

Immel, S.C. (2002). "Facsimile Edition." *Grove Music Online*. Root, D. (ed.). https://doi.org/10.1093/gmo/9781561592630.article.28104 (accessed 8/7/26).

Ingold, T. (2013). *Making: Anthropology, archaeology, art and architecture* (New York: Routledge).

Jackendoff, R. (2009). "Parallels and nonparallels between language and music." *Music Perception: An Interdisciplinary Journal* 26(3), 195–204.

James, J. (1995). *The Music of the Spheres: Music, Science, and the Natural Order of the Universe* (New York: Grove Press).

Jardine, B. (2016). "Charles Darwin's microscopes." *Explore Whipple Collections, Whipple Museum of the History of Science.* University of Cambridge. http://www.hps.cam.ac.uk/whipple/explore/microscopes/darwinsmicroscopes/ (accessed 7/9/26).

Jensen, E.C. (2013). "Overview of Live-Cell Imaging: Requirements and Methods Used." *The Anatomical Record* 296, 1–8.

Jones, D. and Sneath, P.H.A. (1970). "Genetic transfer and bacterial taxonomy." *Bacteriological Reviews* 34(1), 40–81.

Kalin, A. (2019). "Evans In England." *Downbeat.* https://downbeat.com/reviews/detail/evans-in-england (accessed 12/13/23).

Kallberg, J. (1996). *Chopin at the Boundaries: Sex, History and Musical Genre* (Cambridge: Harvard University Press).

Kauffman, S. (1995). *At Home in the Universe: The Search for Laws of Self-Organization and Complexity* (New York: Oxford University Press).

Khan, K. et al. (2004). "Significance of chimerism in hematopoietic stem cell transplantation: new variations on an old theme." Review article. *Bone Marrow Transplantation* (Nature Publishing Group), 1–12.

Kitcher, P. (2001). "Battling the undead: How (and how not) to resist genetic determinism." In Singh, R.S., Krimbas, C. B., Paul, D. B. and Beatty J. (eds). *Thinking about evolution: Historical, philosophical and political perspectives* 2 (New York: Cambridge University Press), 396–414.

Kivy, P. (1959). "Charles Darwin on music." *Journal of the American Musicological Society* 12(1), 42–48.

— (1995). *Authenticities: Philosophical Reflections on Musical Performance* (Ithaca: Cornell University Press).

Knappett, C. and Malafouris, L. (2008). *Material Agency* (Boston: Springer).

Koutsomichalis, M. (2017). "Ad-hoc aesthetics: Context-dependent composition strategies in music and sound art." *Organised Sound* 23(1), 12–19.

Kramer, R. (1994). *Distant Cycles: Schubert and the Conceiving of Song* (Chicago: University of Chicago Press).

Kuljurgis, C. (2010). *Schubert's 1826–1827 strophic songs: Innovations in musical coherence within a folk setting.* Ph.D dissertation, California State University.

Labaree, R. (1989). *"Finding" troubadour song: Melodic variability and melodic idiom in three monophonic traditions.* Ph.D dissertation, Wesleyan University. Publication No. 8919962.

— (2013). "Living with the I-word: Improvisation and its alternates." *Critical Studies in Improvisation* 9(2), 1–12.

— (2019). "European Travelers on Music of the East: Religion, Musical Works and the Otherness of the Ottomans." *The Musical Quarterly* 102(4), 402–439.

— (2023). "Biologists, musicians, and the ecology of variation." In Allen, A.S. and Titon, J.T. (eds). *Sounds, ecologies, musics* (New York: Oxford University Press), 201–220.

Lamarck, J.-B. (1809). *Philosophie zoologique, ou exposition des considérations relatives à l'histoire naturelle des animaux* (Paris: Dentu et l'Auteur).

Lamm, E. (2011). "The Metastable Genome: A Lamarckian Organ in a Darwinian World?" In Gissis, S.B. and Jablonka, E. (eds). *Transformations of Lamarckism: From subtle fluids to molecular biology* (Cambridge: The MIT Press), 345–356.

Landgraf, E. (2011). *Improvisation as Art: Conceptual Challenges, Historical Perspectives. New Directions in German Studies* (New York: Bloomsbury Publishing).

Lane, N. (2009). *Life Ascending: The Ten Great Inventions of Evolution* (New York and London: W.W. Norton), 13–14.

— (2010). "Why are cells powered by proton gradients?" *Nature Education* 3(9), Article 18.

— (2015). "The unseen world: Reflections on Leeuwenhoek (1677) 'Concerning little animals.'" *Philosophical Transactions of the Royal Society B: Biological Sciences* 370(1666), Article 20140344.

Lange, R. and Fugelli, K. (1965). "The osmotic adjustment in the euryhaline teleosts, the flounder, *Pleuronectes flesus L.* and the three-spined stickleback, *Gasterosteus aculeatus L.*" *Comparative Biochemistry and Physiology* 15(3), 283–292.

Latour, B. (2005). *Reassembling the social: An introduction to actor-network theory* (Oxford: Oxford University Press).

Laurentino, T.G., Moser, D., Roesti, M., Ammann, M., Frey, A., Ronco, F., Kueng, B. and Berner, D. (2020). "Genomic release-recapture experiment in the wild reveals within-generation polygenic selection in stickleback fish." *Nature Communication* 11, Article 1928.

Lenneberg, H. (ed. and trans.) (1958). "Johann Mattheson on Affect and Rhetoric in Music." *Journal of Music Theory* 2(1), April.

Lewes, G.H. (1874). *Problems of Life and Mind* II (London: Trübner).

The Liber Usualis with introduction and rubrics in English (1961). Edited by the Benedictines of Solesmes. (Tournai, Belgium. Desclée Company, Printers to the Holy See and The Sacred Congregation of Rites).

Lockwood, L. (2008). *Inside Beethoven's Quartets: History, Interpretation, Performance* (Cambridge: Harvard University Press).

Lohr, N.J. (1993). *Preluding on the harpsichord and pianoforte, circa 1770 to circa 1850.* Ph.D. dissertation, University of Iowa.

London, J. (2022). "A Bevy of Biases: How Music Theory's Methodological Problems Hinder Diversity, Equity, and Inclusion." *Music Theory Online* 28(1), March.

Lord, A.B. (1960). *Singer of Tales* (Cambridge: Harvard University Press).

Lovelock, J. and Epton, S. (1975). "The Quest for Gaia." *New Scientist* 65, 304–306.

Lovelock, J. (1988). *The Ages of Gaia: A Biography of Our Living Earth* (New York: W.W. Norton).

Lubenau, R. (1573–1589). "On the beautiful music of the Turks, and its instruments." Ch. 6 in Martin, R. (trans.) (1990). *Description of the Travels of the most Honorable, Renowned and most Wise Herren Reinholtt Lubenauen, Burger of the great capitol Königsberg in Prussia: How he began in August 5, 1573, and successfully finished in October of 1589. Turkish Music Quarterly*, Autumn, 7.

Luhmann, N. and Roberts, D. (1987). "The medium of art." *Thesis Eleven* 18–19(1), 101–113.

Lutz, P. (2023). "Goodbye, Mr. Bennett: A Tribute." *Downbeat.*

Machado, A, (1912). "Travelors, your footprints." Berg, M.G. and Maloney, D. (trans.) (2003). *Poetry Foundation.* https://www.poetryfoundation.org/poems/58815/traveler-your-footprints (accessed 7/12/26).

Maier, E. (2022). "A Fish at Home in All Waters." *Max Planck Research*, Issue 1.

Margulis [Sagan], L. (1967). "On the Origin of Mitosing Cells," *Journal of Theoretical Biology* 14(3), 255–274.

— (1984/2019). "Gaian Systems." *Lynn Margulis presents the Gaia hypothesis at NASA* [Video]. https://www.youtube.com/watch?v=c5m1pXX8NBM (accessed 7/11/26).

Margulis, L. and Fester, R. (1991). *Symbiosis as a Source of Evolutionary Innovation* (Cambridge: MIT Press).
Margulis, L. and Sagan, D. (1995). *What is Life?* (Berkeley: University of California Press).
Margulis, L. (2000). *Symbiotic planet: a new look at evolution* (New York: Basic Books).
— (2010). "Why are cells powered by proton gradients?" *Nature Education* 3(9), 18.
Mattheson, J. (1737). *Kern melodischer Wissenschaft* (Hamburg: Christian Herald).
— (1739/1958). "Der vollkommene Capellmeister" ["The Perfect Chapelmaster"]. In Lenneberg, H. (ed. and trans.) (1958). "Johann Mattheson on Affect and Rhetoric in Music." *Journal of Music Theory* 2(1), April.
Maturana, H. and Varela, F. (1972). *Autopoiesis and Cognition: The Realization of the Living* (Boston: D. Reidel) (later editions in 1980 and 1991).
Mayr, E. and Provine, W.B. (eds) (1980). *The Evolutionary Synthesis: Perspectives on the Unification of Biology.* First Edition (Cambridge: Harvard University Press).
Mayr, E. (1988). *Toward a new philosophy of biology: observations of an evolutionist* (Cambridge: Belknap Press of Harvard University Press).
— (1991). *One long argument: Charles Darwin and the genesis of modern evolutionary thought* (Cambridge: Harvard University Press).
McDermott, J. and Hauser, M.D, (2001). "The Music of Nature and the Nature of Music." *Science*, v.291, n.5501.
McFall-Ngaia, M., Hadfield, M.G., Bosch, T.C.G. and Wernegreen, J.J. (2013). "Animals in a bacterial world, a new imperative for the life sciences." *Proceedings of the National Academy of Sciences* 110(9), 3229–3236.
"Microbiology by numbers." Editorial. (2011). *Nature Reviews Microbiology* 9, 628.
Miller, J.H. (2015). *A Crude Look at the Whole: The Science of Complex Systems in Business, Life and Society* (New York: Basic Books).
Mitchell, P. (1894). "The Spencer-Weismann Controversy." *Nature* 49, 373–374.
Mocquereau, A. Dom (1920). "La pensée pontificale et la restauration grégorienne." *Revue grégorienne* 5.
Mol, A. (2010). "Actor-network theory: Sensitive terms and enduring tensions." *Kölner Zeitschrift für Soziologie und Sozialpsychologie. Sonderheft* 50, 253–269.
Montague, Lady M.W. (1717). "Letter to Alexander Pope, from Adrianople, April 1, 1717." *The Complete Letters, Vol. 1, 1708–20.* Excerpted in *Turkish Music Quarterly,* fall/winter 1992, 25.
Montévil, M. and Mossio, M. (2015). "Biological organisation as closure of constraints." *Journal of Theoretical Biology* 372, 179–191.
Montgomery, D.L. (1992). "The myth of organicism: From bad science to great art." *The Musical Quarterly* 76(1), 17–66.
Morowitz, H.J. (2004). *Beginnings of Cellular Life: Metabolism Recapitulates Biogenesis* (New Haven: Yale University Press), 8.
Morris, S.C. (2003). "The Cambrian 'Explosion' of Metazoans." In Müller, G.B. and Newman, S.A. (eds). *Origination of organismal form. Beyond the gene in developmental and devolutionary biology* (Vol 2) (Cambridge: MIT Press), 27–33.
Morrison, S.J., Shah, N.M. and Anderson. D.J. (1997). "Regulatory Mechanisms in Stem Cell Biology." *Cell* 88.
Morton, R. (1973). *Come day, go day, God send Sunday: The songs and life story, told in his own words, of John Maguire, traditional singer and farmer from Co. Fermanagh* (Vol. 6) (New York: Routledge and Kegan Paul).

Müller, G.B. and Newman, S.A. (1999). "Generation, integration, autonomy: Three steps in the evolution of homology." In Bock, G.R. and Cardew, G. (eds). *Novartis Foundation Symposium* 222 (Chichester: Wiley), 65–79.

— (eds) (2003). *Origination of organismal form: Beyond the gene in developmental and evolutionary biology* (Vol. 2) (Cambridge: The MIT Pres).

Nelsen, M.P., Lücking, R.C., Boyce, K., Lumbsch, H.T. and Ree, R.H. (2019). "No support for the emergence of lichens prior to the evolution of vascular plants." *Geobiology* 18(1), 3–13.

Nelson, K. (1986). *The Art of Reciting the Qur'an* (Austin: University of Texas Press).

Nelson, T.C. and Cresko, W.A. (2018). "Ancient genomic variation underlies repeated ecological adaptation in young stickleback populations." *Evolution Letters* 2(1).

Nettl, B. (2006). "Response to Victor Grauer: On the concept of evolution in the history of ethnomusicology." *The World of Music* 48(2), 59–72.

Newman, S.A. (1995). "Carnal boundaries: The commingling of flesh in theory and practice." In Birke, L. and Hubbard, R. (eds), *Reinventing biology: Respect for Life and the creation of knowledge* (Bloomington: Indiana University Press), 191–227.

— (2003a) "The Fall and Rise of Systems Biology: Recovering from a half-century gene binge." *GeneWatch,* July–August, 8–12.

— (2003b) "From Physics to Development: The Evolution of Morphogenetic Mechanisms". In Müller, G.B. and Newman, S.A. (eds). *Origination of Organismal Form: Beyond the Gene in Development and Evolutionary Biology* (Cambridge: MIT Press), 221–240.

— (2011). "How Not to Think About Evolution." *The Evolutionary Review* 3, 196–205.

— (2016). "'Biogeneric' developmental processes: Drivers of major transitions in animal evolution." *Philosophical Transactions of the Royal Society B: Biological Sciences* 371(1701), Article 20150443.

— (2019a). "Cell differentiation: What have we learned in 50 years?" *Journal of Theoretical Biology* 485, Article 110031.

— (2019b). "Inherency of form and function in animal development and evolution." *Frontiers in Physiology* 10, Article 702.

— (2021). "Inherency." In Nuño de la Rosa, L. and Müller, G.B. (eds). *Evolutionary Developmental Biology* (Boston: Springer).

Newman, S.A. and Bhat, R. (2009). "Dynamical patterning modules: A 'pattern language' for development and evolution of multicellular form." *The International Journal of Developmental Biology* 53, 693–705.

— (2011). "Lamarck's Dangerous Idea." In Gissis, S.B. and Jablonka, E. (eds). *Transformations of Lamarckism: From Subtle Fluids to Molecular Biology* (Cambridge: The MIT Press), 157–169.

Niklas, K.J. and Kutschera, U. (2017). "From Goethe's plant archetype via Haeckel's biogenetic law to plant evo-devo." *Theory of Biosciences* 136, 49–57.

Noble, D. (2006). *The Music of Life: Biology beyond genes* (New York: Oxford University Press).

— (2017). *Dance to the Tune of Life: Biological Relativity* (Cambridge and New York: Cambridge University Press).

Noble, R. and Noble, D. (2018). "Harnessing stochasticity: How do organisms make choices?" *Chaos* 28(10), Article 106309.

Nooshin, L. (2003). "Improvisation as 'Other': Creativity, Knowledge and Power—the Case of Iranian Classical Music." *Journal of the Royal Musicological Association* 128.

NTU Complexity. (2013). "A Crude Look at the Whole: A Reflection on Complexity (Part 2 of 14)." YouTube. https://www.youtube.com/watch?v=_e6bpLVBwME (accessed 8/13/26).

O'Connell, J.M. (2000). "Fine art, fine music: Controlling Turkish taste at the Fine Arts Academy in 1926." *Yearbook for Traditional Music* 32, 117–142.

Oliver, P. (2001). "Race record." *Grove Music Online*. https://doi.org/10.1093/gmo/9781561592630.article.22778 (accessed 8/9/26).

Oparin, A.I. (1938). *The Origin of Life* (New York: Macmillan).

Oyama, S. (2000). *Evolution's Eye: A Systems View of the Biology-Culture Divide* (Durham: Duke University Press).

Pagel, M. (ed.) (2002). *Encyclopedia of Evolution* (New York: Oxford University Press).

Palisca, C. (1978). *Hucbald, Guido, and John on music: three medieval treatises* (New Haven: Yale University Press).

Paterson, J.R., Edgecombe, G.D. and Yee, M.S. (2019). "Trilobite evolutionary rates constrain the duration of the Cambrian explosion." *Proceedings of the National Academy of Sciences* 116(10), 4394–4399.

Pearce, J.M.S. (2022). "The beginnings of cell theory: Schleiden, Schwann, and Virchow." *Hektoen International. A Journal of Medical Humanities.*

Perlin, J. (1989/2023). *A Forest Journey: The Role of Trees in the Fate of Civilization, rev. ed.* (Ventura, California: Patagonia Press).

Perry, J. (2000). "Music, Evolution and the Ladder of Progress." *Music Theory Online*. https://mtosmt.org/issues/mto.00.6.5/mto.00.6.5.perry.pdf (accessed 8/8/26).

Petrie, G. (1855/1882/1905). *The Ancient Music of Ireland* (Dublin: Printed at the University Press for the Society for the Preservation and Publication of the Melodies of Ireland by M.H. Gill).

Piekut, B. (2014). "Actor-networks in music history: Clarifications and critiques." *Twentieth-Century Music* 11(2), 191–215.

Pietri, J.E., DeBruhl, H. and Sullivan, W. (2016). "The rich somatic life of Wolbachia." *Microbiologyopen* 5(6), 923–936.

Power, R. (2018). *The Overstory* (New York: W.W. Norton).

Powers, H.S., Wiering, F., Porter, J., Cowdery, J., Widdess, R., Davis, R., Perlman, M., Jones, S. and Marett, A. (2001). "Mode." *Grove Music Online*. https://www.oxfordmusiconline.com/grovemusic/display/10.1093/gmo/9781561592630.001.0001/omo-9781561592630-e-0000043718 (accessed 8/8/26).

Pradeu, T. (2012). *The Limits of the Self: Immunology and Biological Identity* (New York: Oxford University Press).

Prigogine, I. (1996). *The End of Certainty: Time, Chaos, and the New Laws of Nature* (New York: Free Press).

Quammen, D. (2018). *The Tangled Tree: A Radical New History of Life New* (York: Simon and Schuster).

Quarto, A. (2025). "Lessons from successful mangrove forest restoration in El Salvador (analysis)." *Mongabay*. https://news.mongabay.com/2025/01/lessons-from-successful-mangrove-forest-restoration-in-el-salvador-analysis/ (accessed 7/13/2026).

Quinton, S. (2017). "Don't Blink: The science of a 100mph fastball." *The Seattle Times*. https://projects.seattletimes.com/2017/mariners-preview/science/ (accessed 7/7/26).

Qureshi, R.B. (1990). "Musical Gesture and Extra-Musical Meaning: Words and Music in the Urdu Ghazal." *Journal of the American Musicological Society* 43(3), 457–497.

Rabelais, F. (1552). "Chapter 1.V. The discourse of the drinkers." In Urquhart, T. and Motteux, P.A. (trans.) *Five books of the lives, heroic deeds and sayings of Gargantua and his son Pantagruel*. https://www.gutenberg.org/files/1200/1200-h/1200-h.htm (accessed 8/11/26).

Rahaim, M. (2006). "What else do we say when we say 'music evolves?'" *The World of Music* 48(3), 29–41.

Raina, J.-P., Eme, L., Pollock, F.J., Spang, A., Archibald, J.M. and Williams, T.A. (2018). "Symbiosis in the microbial world: from ecology to genome evolution." *Biology Open* 7(2), bio032524.

Ravenhall, M., Škunca, N., Lassalle, F. and Dessimoz, C. (2015). "Inferring horizontal gene transfer." *Public Library of Science Computational Biology* 11(5), Article e1004095.

Rehding, A. (2000). "The quest for the origins of music in Germany circa 1900." *Journal of the American Musicological Society* 53(2), 345–385.

— (2011). "Ecomusicology between apocalypse and nostalgia." *Journal of the American Musicological Society* 64(2), 409–414.

— (2021). "Music theory's other nature: Reflections on Gaia, humans and music in the anthropocene." *19th-Century Music* 45(1), 7–22.

Reynolds, C.W. (1987). "Flocks, Herds, and Schools: A Distributed Behavioral Model." *Computer Graphics* 21(4), 25–34.

Ribatti, D. (2009). "William Harvey and the discovery of the circulation of the blood." *Journal of Andiogenesis Research* 1, 3.

Rigg, M.G. (1964). "The Mood Effects of Music: A Comparison of Data from Four Investigators." *The Journal of Psychology* 58(2), 427–438.

Ritterman, J. (1992). "Piano Music and the Public Concert, 1800–1850." In Samson, J. (ed.). *The Cambridge Companion to Chopin* (Cambridge: Cambridge University Press), 11–31.

The Rolling Stones (1969). *Let it Bleed* (Decca Records/London Records).

— (2020). "The Rolling Stones - The Singer Not The Song (Official Lyric Video)." *ABKCOVEVO*. https://www.youtube.com/watch?v=wtlenQW63NM (accessed 7/13/26).

Rosenberg, Y., Bar-On, Y.M., Fromm A. et al. (2023). "The global biomass and number of terrestrial arthropods." *Science Advances* 9(5). https://www.science.org/doi/10.1126/sciadv.abq4049 (accessed 7/9/26).

Royal Society of Chemistry (2026). "Platinum." *Periodic Table*. https://www.rsc.org/periodic-table/element/78/platinum (accessed 7/8/26).

Ruiz, M.F. and Umerez, J. (2021). "Interactionism, Post-interactionism, and Causal Complexity: Lessons From the Philosophy of Causation." *Frontiers of Psychology* 12.

Russell, C.T. (1966). *The Southern French Tonary in the Tenth and Eleventh Centuries*. Ph.D. dissertation, Princeton University.

Saifullah, A.S.M., Kamal, A.H.M., Idris, M.H., Rajaee, A.H. and Bhuiyan, Md. K.A. (2015). "Phytoplankton in tropical mangrove estuaries: role and interdependency." *Forest Science and Technology* 12(2), 104–113.

Sanguinetti, G. (2012). *The Art of Partimento: History, Theory, and Practice* (New York: Oxford University Press).

Sanlıkol, M.A. (2023). *Reform, Notation and Ottoman Music in Early 19th Century Istanbul: EUTERPE. SOAS Studies in Music* (New York: Routledge).

Saygun, A. (1987). *Yunus Emre Oratoryosu* (Budapest: Hungaroton Records).

Schiltz, M. and Verschraegen, G. (2002). "Spencer-Brown, Luhmann and autology." *Cybernetics* and *Human Knowing* 9(3–4), 55–78.

Schippers, H. (2017). "Sound Futures: Understanding Musical Ecosystems." *Folklife*. https://folklife.si.edu/talkstory/sound-futures-understanding-musical-ecosystems-as-a-key-to-sustainability (accessed 7/10/26).

Schmidt, A.R., Steuernagel, L., Behling, I.H., Seyfullah, L.J., Beimforde, C., Sadowski, E-M., Rikkinen, J. and Kaasalainen, U. (2022). "Fossil evidence of lichen grazing from Palaeogene amber," *Review of Palaeobotany and Palynology* 302.

Schmidt, C.E. (1984). "Interlocking techniques In Kpelle music." *Selected Reports In Ethnomusicology* 5, 194–216.

Schmitz, H.-P. (1979). "Jazz and Alte Muzik." *Black Perspectives in Music* 7(1), Spring, 75–78.

Schrödinger, E. (1945). *What is life? The physical aspect of the living cell* (New York: Macmillan).

Schubert, P. (2010). "Musical commonplaces in the Renaissance." In Murray, R.E., Weiss, S.F. and Cyrus, C.J. (eds). *Music Education in the Middle Ages and the Renaissance* (Bloomington: Indiana University Press), 161–192.

— (2014). "From improvisation to composition: Three 16th century case studies." In Moelants, D. (ed.). *Improvising early music* (Leuven: Leuven University Press), 69–130.

[Franz] Schubert's Werke: Kritisch durchgesehene Gesammtausgabe (1884–97) (Leipzig: Verlag Breitkopf & Härtel).

Schuller, K. (2018). "The Microbial Self: Sensation and Sympoiesis." *Resilience: A Journal of the Environmental Humanities.*

Schulz, M., Schmitt, I., Weber, D. and Dal Grande, F. (2022). "Fungal Host Affects Photosynthesis in a Lichen Holobiont." *Journal of fungi* (Basel) 8(12), 1267.

Schweiger, S. (1608). *A Travel Diary* (excerpt). In Martin, R. (trans.) (1990). *Turkish Music Quarterly*, Autumn, 8.

Shanken, E.A. (2002). "Cybernetics and art: Cultural convergence in the 1960s." In Clark, B. and Henderson, L.D. (eds). *From energy to information* (Palo Alto: Stanford University Press), 255–277.

Shapiro, A.D. (1975). *The Tune-Family Concept in British-American Folk-Song Scholarship.* Ph.D dissertation, Harvard University.

Sharp, C. (1907). *English folk-song: Some conclusions* (London: Novello).

Sheldrake, M. (2020). *Entangled Life: How Fungi Make Our Worlds, Change Our Minds and Shape Our Futures* (New York: Random House).

Shultz, T.R. (2001). "Connectionist Models of Development." In Smelser, N.J. and Baltes, P.B. (eds). *International Encyclopedia of the Social and Behavioral Sciences* (Pergamon), 2577–2580.

Simard, S. (2022). *Finding the Mother Tree: discovering the wisdom of the forest* (New York: Vintage Books, a division of Penguin Random House LLC).

Şimşek, H. (1987). *A. Adnan Saygun, 'Yunus Emre Oratoryo'.* Hikmet Şimşek, conductor. Budapest Symphony Orchestra and Choir of the Hungarian Radio and TV. Liner notes by H. Şimşek (Budapest: Hungaraton).

Skelton, G. (1975). *Paul Hindemith: the man behind the music: a biography* (London: V. Gollancz Ltd.).

Small, C. (1998). *Musicking: The meanings of performing and listening* (Middletown: Wesleyan University Press.)

Smith, E. (2015). "McClosky speaker series: New theories of the origin of life with Dr. Eric Smith." *Aspen Institute.* https://www.aspeninstitute.org/blog-posts/mccloskey-speaker-series-new-theories-origin-life-dr-eric-smith/ (accessed 7/10/26).

Smith, E. and Morowitz, H.J. (2016). "The emergence of a biosphere from geochemistry." In Smith, E. and Morowitz, H.J. (eds). *The origin and nature of life on Earth: The emergence of the fourth geosphere* (New York: Cambridge University Press), 340–423.

Solie, R.A. (1980). "The Living Work: Organicism and Musical Analysis." *19th-Century Music* 4(2), 147–156.

— (1982). "Melody and the Historiography of Music." *Journal of the History of Ideas* 43(2), 297–308.

Spenser, H. (1857). "Progress: Its Law and Cause." *The Westminster Review* 67. Revised version (1881) in *Popular Science Literature*.

Steinhardt, A. (1998). *Indivisible by Four: A String Quartet in Pursuit of Harmony* (New York: Farrar, Straus, Giroux).

Stenhouse, W. (1853). *Illustrations of the Lyric Poetry and Music of Scotland* (London: W. Blackwood and Sons).

Stewart, I. (2002). *Does God Play Dice? The new mathematics of chaos* (Malden: Blackwell).

Stravinsky, I. (1936). *Stravinsky, an autobiography* (New York: Simon and Schuster).

— (1947/1970). *Poetics of Music: In the Form of Six Lessons* (Cambridge: Harvard University Press).

Stravinsky, I. and Craft, R. (1959). *Conversations with Igor Stravinsky* (Garden City: Doubleday).

Stravinsky, V. and Craft, R. (1978). *Stravinsky in Pictures and Documents* (New York: Simon and Schuster).

Stubbs, F.W. (1994). *The Art and Science of Taksim: An Empirical Analysis of Traditional Improvisation from 20th Century Istanbul*. Ph.D Dissertation, Wesleyan University.

Sudnow, D. (1972). *Ways of the Hand* (Cambridge: MIT Press).

Talbot, M. (ed.) (2000). *The musical work: Reality or invention?* (Liverpool: Liverpool University Press).

Taruskin, R. (2014). "Agents and causes and ends, oh my." *The Journal of Musicology* 31(2), 272–293.

Tauber, A.I. (2016). "Immunity in context: Science and society in dialogue." *Theoria: An International Journal for Theory, History and Foundations of Science* 31(2), 207–224.

Tennyson, A. Lord. (1850). "In Memoriam, A.H.H" (London: Edward Moxon).

Thévenot, M. de. (1686). *The Travels of Monsieur de Thévenot into the Levant* (London). In Martin, R. (trans.) (1992). *Turkish Music Quarterly*, Spring-Summer, 22.

Thompson, S. (1950). "Tale-Type." In Maria Leach (ed.). *Funk and Wagnalls Standard Dictionary of Folklore* (New York: Funk and Wagnalls), 1137–1138.

Titon, J.T. (2012). "A sound commons for all living creatures." *Smithsonian Folkways Magazine*, Fall-Winter

Tolbert, E. (1994). Review of *Biomusicology: Neurophysiological, neuropsychological and evolutionary perspectives on the origins and purposes of music*, by Wallin, N.L. *Music Library Association: Notes* 50(4), 1354–1359.

Tomlinson, G. (2015). *A Million Years of Music: The Emergence of Human Modernity* (Princeton: Princeton University Press).

Trainor, L.J. (2015). "The origins of music in auditory scene analysis and the roles of evolution and culture in musical creation." *Philosophical Transactions of the Royal Society B: Biological Sciences* 370(1664), Article 20140089.

Treitler, L. (1974). "Homer and Gregory: The transmission of epic poetry and plainchant." *The Musical Quarterly* 60(3), 333–372.

— (1975). "'Centonate' Chant: 'Übles Flickwerk' or 'E pluribus unus'". *Journal of the American Musicological Society* 28(1), Spring, 1–23.

— (1977). "Transmission and the Study of Music History", part of a symposium on "Transmission and Form in oral Tradition". *International Musicological Society, 12th Congress. Proceedings.*

— (1981). "Oral, Written, and Literate Process in the Transmission of Medieval Music." *Speculum* 56(3).

— (2002). "Early Recorded Performances of Chopin Waltzes and Mazurkas: the Relation to the Text". *Journal of the American Liszt Society* 51, 57–77.

— (2015). "Speaking of the I-word." *Archiv für Musikwissenschaft* 72(1), 1–18.

Tsernjavski, N. (2001). *A.B. sea: a cross-curriculum coastal & marine studies planning guide. Level 1–5* (prep-year 8) (Victoria: Marine Discovery Centre).

Van der Werf, H. (1984). *The Extant Troubadour Melodies: Transcriptions and Essays for Performers and Scholars* (Rochester: Published by the author).

Varela, F.J. (2017). "Laying Down a Path in Walking." Ch. 11 in Varela, F.J. with Thompson, E. and Rosch, E. (eds). *The Embodied Mind: Cognitive Science and Human Experience* (New York: Oxford University Press).

Viaggi di Pietro Della Valle Il Pellegrino (1614). In Martin, R. (trans.) (1989). *Turkish Music Quarterly*, Autumn, 11.

Vitak, S. (2021). "Redo of a famous experiment on the origins of life reveals critical detail missed for decades." *Scientific American*. https://www.scientificamerican.com/podcast/episode/redo-of-a-famous-experiment-on-the-origins-of-life-reveals-critical-detail-missed-for-decades/(accessed 7/13/26).

Von Foerster, H. (2003). *Understanding understanding: Essays on cybernetics and cognition* (New York: Springer-Verlag).

Von Herder, W.G. (n.d.). *Stimmen der Völker in Liedern* (Halle: Hendel).

Von Kleist, J. (1805–1806). "On the gradual construction of thoughts during speech." Hamburger, M. (trans.) (1951). Broadcast on B.B.C. Third Program, January 12.

Voros, J. (2018). "Big history as a scaffold for futures education." *World Futures Review* 10(4), 263–278.

Wagner, P. (1901). *Introduction to the Gregorian melodies; a handbook of plainsong* (London: The Plainsong and Mediaeval Music Society).

Wallin, N.L. (1991). *Biomusicology: Neurophysiological, neuropsychological and evolutionary perspectives on the origins and purposes of music* (Stuyvesant: Pendragon Press).

Walter, T. (1721). *The Ground and Rules of Music Explained: or, An introduction to the art of singing by note. Fitted to the meanest capacities* (Boston: Thomas Johnston).

Watkins, H. (2017). "Toward a post-humanist organicism." *Nineteenth-Century Music Review* 14(1), 93–114.

— (2018). *Musical vitalities: Ventures in a biotic aesthetics of music* (Chicago: University of Chicago Press).

Watson, A.J. and Lovelock, J.E. (1983). "Biological homeostasis of the global environment: the parable of Daisyworld." *Tellus B: Chemical and Physical Meteorology* 35(4), 284–289.

Wegman, R.C. (1994). "From maker to composer: Improvisation and musical authorship in the low countries, 1450–1500." *Journal of the American Musicological Society* 49(3), 409–479.

— (1998). "'Das Musikalische Hören' in the Middle Ages and Renaissance: Perspectives from Pre-War Germany." *Musical Quarterly* 82(3–4), 434–454.

— (2014). "What is counterpoint?" In Moelants, D. (ed.). *Improvising early music* (Leuven: Leuven University Press), 9–68.

Wegner, D.M. (2003). "The mind's best trick: how we experience conscious will." *Trends in Cognitive Sciences* 7(2), 65–69.

Weismann A. (1893). "The all-sufficiency of natural selection: a reply to Herbert Spencer." *Contemporary Review* 64, 309–338.

Weiss, P. (1955). "Beauty and the beast: Life and the rule of order." *The Scientific Monthly* 81(6), 286–299.

Wellesz, E. (ed.) (1957). *The New Oxford History Of Music Volume I: Ancient And Oriental Music* (Oxford: Oxford University Press).

West, G. (2017). *Scale: The universal laws of life, growth, and death in organisms, cities and companies* (New York: Penguin Random House).

Whitman, W.B., Coleman, D.C. and Wiebe, W.J. (1998). "Prokaryotes: The unseen majority." *Proceedings of the National Academy of Sciences* 95(12), 6578–6583.

Whitmore, P. (1991). *Unpremeditated Art: The Cadenza in the Classical Keyboard Concerto. Oxford monographs on music* (Oxford: Clarendon).

Wilgus, D.K. (1959). *Anglo-American Folksong Scholarship since 1898* (New Brunswick, Rutgers University Press).

Wilson, D.S. (2007). "The Vital Arts." *Evolution for Everyone: How Darwin's Theory Can Change the Way We Think About Our Lives*, 182–193.

Wiora, W. (1966). *European Folk Song: Common Forms in Characteristic Modification.* Kolben, R. (trans.) (Cologne: Arno Volk Verlag H. Gerig).

Woese, C.R. and Fox, G.F. (1977). "Phylogenetic structure of the prokaryotic domain: The primary kingdom." *Proceedings of the National Association for the Advancement of Science* 74(11), 5088–5090.

Woese, C.R., Kandler. O. and Wheelis, M.L. (1990). "Towards a natural system of organisms: proposal for the domains Archaea, Bacteria, and Eucarya." *Proceedings of the National Academy of Sciences USA* 87(12), 4576–4579.

Wohlleben, J. (1967). "Goethe and the Homeric Question." *The Germanic Review: Literature, Culture, Theory* 42(4), 251–275.

Wolfe, C. (2003). *Animal rites: American culture, discourse of species, posthumanist theory* (Chicago: University of Chicago Press).

Wolkowicz, T. (2020). "The Orchestra as Ecosystem." *New Bedford Symphony Orchestra.* https://nbsymphony.org/the-orchestra-as-an-ecosystem/ (accessed 7/10/26).

Wood, J.A. and Beale, C. (2019). "Starling murmurations: The science behind one of nature's greatest displays." *The Conversation.* https://theconversation.com/starling-murmurations-the-science-behind-one-of-natures-greatest-displays-110951 (accessed 7/13/26).

Wright, O. (2000). *Demetrius Cantemir: The Collection of Notations, Vol. 1, Text, Vol. 2, Commentary* (London: SOAS).

Xie, X., Backman, D., Lebedev, A. et al. (2015). "Primordial soup was edible: abiotically produced Miller-Urey mixture supports bacterial growth." *Science Reports* 5, 14338.

Yates, M. (1979). "Cecil Sharp in America: collecting in the Appalachians." https://mustrad.mainlynorfolk.info/articles/sharp.htm (accessed 8/6/26).

Yavuz, M. (2018). "A Comparative Study of Epiphytic Lichens Mentioned by Ibn Sina and Ibn al-Baytar." *Medicina Nei Secoli* 30(2), 617–640.

Yeats, W.B. (1937). "The long-legged fly." *The Poetry Archive.* https://poetryarchive.org/poem/long-legged-fly/ (accessed 5/8/26).

Yekta Bey, R. (1922). *Türk Musikisi.* Originally published in French as "La Musique Turque," in Lavignac, A. *Encyclopédie de la Musique et Dictionnaire du Conservatoire* (Paris: Librairie Charles Delagrave).

Young, G.F., Scardovi, L., Cavagna, A., Giardina, I. and Leonard, N.E. (2013). "Starling flock networks manage uncertainty in consensus at low cost." *Public Library of Science Computational Biology* 9(1), Article e1002894.

Zallinger, R. (1968). "The March of Progress." In Howell, F.C. *Early Man* (New York: Time-Life Books), 41–45.

Zarlino, G., Marco, G.A. and Palisca, C.V. (1968). *The Art of Counterpoint: Part Three of Le Istitutioni Harmoniche, 1558* (New York: Da Capo Press), 180–181.

Žukauskaitė, A. (2020). "Gaia theory: Between autopoiesis and sympoiesis." *Problemos* 98, 141–153.

Zumthor, P. (1972). *Essai de poetique médiévale* (Paris: Edition du Seuil).

INDEX

f = Figure; *n* = page note